The Biology of Caves and Other Subterranean Habitats

T0202199

THE BIOLOGY OF HABITATS SERIES

This attractive series of concise, affordable texts provides an integrated overview of the design, physiology, and ecology of the biota in a given habitat, set in the context of the physical environment. Each book describes practical aspects of working within the habitat, detailing the sorts of studies which are possible. Management and conservation issues are also included. The series is intended for naturalists, students studying biological or environmental science, those beginning independent research, and professional biologists embarking on research in a new habitat.

The Biology of Streams and Rivers
Paul S. Giller and Björn Malmqvist

The Biology of Soft Shores and Estuaries
Colin Little

The Biology of the Deep Ocean
Peter Herring

The Biology of Soil
Richard D. Bardgett

The Biology of Polar Regions, 2nd Edition
David N. Thomas et al.

The Biology of Caves and Other Subterranean Habitats, 2nd Edition
David C. Culver and Tanja Pipan

The Biology of Alpine Habitats
Laszlo Nagy and Georg Grabherr

The Biology of Rocky Shores, 2nd Edition
Colin Little, Gray A. Williams, and Cynthia D. Trowbridge

The Biology of Disturbed Habitats
Lawrence R. Walker

The Biology of Freshwater Wetlands, 2nd Edition
Arnold G. van der Valk

The Biology of Peatlands, 2nd Edition
Håkan Rydin and John K. Jeglum

The Biology of African Savannahs, 2nd Edition
Bryan Shorrocks and William Bates

The Biology of Mangroves and Seagrasses, 3rd Edition
Peter J. Hogarth

The Biology of Deserts, 2nd Edition
David Ward

The Biology of Lakes and Ponds, 3rd Edition
Christer Brönmark and Lars-Anders Hansson

The Biology of Coral Reefs, 2nd Edition
Charles R.C. Sheppard, Simon K. Davy, Graham M. Pilling, and Nicholas A.J. Graham

The Biology of Grasslands
Brian J. Wilsey

The Biology of Urban Environments
Philip James

The Biology of Caves and Other Subterranean Habitats

SECOND EDITION

David C. Culver and Tanja Pipan

OXFORD

UNIVERSITY PRESS

Great Clarendon Street, Oxford, OX2 6DP,
United Kingdom

Oxford University Press is a department of the University of Oxford.
It furthers the University's objective of excellence in research, scholarship,
and education by publishing worldwide. Oxford is a registered trade mark of
Oxford University Press in the UK and in certain other countries

First Edition published in 2009
Second Edition published in 2019
Impression: 1

Published in the United States of America by Oxford University Press
198 Madison Avenue, New York, NY 10016, United States of America

British Library Cataloguing in Publication Data
Data available

Library of Congress Control Number: 2018964589

ISBN 978–0–19–882076–5 (hbk.)
ISBN 978–0–19–882077–2 (pbk.)

DOI: 10.1093/oso/9780198820765.001.0001

Printed in Great Britain by
Bell & Bain Ltd., Glasgow

Preface to the Second Edition

We are in a golden age of the study of subterranean biology. Thirty-five years ago, when one of us (DCC) wrote a book on the biology of caves, it was easy to read and discuss all the non-taxonomic literature on cave biology written in English. Today, there are well over 200 papers per year published on the biology of subterranean habitats. Thirty-five years ago, for American speleobiologists, but much less so for European biologists, speleobiology meant the biology of caves. There was scarcely any recognition or awareness of non-cave subterranean environments among American speleobiologists. The scope of speleobiology has expanded to include those subterranean[1] habitats whose inhabitants include blind, depigmented species with compensatory increases in other sensory structures. In fact, there has been so much research on these non-cave subterranean habitats, that five years ago we wrote a book devoted to these habitats that are very close to the Earth's surface—*Shallow Subterranean Habitats: Ecology, Evolution, and Conservation.*

Since the publication of the first edition, a number of books on various aspects of speleobiology have been published, including general treatments by Romero (2009) and Fenolio (2016), a book on the microbiology of caves (Engel 2016), four! books on cavefish—Keene *et al.* (2016), Trajano *et al.* (2010), Wilkens and Strecker (2017), and Zhao and Zhang (2009), and a second edition of the *Encyclopedia of Caves* (White and Culver 2012). In addition, there have been more than 1000 research papers on speleobiology published since the first edition of this book.

Our strategy in the second edition has been to update information in the first edition, while still focusing on a relatively small number of well-studied cases. We have replaced some of the case studies from the first edition but have not changed others just because there have been more recent publications on the topic. The growth of information has, of course, not been uniform across subdisciplines. Phylogeography, biodiversity, and evo-devo have in particular experienced a growth spurt. All in all, we have added approximately 125 references published since the first edition.

---

[1] We use subterranean in the sense of organisms living in natural spaces. The word subterranean is also frequently applied to organisms that create their own spaces—especially mammals such as mole rats, termites, and plant roots. The word hypogean is sometimes used in the sense we use subterranean, but its use is uncommon, and we use enough uncommon words as it is. There are many precedents for the way we use the word, such as the International Society for Subterranean Biology and its journal *Subterranean Biology.*

We hope that this book is accessible to a wide variety of readers. We have assumed no training in biology beyond a standard university year-long course, and we have tried to make the geological and chemical incursions self-contained. An extensive glossary should help the readers through any terminological rough spots.

We have organized this book around what seem to us to be the major research areas and research questions in the field. To provide a context for these questions, we review the different subterranean environments (Chapter 1), what the energy sources are for subterranean environments given that the main energy source in surface environments—photosynthesis—is missing (Chapter 2), and the main inhabitants of these underground domains (Chapter 3). The research areas that we focus on are as follows:

- How are subterranean ecosystems defined and organized, and how in particular does organic carbon move through the system (Chapter 4)?
- How do species interact and how do these interactions, such as competition and predation, organize and constrain subterranean communities (Chapter 5)?
- How did subterranean organisms evolve the bizarre morphology of elongated appendages, no pigment, and no eyes (Chapter 6)?
- What is the evolutionary and biogeographical history of subterranean species? Are they in old, relic lineages (Chapter 7)? How does their distribution relate to past geological events?
- What is the pattern of diversity of subterranean faunas over the face of the Earth (Chapter 8)?

We close by 'putting the pieces together' and examining some representative and exemplary subterranean communities (Chapter 9), and how to conserve and protect them (Chapter 10).

With the exception of Chapters 1–3, where we have attempted to provide a comprehensive geographical and taxonomic review of the basics, we have focused on a few particularly well-studied cases. Although we have provided case studies from throughout the world, readers from Africa, South America, and Asia will no doubt find a North American and European bias. Of this we are certainly guilty, but in part this bias is because of longer traditions of study of subterranean life in Europe and North America. We have added several case studies from Asia and South America. We have provided an extensive bibliography and hope that interested readers will pursue the subjects further. Where English language articles are available, we have highlighted them, but we also have not hesitated to include particularly important or unique papers in other languages.

A cautionary word about place names. Many species are limited to a single cave, well, or underflow of a brook, and, if for no other reason, this makes it important to accurately give place names. Throughout the book we have identified the country and state or province in which a site is located. We have, whenever possible, retained the spelling of the local language. Translation runs the risk of confusing anyone trying to identify a particular cave or site, and also runs the risk of repeating the word cave in different languages, as in Postojnska jama Cave (Postojna Cave Cave). Postojnska jama already has names in three languages (Slovene, Italian, and German) and there is no need to add a fourth. A list of sites mentioned in the text is provided.

Even to us, the field of subterranean biology seems especially burdened with obscure terminology. While there is a temptation to ignore it as much as possible, it is widespread in the literature and some of it is even useful. We have defined many terms in the text when we first use them, and have included an extensive glossary to aid readers.

Besides the fascination of their bizarre morphology (which cannot really be overrated), there are two main reasons for biologists to be interested in subterranean faunas. One is numerical. Nearly all rivers and streams have an underlying alluvial system in which its residents never encounter light. Approximately 15 per cent of the Earth's land surface is honeycombed with caves and springs, part of a landscape called karst that is moulded by the forces of dissolution rather than erosion of rock and sediment. In countries such as Cuba and Slovenia, this is the predominant landform.

But there is a more profound reason for biologists to study subterranean biology. Subterranean species can serve as model systems for several important biological questions. As far as we can determine, it was Poulson and White (1969) who first made this notion explicit but it is implicit in the writings of many subterranean biologists. This is a recurring theme throughout this book, and we list just some of the possibilities here:

- Subterranean ecosystems can serve as models of carbon (rather than nitrogen and phosphorus) limited ecosystems and ones where most inputs are physically separated from the community itself.
- Subterranean communities can serve as a model of species interactions because the number of species is small enough that all pairwise interactions can be analysed and then combined into a community-wide synthesis.
- The universal feature of loss of structures (regressive evolution) is especially obvious in subterranean animals, with a clear basis that in turn can allow for detailed studies of adaptation.
- The possibilities of dispersal of subterranean species are highly constrained and so the species (and lineages) can serve as models for vicariant biogeography.
- The highly restricted ranges and specialized environmental requirements can serve as a model for the protection of rare and endangered species.

Whatever reasons you have for reading this book, we hope it leads you to a fascination with subterranean biology, one that lasts a lifetime.

Acknowledgements

The field of subterranean biology is blessed with a strong, cooperative group of scholars from all over the world, and we could not have written this book without the help of many of them. We especially thank Peter Kozel for reading the entire manuscript of the second edition and making many helpful suggestions. Cene Fišer, Daniel W. Fong, and Mike Slay all read selected chapters and helped us avoid many mistakes. Several colleagues provided photographs and drawings for the first and/or second editions—Gregor Aljančič, Magda Aljančič, Matej Blatnik, Marie-Jose Dole-Olivier, Annette Summers Engel, the late Horton H. Hobbs III, Hannelore Hoch, William R. Jeffery, Arthur N. Palmer, Borut Peric, Slavko Polak, Megan Porter, Mitja Prelovšek, Nataša Ravbar, Andreas Wessel, Jill Yager, and Maja Zagmajster. Colleagues also provided us with preprints and answered sometimes naive questions—Gergely Balázs, Louis Deharveng, Marie-Jose Dole- Olivier, Stefan Eberhard, David Eme, Annette Summers Engel, Daniel W. Fong, Franci Gabrovšek, Janine Gibert, Benjamin Hutchins, Lee Knight, Florian Malard, Georges Michel, Pedro Oromí, Metka Petrič, Megan Porter, Graham Proudlove, Katie Schneider, Trevor Shaw, Boris Sket, Peter Trontelj, Rudi Verovnik, and Maja Zagmajster. Franjo Drole of the Karst Research Institute ZRC SAZU devoted many hours to scanning and producing diagrams. Maja Kranjc, and later Janez Mulec, in charge of the magnificent library at the Karst Research Institute, have helped with our requests for books and journals.

We are especially grateful to the Karst Research Institute ZRC SAZU, especially the head of the institute, Dr Tadej Slabe and the administrative assistant, Sonja Stamenković, for making the writing go as smoothly as possible. Tadej Slabe provided time for TP to work, space for DCC to work, and an appointment to DCC as Associate Researcher. Financial support was provided by Ad Futura (Javni sklad Republike Slovenije za razvoj kadrov in štipendije) to DCC during his stay in Slovenia.

We thank Ian Sherman of Oxford University Press for providing us with the opportunity to write both editions of the book, Helen Eaton for shepherding through the first edition, and Bethany Kershaw for shepherding through the second edition.

A project of this magnitude was a burden on both of our families, and we are especially grateful to our spouses, Gloria Chepko and Miran Pipan, for providing both understanding and support.

Postojna, Slovenia
April, 2018

Contents

List of Sites

A list of sites from 29 countries mentioned in the text. General references to countries or large regions are omitted. Specific reference to the sites can be found in the appendix. While most of the sites are caves, there are a number of specific regions, e.g. Cape Range in Australia, that are listed.

Abkhazia (Georgia)
- Veryovkina

Australia
- Bayliss Cave (Queensland)
- Bundera sinkhole (Western Australia)
- Bungonia Cave (New South Wales)
- Cape Range (Western Australia)
- Pilbara (Western Australia)
- Robe River (Western Australia)
- Sturt Meadows (Western Australia)
- Tantabiddi well (Western Australia)
- Yilgarn (Western Australia)

Austria
- Danube Flood Plain National Park
- Lobau wetlands

Belgium
- Walloon

Bermuda
- Walsingham Cave

Bosnia & Hercegovina
- Dabarska pećina
- Popovo polje
- Trebišnjica River System
- Vjetrenica

Brazil

- Atlantic Coastal Rain Forest
- Lapa da Fazienda Extrema I (Goiás)
- São Mateus Cave (Goiás)

Burma

- Farm caves

China

- Ma San Dong Cave
- Shihua Cave

Croatia

- Krk Island
- Markarova pećina
- Miljacka pećina
- Pincinova jama
- Rupečiča
- Šipun

France

- Ardèche
- Ariège
- Baget basin
- Bellissens
- Canal di Mirabel
- Col des Marrous
- Dorvan-Cleyzieu
- Grand Gravier
- Grotte du Cormoran
- Grotte du Pissoir
- Grotte de Sainte-Catherine
- Heraut
- Isère River
- Jura Mountains
- Lachein Creek
- Las Hountas
- Rhône River at Lyon
- Roussilon
- Tour Laffont

Germany

- Aach Spring
- Segeberger Höhle

Indonesia

- Batu Lubang (Halmahera)
- Gua Sulukkan (Sulawesi)
- Niah Caves (Sarawak)

Israel

- Ayalon Cave

Italy

- Abisso di Trebiciano
- Grotta Azzurra
- Grotta di Frasassi
- Lessinian Mountains

Malaysia

- Lubang Nasib Bagus (Sarawak)
- Sarang karst (Borneo)
- Subis karst (Borneo)

Mexico

- Cueva Chica
- Cueva de la Curva
- Cueva de Los Sabinos
- Cueva da Villa Luz
- Cueva de El Pachón
- Sierra del Abra
- Sistema Zacatón
- Sotano de al Tinaja
- Sotano de las Golandrinas

Montenegro

- Dormitor National Park

Papua New Guinea

- Kavakuna Matali system

Phillipines

- Bohol Island caves
- Montalban caves
- Taninthayri caves

Portugal

- Azores lava tubes
- Serra da Estrela

Romania
- Peştera Movile
- Peştera Urşilor

Slovakia/Hungary
- Baradla/Domica

Slovenia
- Črna jama
- Huda luknja
- Jelševnik
- Kompolska jama
- Krim
- Križna jama
- Mejama
- Otovski breg
- Logarček
- Malo okence
- Paka karst
- Pivka jama
- Pivka River
- Planinska jama
- Postojna–Planina Cave System
- Postojnska jama
- Rak River
- Šica–Krka system
- Škocjanske jame
- Tular

Spain
- Cantabrica
- Cueva del Felipe Reventón (Canary Islands)
- Cueva del Mattravies
- Cueva del Naciemento del Arroyo de San Blas
- Cueva del Viento (Canary Islands)
- Jameos del Agua (Canary Islands)

Thailand
- Central Plain
- Tham Chiang Dao
- Tham Phulu
- Tham Thon

Ukraine
- Zoloushka Cave

United Kingdom

- Otter Hole Cave (Wales)
- Swildon's Hole (England)

United States

- Alpena Cave (West Virginia)
- Blue Lake Rhino Cave (Oregon)
- Bracken Cave (Texas)
- Carlsbad Caverns (New Mexico)
- Cave Spring Cave (Arkansas)
- Cesspool Cave (Virginia)
- Classic Cave (New Mexico)
- Coldwater Cave (Iowa)
- Columbia River basalt (Washington)
- Devil's Hole (Nevada)
- Dillion Cave (Indiana)
- Edwards Aquifer (Texas)
- Fern Cave (Alabama)
- Fisher Cave (Missouri)
- Flathead River (Montana)
- Floridan Aquifer (Florida)
- Glenwood Caverns (Colorado)
- Greenbrier Valley (West Virginia)
- Hellhole (West Virginia)
- Hering Cave (Alabama)
- Hidden River Cave (Kentucky)
- Howe Caverns (New York)
- Inner Space Caverns (Texas)
- Kartchner Caverns (Arizona)
- Kazumura Cave (Hawaii)
- Lanikai Cave (Hawaii)
- Lava Tubes National Monument (California)
- Lechuguilla Cave (New Mexico)
- Limrock Cave (Alabama)
- Logan Cave (Arkansas)
- Lower Kane Cave (Wyoming)
- Lower Potomac (District of Columbia)
- Mammoth Cave (Kentucky)
- McClean's Cave (California)
- Ogalalla Aquifer (Nebraska)
- Old Mill Cave (Virginia)
- Organ Cave (West Virginia)
- Parker Cave (Kentucky)
- Pless Cave (Indiana)
- Robber Baron Cave (Texas)

- Scott Hollow Cave (West Virginia)
- Scotts Run Park (Virginia)
- Shelta Cave (Alabama)
- Shenandoah Valley (Virginia)
- Silver Spring (Florida)
- South Platte River (Colorado)
- Sunnyday Pit (West Virginia)
- Thompson Cedar Cave (Virginia)
- Thornhill Cave (Kentucky)
- Tony Sinks (Alabama)
- Ward's Cove (Virginia)
- Young–Fugate Cave (Virginia)

Vietnam

- Halong Bay
- Tan phu lava tubes

1 The Subterranean Domain

1.1 Introduction

Beneath the surface of the earth are many spaces and cavities. These spaces can be very large—some cave chambers such as the Sarawak Chamber, with an area of over 21 000 000 m³ in Lubang Nasib Bagus (Good Luck Cave) in Sarawak, Malaysia (Waltham 2004) can easily accommodate the world's largest aircraft. They can also be very small, such as the spaces between grains of sand on a beach. These spaces can be air-filled, water-filled, or even filled with petroleum. All of these spaces share one very important physical property—the complete absence of sunlight. This is a darkness that is darker than any darkness humans normally encounter, a darkness to which our eyes cannot acclimate no matter how long one waits. There are some habitats that are dark and yet have some light. The ocean abyss is nearly without light but many organisms of the abyss, such as the well-known angler fish, produce their own light with the help of microbes (Fenolio 2016). In addition, the heat of deep sea vents is high enough that light is emitted (Van Dover 2000). In subterranean habitats, with very rare exceptions, this does not happen. The most notable exception is that of glow-worms (actually fungus-gnat larvae) in a few caves in Australia and New Zealand (Broadley and Stringer 2001). But even in these special cases, organisms cannot use light to find their way about, to find food, to find mates, and so on.

Taken together, the water-filled and air-filled cavities are quite common, perhaps more common than surface habitats. Over 94 per cent of the world's unfrozen freshwater is stored underground, compared with only 3.6 per cent found in lakes and reservoirs, with the rest in soil, rivers, and the atmosphere (Heath 1982). Heath estimates that there are 521 000 km³ of subsurface spaces and cavities in the soils and bedrock of the United States, and most of these contain water. Whitman *et al.* (1998) indicate that between 6 per cent and 40 per cent of the total prokaryotic (organisms with no nuclear membrane such as Bacteria) biomass on the planet may be in the terrestrial subsurface. The number of caves is also large—for example the Karst Research

The Biology of Caves and Other Subterranean Habitats. Second Edition. David C. Culver and Tanja Pipan. Published 2019 by Oxford University Press. © David C. Culver and Tanja Pipan 2019. DOI: 10.1093/oso/9780198820765.001.0001

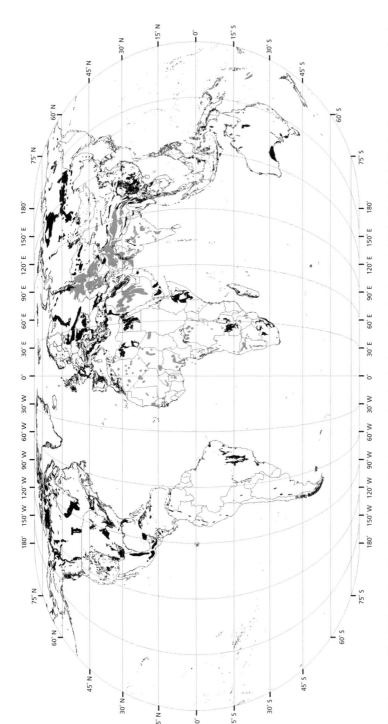

Fig. 1.1 Global distribution of major outcrops of primary cave-bearing (carbonate) rocks, shown in black. Not included in the figure are areas of volcanic rock with lava tubes. Impure or discontinuous carbonate regions are in grey. Map by P.W. Williams, used with permission.

Institute of Slovenia has records of more than 12 000 caves in a country with an area of about 20 000 km². More than 100 000 caves are known from Europe, and nearly 50 000 are known from the United States (Culver and Pipan 2013). All of the continents except Antarctica have caves, as do most countries. A map (Fig. 1.1) of cave regions shows that North America and Eurasia are especially rich in cave-bearing rocks.

The absence of light has profound effects on the organisms living in such habitats. Eyes and the visual apparatus in general have no function there. There are no photons to capture; therefore, no increase in visual acuity will have any benefit to the organisms exclusively living in darkness. Food-finding, mate-finding, and avoidance of competitors and predators all must be accomplished without vision. As is discussed in more detail in Chapter 7, this is a profound barrier that surface-dwelling animals must overcome to successfully colonize subsurface habitats. The absence of light means an absence of both photosynthesis and primary producers (plants, algae, and some bacteria). In some rare but very interesting cases, microorganisms can obtain energy from the chemical bonds of inorganic molecules (Engel 2012), but most subsurface communities rely on food transported in from the surface. This will be taken up in detail in Chapter 2, and we just note in this chapter that the general absence of autotrophy means the amount and variety of resources are usually reduced.

For all subsurface habitats, the amplitude of variation of environmental parameters, especially temperature, is much less than that of the surface habitats. This reduction in amplitude is especially noticeable in regions where variation in surface temperatures is extreme. In Kartchner Caverns, Arizona, USA, the daily average temperature on the surface varies by more than 17°C, whereas temperatures within the cave vary less than 1°C (Fig. 1.2) (Cigna 2002). The range of variation in most spots in Kartchner Caverns was around 1 per cent or 2 per cent of the surface variation. Nevertheless, in Kartchner Caverns, as in nearly all subterranean habitats, there is still an annual temperature cycle. With the possible exception of groundwater aquifers at depths of hundreds of metres, there are no truly constant subsurface environments. In many older references (e.g., Poulson 1963, Vandel 1964), environmental constancy is over-emphasized. With the availability of better monitoring devices, especially ones taking multiple measurements, environmental variability can be detected. Other parameters besides temperature vary, including air currents, water levels, and the amount of food brought into caves. The pulse of spring flooding may be an important cue for reproduction for many cave animals (Hawes 1939). It varies in amplitude, predictability, and seasonality in different caves, but shows the general lack of constancy of the subterranean environment.

Traditionally, subsurface habitats are divided into large cavities (caves) and small cavities (interstitial habitats) (Botosaneanu 1986). We follow this division but add a third category—shallow subterranean habitats, which fit uneasily into the traditional dichotomy (Culver and Pipan 2008a, 2014).

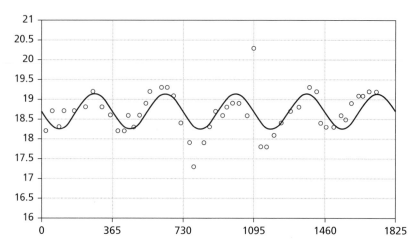

Fig. 1.2   Temperature profiles from Kartchner Caverns, Arizona, USA. Sampling began on 1 January 1996 and continued for 5 years. Solid line is a sinusoidal fit of the data. Time (in days) is shown on the x-axis and temperature (°C) is shown on the y-axis. From Cigna (2002). Used with permission of Karst Research Institute ZRC SAZU.

1.2 Caves

Caves are more difficult to define that one might expect. Geologists (e.g., White 1988) often define caves as natural openings large enough to admit a human being, but this is not an especially useful biological definition. A more useful definition is a natural opening in solid rock with areas of complete darkness, and larger than a few millimetres in diameter. The first criterion excludes spaces among sands, gravels, and stones because they are not openings in solid rock. The second criterion excludes some geographical features that are sometimes called caves, such as rock shelters and natural tunnels, which have no zone of complete darkness. The third is a more technical restriction which eliminates very tiny tubes that are too small to have turbulent water flow. Eventually, many of these tiny tubes will develop into caves but below this critical diameter processes of enlargement and dissolution are very slow indeed, taking up to hundreds of thousands of years (Dreybrodt *et al.* 2005, Ford and Williams 2007, Audra and Palmer 2015).

1.2.1 Caves formed by dissolution of rocks

Landscapes in which the primary agent moulding the landscape is dissolution rather than erosion are called karst landscapes (Fig. 1.3). That is, the features of karst landscape (caves, sinkholes, springs, blind valleys, and the like) result from the action of the hollowing out of rocks by weak acids rather than by erosion, volcanic activity, earthquakes, and so on. Caves are the most biologically interesting part of this landscape, but there are karst landscapes with very few caves (the extreme northern Shenandoah Valley in

Fig. 1.3   Photo of the karst landscape of Halong Bay, Vietnam. Karst landscapes take many different shapes and forms in different regions. Among the most spectacular are the towers and pinnacles of Halong Bay, a UNESCO World Heritage site. The remaining limestone is slowly being dissolved away. See Plate 1.

Virginia, USA, and Krk Island, Croatia, are examples); apparently, the result of the absence of suitable hydrological conditions for caves to form. Klimchouk (2015, 2017) makes a counter-argument, putting speleogenesis itself at the centre of the definition of karst, and points out that there are regions with caves (especially hypogenic caves) with relatively few other karst features. Comprising approximately 15 per cent of the Earth's surface (see Fig. 1.1), karst represents 75 per cent of the land area of Cuba, 45 per cent of Slovenia, 25 per cent of France and Italy, and 40 per cent of the United States east of Tulsa, Oklahoma (White *et al.* 1995). Caves are present in rocks more than 400 million years old to rocks less than 10 000 years old.

Many caves are formed by the action of acidic waters on carbonates (particularly limestone but sometimes dolomite and marble) and evaporites (particularly gypsum but sometimes rock salt). Most caves form in limestone rock, which consists mostly of the mineral calcite ($CaCO_3$). Calcite barely dissolves in pure water but readily dissolves in the presence of an acid (Palmer 2007):

$$CaCO_3 + H^+ \leftrightarrow Ca^{2+} + HCO_3^-$$

The bicarbonate ion is in solution and as a result the calcite is dissolved. The question is where does the hydrogen ion come from? Usually it comes from the action of atmospheric CO_2 and of biological activity in the soil. The metabolism of bacteria and other soil organisms produces CO_2. CO_2 dissolves in water to form hydrogen and bicarbonate ions:

$$CO_2 + H_2O \leftrightarrow H_2CO_2 \leftrightarrow H^+ + HCO_3^-$$

Initially, small fissures in the rock are created in this way. Once they reach a diameter of about 0.2 mm they rapidly enlarge (Dreybrodt *et al.* 2005), forming a network of passages (Fig. 1.4). Some caves may be many millions of years old (Osborne 2007), but significant cave development can occur in tens of thousands of years (Bosák 2002; Dreybrodt and Gaborovšek 2002). In some geological settings, especially those with a protective sandstone cap over the cave, cave development can be extensive. The most spectacular example of this is the Mammoth Cave system in Kentucky, USA, with 650 km of passage (Palmer 2007, Toomey *et al.* 2017) (Fig. 1.5).

Until recently, caves were thought to form almost exclusively in hydrologically open, near-surface systems, in intimate contact with the landscape, an eogenic karst system. But caves can form in more hydrologically deep-seated conditions in a confined, stratified aquifer system (Klimchouk 2007, 2017). Klimchouk (2017) defines hypogene speleogenesis as the formation of voids by upwelling liquids, independent of recharge from the overlying or adjacent surface. Mechanisms include sulfuric acid speleogenesis, hydrothermal speleogenesis, mixing corrosion speleogenesis, and others. Of these, perhaps sulfuric acid speleogenesis is the most common, or at least the best known. In regions where there is considerable underground sulfur, especially in areas of petroleum deposits, sulfuric acid rather than carbonic acid is the source of hydrogen ions in the dissolution of limestone (Egemeier 1981). Since sulfuric acid is a much stronger acid than carbonic acid, large caves can form, and they form more rapidly. The best examples of caves formed by sulfuric acid are those in the Guadalupe Mountains of Mexico, USA, including Carlsbad Caverns and Lechuguilla Cave. Both extend many tens of kilometres. Gypsum caves, including the giant gypsum caves of Ukraine, some of which extend up to 250 km (Klimchouk and Andreychouk 2017) are also formed by hypogenic processes.

The ongoing discussion among geologists about hypogene is biologically relevant for several reasons. Sendra *et al.* (2014) found that hypogenic caves in the Iberian Peninsula were less biologically diverse than eogenic caves, presumably because they are more isolated from surface waters. Gypsum caves appear to be generally less diverse, probably because they form quickly, and decay quickly (see Cokendolpher and Polyak (1996) for species lists

Fig. 1.4   (a–d) Example of a two-dimensional computer model of conduit growth. In the simulation, water and CO_2 are injected along the left side with outflow along the right side. An additional input is located in the centre. The length of the model system is 2 km and its width is 500 m. The hydraulic head is 50 m on the left edge, and 0 m on the right edge. The result resembles the development of a cave by sinking streams, particularly in (d), in which the streams are integrated into a single system. The thickness of the conduit lines indicates passage width (scale at bottom). From Dreybrodt and Gabrovšek (2002). Used with permission of ZRC SAZU, Založba ZRC.

t = 10000y

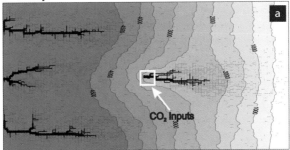

t = 12000y

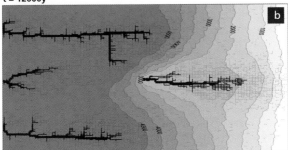

t = 12105y

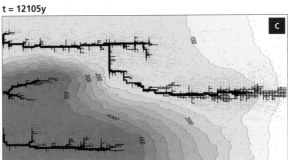

t = 12173y

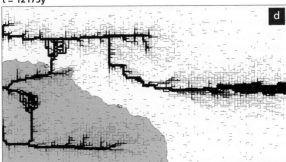

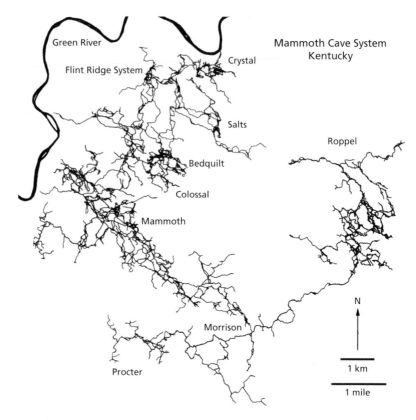

Fig. 1.5   Map of Mammoth Cave, Kentucky, USA. From Palmer (2007). Used with permission of Cave Books. For more detailed maps, see Toomey *et al.* (2017).

from gypsum caves). On the other hand, some hypogenic caves are chemo-autotrophic and, because of this energy source, harbour a very diverse fauna. Such caves include Peştera Movile in Romania, one of the richest caves in the world (Culver and Pipan 2013) and Ayallon Cave in Israel (Por *et al.* 2013).

All eogenic caves have a few basic components (Ravbar 2007). Water enters the subterranean system at the rock–soil interface, which typically has many small solution pockets and cavities with complex horizontal and vertical pathways—the epikarst. Eventually, water percolating through the epikarst reaches a cave stream. The cave stream may be entirely fed from epikarst flow or it may also be fed by a surface stream that sinks into the porous lime-stone. Cave water exits at a spring and flows into a surface river. Beneath the cave stream is a permanent saturated (phreatic) zone that itself often con-tains large cavities (Fig. 1.6).

Epikarst retains water considerably above the water table, and is an ecotone between surface water and subsurface water (Pipan 2005). The principal

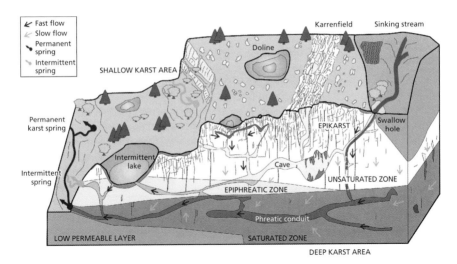

Fig. 1.6   Conceptual model of water flow in a karst aquifer system. Arrows indicate direction of water flow. See glossary for an explanation of the karst features. From Ravbar (2007). Used with permission of ZRC SAZU, Založba ZRC.

characteristic of epikarst is its heterogeneity, with many semi-isolated solution pockets whose water chemistry is also quite variable (Musgrove and Banner 2004; Williams 2008). It forms a more or less permanently saturated zone with a considerable volume of water close under the surface, and is the reason that cave streams rarely dry up. Epikarst water is transmitted vertically through tubes and small fissures to the vadose zone, the region of air-filled passages. In addition to vertical transmission, lateral transmission occurs through poorly integrated cracks and solution tubes (Figs 1.7 and 1.8).

The vadose (unsaturated) zone is a subterranean terrestrial realm. Some terrestrial habitats are perpetually dry, especially those close to the epikarst or in caves well above the water table. Some of these habitats have a substrate of sand or clay while others are largely without sediment. Other terrestrial habitats periodically flood, such as those near streams—riparian habitats. Floods are also an important source of food because of the organic material deposited (Hawes 1939; see Chapter 2). Additional habitats also occur in the vadose zone. Piles of bat guano are themselves habitats as well as sources of food, and are especially common habitats in tropical caves (Deharveng and Bedos 2000; Gnaspini and Trajano 2000).

From the vadose zone water goes to the epiphreatic and phreatic zone, the area permanently saturated with water (Fig. 1.6). The water in cave streams, which typically are at the boundary between the vadose and phreatic zones, not only comes from infiltration from the epikarst but also may come from sinking streams. Cave streams supplied largely from percolating epikarst water tend to be very stable temporally, while cave streams fed largely from

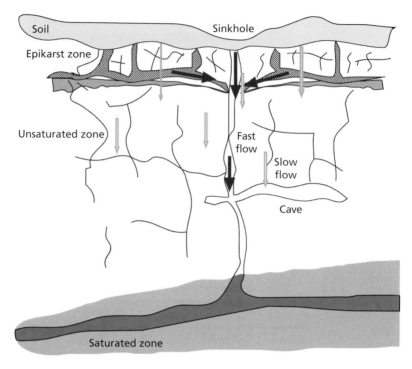

Fig. 1.7   Conceptual model of epikarst. Arrows indicate direction of water flow and black arrows are faster flow paths. From Pipan (2005). Used with permission of ZRC SAZU, Založba ZRC.

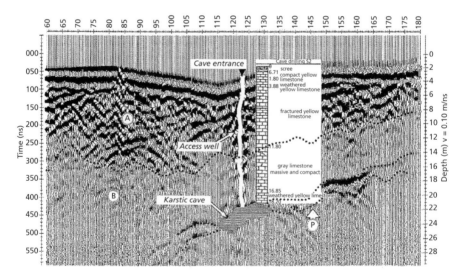

Fig. 1.8   Ground penetrating radar profile through the Hortus field test site (Herault, France). A cave, dipping rock, a local fault, and the epikarst are shown. From Al-fares *et al.* (2002). Used with permission of Elsevier Ltd.

sinking surface streams show much greater temporal variability (Culver *et al.* 1995). Cave streams formed by sinking surface streams can be quite large. In some karst areas, rivers with flows of tens of cubic metres per minute sink to form large river caves (Fig. 1.9).

Phreatic water (permanent ground water) in karst areas is in many ways similar to phreatic habitats in non-karst areas (see section 1.3) except that phreatic habitats in karst regions occupy more space. This is because of the hypogenic dissolution of limestone that results in large cavities deep underground. In non-karst regions water occupies the interstices of spaces in the rock, such as basalt or conglomerates. Wells in karst areas often connect

Fig. 1.9   Photo of Pivka River sinking at the entrance to Postojnska jama, Slovenia. Photo by M. Petrič, used with permission. See Plate 2.

with water-filled underground voids tens to hundreds of times—over decades or even centuries. When there is sufficient water for extraction, the term aquifer is given to these subterranean bodies of water.

Water emerges from caves at springs, which are boundaries (ecotones) between surface and subsurface waters (Gibert *et al.* 1990). Springs come in a wide variety of sizes and shapes (Fig. 1.10). In countries with extensive karst development such as Slovenia and Croatia, there are multiple words that subdivide the English word 'spring' based on geological setting, size, and periodicity. Even though springs are the boundary between surface and subsurface waters, the chemistry of the water at the spring is often more stable. This is because the water emerging from the spring has been underground longer than at any other point in the cave and has had more time to equilibrate with the surrounding rock. Many springs harbour a unique set of species, differing both from the surface stream and the underground water courses (Botosaneanu 1998).

While the basic components of a cave are relatively few (percolating water, streams, and resurgences), the actual geometry of caves varies widely. Lengths vary from a few metres to the 650-km-long Mammoth Cave. Depths vary from less than a metre to 2204 metres in Veryovkina in the Arabika Massif of Abkhazia (Gulden 2017). Several factors are at play in determining shape and geometry of caves (Palmer 2012). In areas with sinkholes, caves tend to be branchwork types, with the details depending on whether the dominant geological structures are vertical fractures or horizontal bedding-plane partings (Fig. 1.11). In contrast, caves in areas without sinking streams

Fig. 1.10   Photo of Unica Spring, the resurgence of the Postojna–Planina Cave System, Slovenia. Photo by M. Blatnik, used with permission. See Plate 3.

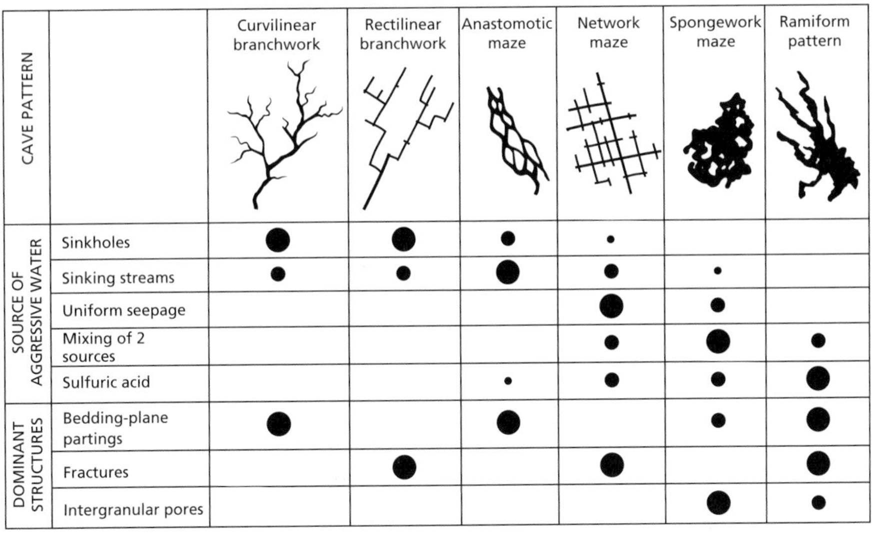

CAVE PATTERN		Curvilinear branchwork	Rectilinear branchwork	Anastomotic maze	Network maze	Spongework maze	Ramiform pattern
SOURCE OF AGGRESSIVE WATER	Sinkholes	●	●	●	•		
	Sinking streams	•	●	●	•	·	
	Uniform seepage				●	•	
	Mixing of 2 sources				•	●	•
	Sulfuric acid			·	•	•	●
DOMINANT STRUCTURES	Bedding-plane partings	●		●		•	●
	Fractures		●		●		●
	Intergranular pores					●	•

Fig. 1.11 Relation of cave patterns to mode of source of water and geological structure. Dot sizes show the relative abundance of each pattern within the various categories. Categories across the top of the figure are the different geometries of cave passages. From Palmer (2012). Used with permission of Elsevier Ltd.

or sinkholes tend to have maze shapes. Caves formed by sulfuric acid waters rising under pressure have a branching, ramiform shape.

There are also interesting regional differences in cave size and geometry. Many of the world's deepest caves are in the Caucasus Mountains in Abkhazia, the Pyrenees Mountains in France and Spain, the Julian Alps in Slovenia, and the Sierra Madre Oriental in Mexico. Many of the world's longest caves are in Interior Low Plateau (eastern USA), the Black Hills of South Dakota (USA), the gypsum karst of Ukraine, and the Yucatan Peninsula of Mexico.

While cave entrances are of course critical for human access, they typically occur independently of cave development. Usually, an entrance is the result of the chance intersection of the developing cave with the surface. Some caves have upwards of one hundred entrances and multiple entrances to large cave systems are common (Curl 1966). More interestingly, there are caves without any entrance. Some large caves, such as Scott Hollow Cave in West Virginia, USA, with over 30 km of passage, have no natural entrances and it was entered only by digging an artificial entrance. Peștera Movile, a chemoautotrophic cave in Romania, likewise has no natural entrances, which is typical of hypogenic caves (Klimchouk 2017). Using a statistical approach, modelling the number of caves with entrances 0, 1, 2, 3, and so on, Curl (1966) estimates that in most karst areas, the majority of caves have no entrances although these caves without entrances are generally shorter in length.

A very interesting kind of cave and habitat occurs at the interface between freshwater and saltwater. Anchialine habitats (haline water, usually with restricted exposure to open air, and always with subterranean connections to the sea) were first carefully investigated by Sket (1986) in a cave along the coast of Croatia. It is a complex-structured habitat with a freshwater lens, the Ghyben–Herzberg lens, on top, underlain by salt water. The boundaries are maintained by differences in densities of water. Anchialine habitats occur in many coastal areas with either carbonate or volcanic rock throughout the world.

Intermediate between an aquatic and terrestrial habitat, is the cave hygropetric (Sket 2004a). It is where there is a thin layer of water moving over vertical rock surfaces, resulting from water percolating from above. The flow of the water is usually laminar rather than turbulent, and it is well oxygenated and often relatively rich in organic matter. It is well developed in the cave Vjetrenica, a biologically diverse cave in Bosnia & Herzegovina, which also harbours several beetles and centipedes specialized for this habitat.

The vertical range of terrestrial cave habitats is much more restricted than that of aquatic cave habitats. Terrestrial habitats are of course above the water table. A wide variety of substrates, including sand, bare rock, and mud, occur in the vadose zone. Some habitats are perpetually dry, but typically relative humidity is high, and in many cave passages the air is at or near saturation. The distribution of food in the terrestrial part of caves largely determines where animals will be found (see Chapter 2 for more details on food sources). Sinking streams bring in food as do the organisms that periodically enter and leave caves. The most familiar examples are bats that routinely leave caves during the night in summer in search of food. On return, their guano routinely supports a variety of organisms. Bats are only the most familiar of the 'regular visitors' to caves. In many regions cave crickets leave caves in the evenings and return with food. Other less regular visitors enter caves, sometimes dying in the process. In wooded north temperate areas, it is rare to find an open air pit without at least one dead wild mammal at the bottom. The flooding or rise of cave streams also brings in food and makes the riparian habitat an important one.

1.2.2 Caves formed by lava

Caves in volcanic rock are often called lava tubes. Their origin is entirely different from those in soluble rock. Most lava tubes are byproducts of volcanic processes and caves are the same age as the rock (Palmer 2007). Lava tubes are most common in panhoehoe basalt flows, which have smooth surfaces with ropy wrinkles. They are most commonly formed by very thin sheets of very fluid lava. Lava tubes are formed by the outflow of fluid lava beneath a hardened and cooler crust (Fig. 1.12). Typically, lava tubes are quite shallow, often less than 5 m below the surface. Even Kazumura Cave, the world's longest known lava tube at 65 km, is less than 20 m below the surface throughout its

Fig. 1.12   Arched ceiling and rough floor of a typical lava-tube cave (Classic Cave, New Mexico, USA). From Palmer (2007). Used with permission of Cave Books.

length (Allred 2012). Limestone caves are rarely this shallow. Because of their shallow character, we include lava tubes with other shallow subterranean habitats (see section 1.4.7). There are some features of lava flows that are deeper, including pit craters, volcanic fissure caves, and open vertical volcanic conduits that are deeper (Palmer 2007), but they have rarely been studied biologically.

1.2.3 **Other caves**

Nearly all caves are formed as the result of solution or lava flows. Solution caves can occasionally occur in other kinds of rock, such as quartzite (Auler 2012). This is the result of the existence of some unusual water chemistry, such as extremely low pH. Occasionally, caves develop in ice, talus slopes, crevices, or as the result of wave action (Palmer 2007). Species specially adapted to subterranean life are rare in such caves. Small caves, with a significant fauna, are common in iron ore regions of Brazil (Parker *et al.* 2013), but since they are all shallow, we defer discussion of them to section 1.4.

1.3 Small-cavity subterranean habitats

To understand interstitial habitats, we need to understand a bit about aquifers and groundwater. The general definition of an aquifer is a water-bearing stratum but in American usage it has come to mean a water-bearing stratum capable of delivering abundant water to a well. We use it here in the more general sense. Interstitial habitats can be relatively shallow with regular interchanges with the surface, or deep with no interchange with the surface. Most biological research has been done on shallow interstitial habitats, especially the hyporheic, the underflow of rivers and streams (Krause *et al.* 2011). We defer discussion of shallow interstitial habitats to section 1.4.

Deep interstitial habitats have permanent groundwater in the fractures and pore spaces within consolidated rock aquifers, such as basalt. The deepest aquifers have no recharge and no connection with the surface. Other aquifers are slowly recharged by water percolating down from the surface. Even groundwater that is a kilometre or more below the surface still harbours life in the form of Bacteria and Archaea (Frederickson *et al.* 1989). In some cases, the microbes most probably colonized the habitat as it was being formed. Multicellular animals, especially Crustacea, have been found in aquifers that are nearly 500 m deep (Longley 2004), although these aquifers have connections to the surface.

Aquifers come in a bewildering array of sizes and shapes. The Ogalalla Aquifer extends over 500 km in the western United States. Others are only a few tens of metres in extent, and even water in seeps (see section 1.4.1) can be considered as an aquifer. The interstitial habitat is a very ancient one and so organisms in interstitial habitats may be from ancient lineages (see Chapter 6).

Stability of environmental variables in interstitial habitats generally increases with depth in these habitats. Grain size determines interstitial space as well as many chemical parameters, such as oxygen concentration. At a microscale, interstices are interconnected, and the porous medium is heterogeneous and constitutes a mosaic of microenvironments. Three factors are especially important in determining the biological composition of interstitial aquifers. The first is permeability, the ability of a rock to transmit water. For example, sand has high permeability because water can easily move through it. In contrast, clay has low permeability because water cannot easily move through it even though clay contains a great amount of water. The second is pore size—the space between particles. If pore size is very small, then many invertebrates, let alone vertebrates, cannot occur there. The third factor is degree of connection with the surface. Aquifers isolated from the surface tend to have low oxygen content because there is no way to replenish oxygen used up by aerobic respiration of the organisms living in the aquifer and relatively little food unless chemoautotrophy occurs. Aquifers with multiple sources of water, including infiltration of river water and infiltration from precipitation, have close connections to surface water. Aquifers that are only fed by infiltration of precipitation have a less intimate connection with surface waters because infiltration takes days and weeks if not longer, and there is no reciprocal exchange with the surface. Aquifers that are 'confined' have the least connection with surface waters. Confined aquifers are ones that are sandwiched between two impermeable layers such as clay or sandstone.

1.4 Shallow subterranean habitats

What do wet spots in the woods, talus slopes in mountains, and fissures and cracks in the ceiling of shallow caves have in common? They are all shallow subterranean habitats (< 10 m from the surface), habitats in constant darkness and dependent on food produced elsewhere, directly or indirectly by photosynthesis. Not all subterranean habitats fit conveniently into the subdivision of karst and interstitial habitats. These superficial subterranean habitats, which include fissures and cracks in the ceiling of caves (the epikarst), seeps, and small cavities among rocks and soil on mountain slopes, share (1) the absence of light, (2) a close connection with the surface, and (3) the presence of species highly modified for subterranean life (Culver and Pipan 2008a, 2014).

Culver and Pipan (2014) provide an extensive treatment of seven shallow subterranean habitats:

1. Seepage springs and the hypotelminorheic habitat
2. Epikarst: the soil–rock interface in karst
3. Intermediate-sized terrestrial shallow subterranean habitats
4. Calcrete aquifers
5. Hyporheic: interstitial habitats along rivers and streams
6. Soil
7. Lava tubes

More recently, iron-ore caves have been shown to also harbor a fauna modified for subterranean life (Souza-Silva *et al.* 2011).

1.4.1 Seepage springs and the hypotelminorheic habitat

The first kind of shallow subterranean habitat is small seeps, really little more than wet spots in the woods (Fig. 1.13). Given the tongue-twisting name of 'hypotelminorheic' by Meštrov (1962), this habitat has proven to be much more extensive and important than its name would suggest (Culver *et al.* 2006a). The habitat has the following characteristics:

1. a persistent wet spot, a kind of perched aquifer;
2. fed by subsurface water in a slight depression in an area of low to moderate slope;
3. rich in organic matter;
4. underlain by a clay layer typically 5 to 50 cm beneath the surface;
5. with a drainage area typically of less than 10 000 m^2; and
6. with a characteristic dark colour derived from decaying leaves which are usually not skeletonized.

What is remarkable about it is the richness of its fauna, a fauna that shows many of the characteristics of the animals that live in interstitial habitats and caves (Culver and Pipan 2011).

Fig. 1.13   Photograph of the authors at a hypotelminorheic site at Scotts Run Park, near Washington, DC, USA. Photo by W.K. Jones, with permission. See Plate 4.

1.4.2 Epikarst: the soil–rock interface in karst

Epikarst is the uppermost layer of karst, often lying only a few metres beneath the land surface (Fig. 1.7). In lava tubes, the cracks and fissures that extend from the surface to the cave ceiling have a similar function although they are formed in very different ways. Lava tubes have an equivalent to epikarst—cracks and fissures in the lava for which Howarth (1983) coined the term mesocavern. It is especially important because roots penetrate through the cracks and fissures to provide an important energy resource in lava tubes (Stone *et al.* 2012; see Chapter 2). Although the amount of water entering a cave through epikarst drips may be much smaller than the amount of water entering from a sinking stream, epikarst water is especially important both for the organic carbon (Simon *et al.* 2007a) and the rich, unique fauna it contains (Pipan 2005). Not all epikarst is water-filled, and the terrestrial also harbours a resident fauna (Culver and Pipan 2014).

1.4.3 Intermediate-sized terrestrial shallow subterranean habitats

Terrestrial subsurface habitats also occur in places besides caves. In addition to mesocaverns in lava, they include the superficial zone of rock fissures and debris slopes in schists, gneiss, and granite. Generally occurring at a depth of a few metres, this habitat was called *milieu souterrain superficiel* (MSS) in French, or mesovoid shallow substratum in English (Juberthie *et al.* 1980).

The MSS (Fig. 1.14) is generally found in mountains in temperate zones but apparently not in the tropics, where spaces are usually filled with sediment such as clay. Like seeps, species occurring in MSS habitats have many of the same characteristics as species living in caves. In some cases they have the same species as caves and in other cases there are species unique to the MSS. Mammola *et al.* (2016) distinguish four main types of MSS habitats:

1. alluvial, streambeds of temporary watercourses;
2. bedrock, weathered bedrock at the soil–rock interface;
3. colluvial, fragments of rocks accumulating at the bottom of rocky walls;
4. volcanic.

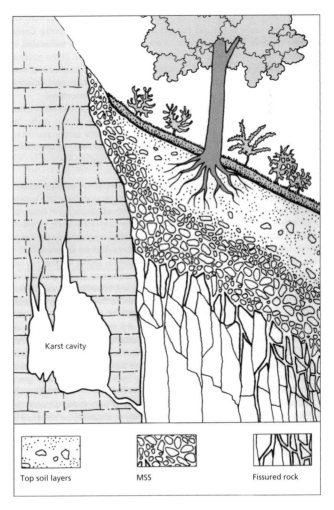

Fig. 1.14   Conceptual model of MSS in a calcareous zone in scree at the base of a limestone cliff. It also occurs in non-carbonate rock, and is probably more common there. From Juberthie (2000). Used with permission of Elsevier Ltd.

1.4.4 Calcrete aquifers

Calcrete aquifers are a feature of arid landscapes and are formed by the pre-cipitation of carbonates from shallow groundwater in climates with precipi-tation of less than 200 mm/year (Mann and Horwitz 1979). In Western Australia, where they occur extensively, they are upflows of salt lakes and are about 10 m thick. They have some functional analogies with epikarst, both being relatively shallow subterranean habitats with capacity for water stor-age, and relatively rich in organic matter. Calcrete aquifers contain a diverse, highly endemic, specialized fauna (Humphreys 2001).

1.4.5 Hyporheic: interstitial habitats along rivers and streams

Shallow aquatic interstitial habitats are comprised of water-filled spaces between grains of unconsolidated sediments. These habitats occur in littoral sea bottoms and beaches, freshwater lake bottoms, river beds, and the hypor-heic zone (the porous aquifer beneath and lateral to streams). The most useful distinction among shallow interstitial aquatic habitats is between the hyporheic zone of rivers and groundwater flowing through unconsolidated sediments.

The hyporheic zone, the best studied of all interstitial habitats (see Buss *et al.* 2009), is the surface–subsurface hydrological exchange zone beneath and alongside the channels of rivers and streams. Although the hyporheic zone is relatively close to the surface compared with most other interstitial habitats, it has already been proved to be relatively easy to monitor and sample. The hyporheic zone of rivers is an ecotone between surface and groundwater. The connection between the hyporheic zone and permanent groundwater (phreatic water) can be very direct or without any direct con-nection at all (Fig. 1.15). In the case of direct connections between the hypor-heic and permanent groundwater, fauna showing adaptation to subterranean life (see Chapters 3 and 6) is often found. Even though the hyporheic zone appears to be highly uniform, it actually has a series of upwellings and downwellings (Fig. 1.16). Downwellings typically have higher oxygen levels and more organic matter. The exact position of these upwelling and down-welling zones along the stream course depends on the relative pressure of the subsurface and surface waters, and other hydrological details. When there are unconsolidated sediments along the stream bank, the hyporheic zone can extend tens of metres from the stream bank.

1.4.6 Soil

Soil forms a thin mantle over the Earth's surface and acts as the interface between the atmosphere and lithosphere, and provides a medium where an outstanding variety of organisms live (Bardgett 2005). For more than 100

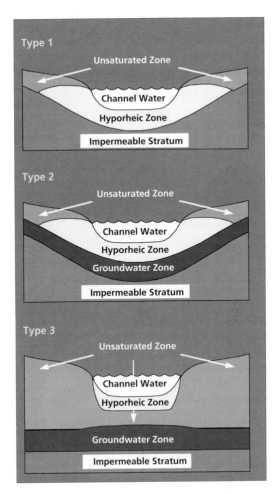

Fig. 1.15   Conceptual cross-sectional models of surface channels and beds showing relationship of channel water to hyporheic, groundwater, and impermeable zones. From Malard *et al.* (2000). Used with permission of E. Schweizerbart'sche Verlagsbuchlanglung (www.schweizerbart.de/9783510540525).

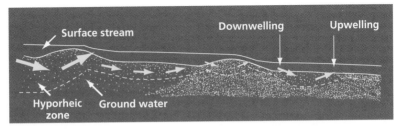

Fig. 1.16   Surface–subsurface hydrological exchanges in the hyporheic zone induced by spatial variation in stream bed topography and sediment permeability. Downwellings bring both organic matter and oxygen to the hyporheic zone. From Malard *et al.* (2002). Used with permission of Blackwell Publishing.

years, soil has been a curious exclusion from the list of subterranean habitats (Racoviță 2006, Sket 2004b), in part because the morphology of soil organisms, particularly from deep soil, is quite different from cave organisms, even though they share lack of eyes and pigment. A notable exception to this exclusion was the work of Coiffait (1958).

1.4.7 Lava tubes

Unlike solution caves which may take millions of years to form, lava tubes form almost instantaneously. Furthermore, lava tubes are quite transient, rapidly being eroded or covered by new lava flows. Even Kazumura Cave in Hawai'i, with over 65 km of passage, a vertical extent of more than 1100 m, and 101 entrances, is only 350–500 years old (Allred 2012). Melting and re-melting of lava can take on quite complicated forms, and some of the details of the geometry of lava tubes are quite bizarre. For example, Blue Lake Rhino Cave, Oregon, USA, is a lava cave formed around the mould of an extinct rhinoceros (Palmer 2007). Some lava tubes are cold air traps. Several lava tubes in Idaho retain ice throughout the summer even though surface temperatures are often in excess of 35°C. The roots of plants on the surface often penetrate the ceiling of the lava tube. The root fauna is both unique and diverse (Stone *et al.* 2012). Streams sometimes secondarily invade lava tubes, and so lava tubes can have roughly equivalent habitats to those in solution caves, sometimes given the prefix 'pseudo' to distinguish those formed by solution, such as 'pseudokarst'.

1.4.8 Iron-ore caves

Over 3000 caves have been reported from iron-ore deposits in Brazil (Auler *et al.* 2014). They tend to be very short and superficial, some so short they lack a dark zone. They are embedded in a matrix of fractured porous rock. While the caves are convenient collecting sites, the primary habitat for the highly endemic fauna (see Souza-Silva *et al.* 2011) is likely the cracks, pores, and fissures of the canga and the Banded Iron Formation (BIF). The primary agent for formation of the caves involves bioreduction of Fe(III) by iron-reducing bacteria that convert insoluble solid Fe(III) into aqueous Fe(II), allowing for the mobilization of iron and generation of voids (Parker *et al.* 2013).

1.4.9 Overview of shallow subterranean habitats

The existence of shallow subterranean habitats forces us to change the paradigm of evolution in subterranean environments, and indeed to rethink the very definition of what subterranean means. The barrier to colonization of subterranean habitats may be much lower than previously thought. The barrier is not one of resource scarcity as implied by many studies of adaptation (see Chapter 6). Of course there may be resource limitation, but this is true for many if not most surface-dwelling species as well. The barrier is not one of environmental sameness, with no cues about day or season. Indeed part of the challenge of survival in shallow subterranean habitats is the ability to cope with environmental fluctuation. The barrier is that of life in complete darkness.

1.5 Summary

The main subterranean habitats are: small cavities—interstitial spaces beneath surface waters; large cavities—caves; and shallow subterranean habitats—voids of various sizes close to the surface. The defining feature of all these habitats is the absence of light. Environmental variation is also reduced relative to surface habitats and most subterranean habitats rely on nutrients transported from the surface.

Most caves result from the dissolution of soluble rocks, especially limestone, but caves in lava are created by the flow of lava beneath a hardened surface. The aquatic component of caves has three main components—water percolating from the surface (including epikarst), streams, and resurgences. Terrestrial habitats include epikarst, and the vadose zone. Cave-bearing regions account for about 15 per cent of the Earth's surface.

The aquatic interstitial habitat is comprised of the water-filled spaces between grains of unconsolidated sediments. They occur in lake bottoms, river beds, littoral sea bottoms, and deep in the earth. Some, especially the hyporheic, are transition zones between surface and subsurface waters. Biological activity has also been found at depths of more than one km beneath the Earth's surface. Important differences among interstitial habitats are permeability, pore size, and degree of connection with the surface.

Shallow subterranean habitats are ones close to the surface but which do not fit conveniently into a classification of caves vs. interstitial. They include the hypotelminorheic, interstitial, epikarst, MSS, soil, lava tubes, calcrete aquifers, and iron-ore caves. They share an absence of light, close surface connections, relatively high nutrient levels compared to other subterranean habitats, and the presence of species highly modified for subterranean life. These habitats may be more important than their areal extent indicates.

2 Sources of Energy in Subterranean Environments

2.1 Introduction

Because of the absence of sunlight, there is no photosynthesis in subterranean environments. Except for a very few caves and possibly most deep interstitial aquifers, all of the energy in subterranean habitats is transferred from surface habitats. One of the most obvious examples of this transfer is the guano left by bats that roost in caves during the day and leave the caves to forage at night. This reliance on external (allochthonous) energy sources generally means that there are fewer energy resources available in subterranean habitats and the diversity of energy resources is low. In surface habitats, insects can specialize in feeding on a particular species of plant. In the subterranean realm, all plants become detritus of very similar composition.

We first consider the sources of energy in subterranean environments, and then summarize what is known about energy movement through the ecosystem by looking at the path of organic carbon.

2.2 Sources of energy

In many ways, the most interesting energy source in the subterranean realm is the energy in inorganic chemical bonds. The utilization of the energy of chemical bonds, chemoautotrophy, rather than the energy of sunlight as the basic source of energy, is common in a few other extreme environments, such as deep sea vents (Van Dover 2000). Chemoautotrophs, primary producers whose energy is derived from light-independent chemical reactions, are also found in nearly every environment on Earth, but it is in subterranean habitats, where there is no competition with photosynthetic organisms, that chemoautotrophs sometimes prevail (Engel 2012).

The Biology of Caves and Other Subterranean Habitats. Second Edition. David C. Culver and Tanja Pipan. Published 2019 by Oxford University Press. © David C. Culver and Tanja Pipan 2019. DOI: 10.1093/oso/9780198820765.001.0001

External energy sources enter subterranean habitats in a variety of ways. *Percolating water* carries with it dissolved organic matter (DOM), some suspended particles of organic matter, and a variety of microbes and minute invertebrates. This seemingly unimportant source of nutrients is actually the most important one in many situations. *Flowing water*, especially streams entering caves, carries with it not only dissolved organic material, but also particulate organic material, in some cases up the size of logs. Flowing water provides nutrients not only to aquatic communities in caves but also to terrestrial communities that live alongside cave streams (riparian communities). *Wind and gravity* bring nutrients into caves when organic material comes into an entrance. Examples include falling leaves as well as animals that fall or wander into a cave, cannot exit, and die. The hallmark of this food source is its unpredictability. *Active movement of animals* is, in some caves, a major source of nutrients, especially in terrestrial cave habitats. The most notable examples of this food source are bats, and in fact distinct communities of organisms specialize on the bat guano of caves (Gnaspini and Trajano 2000). Finally, *roots* penetrate into some shallow caves, especially lava tubes, and species utilize the roots as a food source.

The sources and importance of these six types of nutrients are summarized in Table 2.1. While caves may receive energy from all six sources, interstitial habitats are restricted to at most three—chemoautotrophy, percolating water, and flowing water. In general, shallow subterranean habitats are restricted in terms of sources, even though available sources may provide relatively plentiful energy (Culver and Pipan 2014). For example, seepage springs and the hypotelminorheic are restricted to organic matter from percolating water and water brought to the seepage spring by gravity. Of food sources, only percolating water is universal, or nearly so.

Table 2.1 Classification of sources and origins of energy and their destinations in subterranean environments.

Energy source	Origin of energy	Destination	Subterranean habitats
Chemoautotrophy	Autochthonous (aquatic)	Aquatic and occasionally terrestrial	Deep interstitial and a few caves
Percolating water	Allochthonous (aquatic)	Aquatic	Caves, interstitial, and shallow subterranean habitats
Flowing water	Allochthonous (aquatic)	Aquatic and terrestrial	Caves and hyporheic
Wind and gravity	Allochthonous (terrestrial)	Terrestrial	Caves and shallow subterranean habitats
Active movement of animals	Allochthonous (terrestrial)	Terrestrial	Caves
Roots	Allochthonous (terrestrial)	Terrestrial	Caves and some shallow subterranean habitats

2.2.1 Chemoautotrophy

Probably the first speleobiologist to recognize that chemoautotrophy was a possible energy source in caves was Dudich (1930), commenting about Baradla/Domica Cave, on the Hungarian/Slovakian border.

> The nature of the autotrophy of the Beggiotoa and the Leptothrix is that the energy source is not the light but oxidation. Therefore the organic material is not a result of photosynthesis but of chemosynthesis. They are capable of autotrophy in the absence of light, in the darkness. Nobody has found chemosynthetic, auto-troph producing organisms in caves before, but it seems like they exist and they undoubtedly play a role in the nutrition biology of the cave. Since it is the first such finding it is hard to assess the importance of the phenomenon, and further research is needed. (translation by Gergely Balázs).

The essential aspect of chemoautotrophy is that the energy of inorganic chemical bonds is converted to a biologically useful form, particularly adenosine triphosphate (ATP). The best documented example of chemoautotrophy from a subterranean environment involves the following reaction:

$$H_2S + 2O_2 \rightarrow SO_4^{2-} + 2H^+$$

where hydrogen sulfide is biologically oxidized to form sulfuric acid, and the energy produced is used to form ATP (Fig. 2.1). In Peştera Movile, Romania, a small cave near the Black Sea, is the best studied subterranean

Fig. 2.1   Main trunk passage in Lower Kane Cave, Wyoming, USA. White, filamentous microbial mats dominated by sulfur-oxidizing bacteria are present in shallow sulfidic water, beginning at the lower right corner (water flows from the lower right to upper left). The microbial mat extends for approximately 20 m with an average thickness of 5 cm. From Engel (2012). Photo by A.S. Engel, with permission. See Plate 5.

chemoautotrophic system (Sârbu *et al.* 1996). The reaction is mediated by the bacterium *Thiobacillus thioparus* (Vlăsceanu *et al.* 1997). The reaction is energy releasing (exothermic), with a Gibbs free energy (ΔG) of -798.2 kilojoules (kJ)/mol. It is this energy that is used by *T. thioparus* to make ATP. This sulfur oxidation reaction is also interesting because a strong acid (sulfuric acid) is produced and it readily dissolves limestone. Sulfur oxidation not only supplies energy, it also enlarges the cave. In spite of the considerable interest in this phenomenon, there are only a small number of caves known where it has been documented (Table 2.2). There are other reactions, both anaerobic and aerobic, that are also utilized in chemoautotrophy, but sulfur oxidation is the most commonly reported one (Engel 2012) (Table 2.2).

Many phreatic deep systems are also chemoautotrophic. Chemoautotrophic Bacteria and Archaea have been found at depths of more than 2 km. Their presence in deep sedimentary and basaltic rock may be the result of slow infiltration of water from the surface, or they may have been present at the time of the deposition of the rock, millions of years ago. The possible great phylogenetic age of some of these chemoautotrophic species combined with their ability to survive in the absence of organic matter makes it likely that they are more representative of some of the earliest forms of life on the planet than are surface-dwelling microorganisms. They may be useful analogues for understanding what life on other planets, such as Mars, might be like (Boston *et al.* 2001). Among the energy-producing reactions occurring in deep groundwater is methanogenesis (see also Table 2.2):

$$4H_2 + CO_2 \rightarrow CH_4 + 2H_2O$$

This reaction is exothermic with a Gibbs free energy (ΔG) of -32.5 kJ/mol, and it is a much less efficient source of energy than sulfur oxidation because of the smaller ΔG. Hydrogen serves as the energy source and oxygen acceptor, and inorganic carbon from CO_2 as the carbon source. Methanogenesis thus requires no source of organic carbon from the surface and can persist indefinitely as long as there is a source of hydrogen. The hydrogen derives from the reaction of groundwater with iron-bearing minerals (Stevens and McKinley 1995). Not surprisingly, metabolic rates of these systems are the lowest recorded, indicating that they are the most energy-poor (oligotrophic) environments known.

The demonstration that the entire food web of a cave is dependent on chemoautotrophic production has only been done in a handful of caves. The basic technique used is to look at the relative frequency of stable (i.e., nonradioactive) isotopes of carbon and nitrogen. For example, three isotopes of carbon exist, with six (^{12}C), seven (^{13}C), or eight neutrons (^{14}C). The eight-neutron isotope (^{14}C) is radioactive and decays to one of the stable carbon isotopes. As carbon and nitrogen move up the food chain, the relative frequency of the stable isotopes remains similar but lighter isotopes, such as

Table 2.2 Summary of studies describing chemoautotrophy in caves. From Engel (2012). Used with permission of Elsevier Ltd.

Site	Ammonium oxidation	Anaerobic ammonium oxidation	Denitrification	Hydrogen oxidation	Iron oxidation	Iron reduction	Methanogenesis	Methanotrophy	Nitrite oxidation	Sulfur oxidation	Sulfate reduction
Anchialine caves, Mexico	X							X		X	
Acquasanta Terme Caves, Italy										X	X
Ayalon Cave, Israel							X	X		X	
Bundera Sinkhole, Australia										X	
Bungonia Caves, Australia						X				X	X
Caves in France and Iowa, USA					X						
Caves of Kugitangtou region, Turkmenistan					X					X	X
Cesspool Cave, Virginia										X	X
Coldwater Cave, Iowa and Minnesota	X	X							X		
Cueva de Maltravies, Spain	X										
Cueva de Villa Luz, Mexico										X	X

Edwards Aquifer, Texas	X						X	X		X	X
Florida Aquifer karst	X	X					X	X		X	
Glenwood Caverns and nearby mine adit, Colorado				X						X	
Grotta Azzurra, Italy										X	X
Grotta di Frasassi, Italy										X	X
Karst springs, Spitsbergen, Norway										X	X
Lower Kane Cave, Wyoming					X	X	X			X	X
Mammoth Cave, Kentucky	X								X		
Parker Cave, Kentucky										X	
Peştera Movile, Romania	X							X	X	X	
Sistema Zacatón, Mexico		X					X			X	X
Zoloushka Cave, Ukraine			X							X	
TOTAL	7	3	1	1	4	2	6	5	4	19	11

[12]C, become less common. The graph in Fig. 2.2 shows the situation in Grotta di Frasassi in Italy. There are two food webs—one chemoautotrophically based web present in the sulfidic sections of the cave and one heterotrophically based web present in the cave entrances, and guano rooms in Grotta di Frasassi (Sârbu 2000). The two are easily separated because the isotopic signature of the two energy sources (sulfidic bonds and heterotrophically derived nutrients such as guano and fallen leaves) is very different (Fig. 2.2).

2.2.2 Percolating water

Some precipitation enters lakes and streams, some quickly evaporates, but some also infiltrates into the soil. In turn, some of this infiltrating water moves

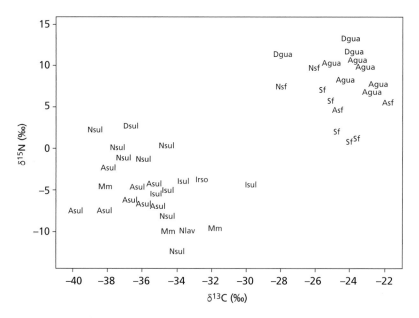

Fig. 2.2   Stable isotope ratios of carbon and nitrogen in organic samples from Grotta di Frasassi. Nearby values are part of the same food web, and two different webs are clearly separated. The sulfidic web is on the lower left and the detrital web is on the upper right. Coding is as follows:

Asf, the isopod *Androniscus dentiger* collected at the surface near the cave entrance;
Agua, *A. dentiger* from cave sections containing deposits of guano;
Asul, *A. dentiger* from sulfidic passages;
Dgua, the beetle *Duvalius bensai lombardi* from guano;
Dsul, *D. bensai lombardi* from sulfidic passages;
Isul, *Niphargus ictus* from sulfidic groundwater;
Mm, microbial mat from sulfidic passage;
Nsf, *Niphargus eremita* collected at the surface near the cave entrance;
Nsul, *N. eremita* from sulfidic passage;
Sf, surface fauna collected under leaf litter near the cave entrance.
From Sârbu (2000). Used with permission of Elsevier Ltd.

vertically into groundwater or caves (Figs 1.7 and 2.3). Hydrogeologists use the phrase recharge to describe this process, and without recharge water tables drop, wells dry up, and in karst regions, cave streams and springs dry up.

Rainwater, just before it reaches the surface of the Earth, is without organic carbon. As it passes through the soil, it accumulates carbon dioxide as a result of the respiration of organisms living in the soil. More importantly from the biological point of view, water accumulates, in solution and suspension, organic carbon in a variety of compounds (Lechleitner *et al.* 2017). The sources of organic carbon include decay products of surface vegetation, excretory products of soil organisms, and decay products of soil organisms and underground parts of vegetation. The organic compounds thus produced range from very easily assimilated compounds such as simple sugars to very difficult to assimilate compounds such as cellulose and lignin. Besides dissolved organic compounds, water moving vertically through the soil carries many microbes, small particles of soil with particulate organic material, and small micro-invertebrates, such as copepods.

Fig. 2.3   Photograph of drip water in Organ Cave, West Virginia, USA which percolates into the cave from the epikarst. Photo by H.H. Hobbs III, with permission. See Plate 6.

In non-karst areas, water is likely to move slowly but more or less uniformly through soil and then rock. As this water moves vertically, the amount of organic carbon declines (Pabich *et al.* 2001). In the absence of chemoautotrophy, which is the usual case, as water moves vertically, organisms utilize organic carbon, and so the total amount of available organic carbon must decline. This is also why many deep aquifers contain high-quality drinking water with little organic content.

The path of percolating water in karst areas is more complicated than in non-karst areas. At the base of the soil layer is a zone of rock, with fissures and cracks as well as solution pockets and channels—the epikarst (Fig. 1.7). Unlike the situation in non-karst areas, epikarst retains significant amounts of water in the voids present in the limestone (Williams 2008). These voids were of course created by the action of carbon dioxide present in soil water, which in turn dissolves the limestone.

In temperate zone caves, the concentration of dissolved organic carbon in epikarst water collected from ceiling drips is typically about 1 mg/L (Simon *et al.* 2007a), but it can reach over 2000 mg/L in some circumstances (Laiz *et al.* 1999). Although the concentration of organic carbon is usually low, it is often the only carbon available in a cave passage. DOC from drips may also be disproportionately important relative to its concentration in forming biofilms, the base of the aquatic invertebrate food web in cave streams (Simon *et al.* 2003). Biofilms in subterranean habitats are coatings on rocks and sediments consisting of microorganisms, extracellular polysaccharides, and particles, both organic and inorganic, trapped in the polysaccharides (Boston 2004).

Other organic matter is carried into subterranean habitats by percolating water, including bacteria (Gerič *et al.* 2004), meiofauna (Pipan 2005), and even terrestrial microarthropods (Pipan and Culver 2005). The amount of particulate organic carbon from these sources is usually much less than that of dissolved organic carbon in percolating water (Gibert 1986, Simon *et al.* 2007a).

The overall flux of organic carbon via percolating water is both spatially and temporally variable. Ban *et al.* (2006) provide a striking example of this from Shihua Cave near Beijing, China. During rain events, DOC content in drips increased to 3 mg/L from a baseline of 1 mg/L and drip rate also increased several fold. There are also spatial 'hotspots' of organic carbon in percolating water, perhaps the result of different residence times of the water in soil and epikarst (Simon *et al.* 2007b).

2.2.3 Flowing water

Streams flowing along the surface of the landscape often sink when the underlying rock changes from an impermeable type such as sandstone, flysch, or shale to a soluble type such as limestone (Fig. 1.9). Such sinking streams are sometimes called swallets. In large sinkholes and dolines, temporary surface streams may develop during periods of precipitation, and these streams also sink underground. All of these streams, both temporary and permanent, bring not only DOM into caves, but also particulate organic matter (POM), sometimes the size of trees. Generalizations about the quantity and size of organic carbon brought in by flowing water are difficult to make because the size of sinking streams varies from tiny brooks to large rivers and because of seasonal variation in flow. In tropical caves during the wet season, flow rates in underground rivers can be enormous and vary enormously as well. Audra and Maire (2004) report that flow rates in the Kavakuna Matali System in Papua New Guinea range from less than 20 m³/s to 1000 m³/s during flooding.

A well-studied example of nutrient transfer from sinking streams is the work of Simon and Benfield (2001) on streams in Organ Cave, West Virginia, USA, a large cave with more than 50 km of passages and several sinking streams. They estimated the amount of POM in two cave streams that originate as sinking streams, how fast the organic matter disappears, and how far it moves as a result of the current before it is utilized by the organisms in the stream (Table 2.3). Rates of consumption of leaves and wood are similar to those of surface streams, and generally there is a considerable amount of organic matter present in the stream, although much of it is fine particulate organic matter (FPOM). In both cave streams, biomass of leaves was much higher than that of wood (sticks and branches). Especially interesting is the fact that neither leaves nor wood travel very far into the cave before they are consumed or converted to fine particulate and dissolved organic matter. Invertebrates are not concentrated near the points where these streams go underground, so those animals not near entrances must have other sources of food. Fine particulate and dissolved organic matter are likely to be more important deeper in the cave. There were no floods during

Table 2.3 Summary of organic matter in two cave streams in Organ Cave, West Virginia, USA.

	Stream 1	Stream 2
Fine Particulate Organic Matter	28.2 (±7.4) g/m²	19.8 (±2.6) g/m²
Coarse Particulate Organic Matter		
Leaves	0.05 (±0.02) g/m²	4.4 (±3.5) g/m²
Wood	0.04 (±0.04) g/m²	2.5 (±2.4) g/m²
Percent Loss of Ash Free Dry Mass/Day		
Leaves	0.0158	0.007
Wood	0.0049	0.0048
Turnover Length		
Leaves	2.3 m	56.4 m
Wood	0.6 m	no data

Numbers in parentheses are standard errors. Stream 1 is C1 and Stream 2 is C4 in Simon and Benfield (2001), the source of the data.

Simon and Benfield's study; leaves and wood would be transported further during floods. When wood and leaves were put in cave streams formed by percolating water, rates of removal were much lower than in cave streams formed by sinking streams. Simon and Benfield conclude that energy was limiting in these sinking streams as it is in many surface streams. All in all, they found remarkably few differences in standing crop of organic carbon and processing rates between cave streams and similar-sized surface streams.

Venarsky *et al.* (2012b) also examined the dynamics of organic carbon in four cave streams in Tennessee and Alabama, USA. They utilized mesh bags of red maple (*Acer rubrum*) and corn (*Zea mays*) and found that breakdown rates were independent of the ambient biomass. Interestingly, they also found that invertebrates which showed no obvious signs of adaptation to subterranean life, such as eye and pigment loss, dominated the processing of organic matter, even though a number of cave-adapted species were present. Overall, processing rates were similar to those in Organ Cave, but they found a greater range of ambient organic matter, up to 850 g/m^2 of ash-free weight (cf. Table 2.3).

Souza-Silva *et al.* (2012) have done a similar study in a tropical cave—Lapa da Fazienda Extrema I in Goiás, Brazil. They found that the input of particulate organic matter was highly seasonal, and dominated by large pieces of wood, followed by leaves and fragments. They examined processing rates of a variety of native Brazilian tree leaves in mesh bags, and found processing rates were slow in riparian habitats (k = 0.0099) and moderate in the stream (k = 0.017–0.025) for the largest mesh size (35 mm^2). Processing was more rapid in the cave stream than in the sinking surface stream.

The quantitative difference between organic carbon from epikarst and from sinking streams seems usually to be strongly in favour of sinking streams. In Organ Cave, West Virginia, at least 75 per cent of the DOC is in water that originated from sinking streams. In the Postojna–Planina Cave System, Slovenia, which has two main sinking streams (Pivka and Rak Rivers), 99 per cent of the DOC is in water that originated from sinking streams (Simon *et al.* 2007a). On several counts, this quantitative picture is misleading. Percolating water from epikarst is more widely distributed throughout a cave than cave streams, because many passages have no stream and many caves have no sinking streams. As indicated earlier, percolating water appears to be especially important in forming biofilms. Finally, there are of course differences in ease with which the organic compounds in DOC are assimilated by heterotrophic organisms. It may be that the carbon in drips is more easily assimilated than the organic carbon from cave streams (Simon *et al.* 2010).

Whatever its origin (sinking streams or percolating water), DOC appears to be a more important quantitative source of carbon than either particulate organic carbon at the cave (Simon *et al.* 2007a) and subterranean stream scale (Gibert 1986) or stream reach (Graening and Brown 2003). Graening and Brown compared fluxes and standing crops of organic matter in two caves in Arkansas, USA, to 19 surface streams of roughly similar size (Table 2.4). The cave streams were within the range of surface streams, but generally had (1) lower lateral movement of organic carbon (however, one of their caves—Cave Spring Cave—had no sinking stream and hence no lateral transport), (2) a higher dependence on DOM, (3) relatively high standing crops of FPOM, and (4) little wood or other CPOM present in the cave stream. Even though both caves had bat colonies numbering in the thousands, bat guano was an unimportant flux compared to DOM, with values less than 1 per cent of the DOM flux. However, bat guano can be an important carbon source for some aquatic cave species (Fenolio *et al.* 2006).

A conceptual model of the role of sinking streams and percolating water is shown in Fig. 2.4. The values for standing crops and fluxes are for Organ Cave, and of course will differ from cave to cave, but the qualitative features should be quite general. In this model, organic carbon enters the subsurface either through the soil or through the sinking streams. In the soil pathway, considerable processing of the organic carbon occurs, through microbial metabolism. It enters epikarst, where it undergoes further microbial processing. From the epikarst it enters into the cave in drips and seeps largely as DOC. DOM from percolating water plays a major role in formation of biofilms. A similar process occurs in interstitial aquifers, but without the presence of epikarst. In interstitial aquifers it is the major source of organic carbon. The movement of water from the surface to the cave stream or aquifer usually takes days to months.

The second source of organic carbon is from sinking streams, which may contain considerable amounts of particulate organic matter as well as DOM.

Table 2.4 Comparison of organic matter budget of Cave Springs Cave, Logan Cave (both in Arkansas, USA) and 19 surface streams worldwide of roughly comparable size.

			19 Surface Streams	
	Cave Springs Cave	Logan Cave	Median	Range
Inputs				
Gross Primary Productivity (g/m²/yr)	0	0	71	0–5400
Leaf-fall (g/m²/yr)	0	0	448	0–736
Lateral Movement (g/m²/yr)	0	56	10	0–1111
DOM Input (g/m²/yr)	4370	3250	95	0–36 037
Bat Guano Deposition (g/m²/yr)	10	3	0	0
Standing Crops				
Fine Benthic Organic Matter (g/m²)	1209	632	333	0–1400
Wood (g/m²)	0	25	2988	0–28 993

All masses are expressed as grams of ash-free dry mass.
Source: modiified from Graening and Brown (2003). Used with permission of Blackwell Publishing.

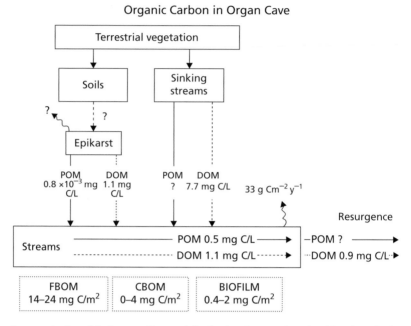

Fig. 2.4 A conceptual model of energy flow and distribution (as organic carbon) in a karst basin with estimates of fluxes and standing crops for Organ Cave, West Virginia, USA. Standing stocks are particulate (POM) and dissolved (DOM) organic matter in the water column and fine (FBOM) and coarse (CBOM) benthic organic matter and microbial films on rocks (epilithon). Solid and dashed arrows represent fluxes. Data are standing stocks of carbon except for respiration flux. Values for FBOM, CBOM, and microbial film are taken from Simon et al. (2003); the whole-stream respiration rate is from Simon and Benfield (2002); and the remaining values are from Simon et al. (2007a). Modified from Simon et al. (2007a). Used with permission of the National Speleological Society (www.caves.org).

After processing in the cave stream, which results in a reduction in the amount of both POM and DOM, organic carbon exits the system at the resurgence (spring).

2.2.4 Wind and gravity

In open air pits, it is common to see piles of leaves and even dead animals at the bottom of the pit (Fig. 2.5). For almost any entrance to a cave, leaves and other dead plant material will be present. The distance it moves will depend on the geometry of the setting, but combined with wind currents, leaves can move tens of metres or more into caves.

In some circumstances, wind-blown organic matter may be nearly the only source of organic carbon. On relatively new lava flows, ones in which there is no vegetative cover, nearly all organic matter present may be the result of wind. Because new lava flows are highly fractured with many small cavities (mesocaverns), it is relatively easy for wind-blown organic matter to enter

Fig. 2.5   Dead raccoon at the base (10 m depth) of Sunnyday Pit, West Virginia, USA. Photo by H.H. Hobbs, with permission. See Plate 7.

subterranean habitats. The nature of the organic matter can vary from place to place, and includes plant fragments and aerially dispersing arthropods (Ashmole and Ashmole 2000). They provide a quantitative dry-weight estimate of 40 mg/m²/day of aerially dispersing arthropods falling on a lava field in Tenerife, Canary Islands.

2.2.5 Active movement of animals

Many animals move in and out of caves on a regular basis. Some are occasional or accidental visitors, including many vertebrates such as raccoons, rattlesnakes, foxes, dogs, and even sheep and cattle. Other animals, such as the extinct European cave bear, bats, and dormice, use caves on a seasonal basis for hibernation. Still others move in and out of caves for food on a much more frequent, sometimes daily basis, including summer roosts of bats and cave-crickets. What all these animals share in common is that they are moving resources from surface habitats into caves. Typically, the resources in caves are in the form of guano, but can also be dead bodies (such as hibernating bats that fall or occasional visitors that become trapped, Fig. 2.5), or in the case of cave-crickets, eggs that are deposited in soft substrates in caves (see Chapter 5). In the case of dead bodies of invertebrates, Novak *et al.* (2013) show it is an unimportant carbon source, based on detailed quantitative measurements in caves in northeast Slovenia. This is a spatial subsidy (Polis and Hurd 1996; Fagan *et al.* 2007) where resources from one system (surface) are transferred to another adjoining system (caves).

The carbon fluxes and standing crops resulting from these activities can be considerable. Graening and Brown (2003) provide a quantitative estimate of the amount of guano produced in Cave Springs Cave in Arkansas by a maternity colony of 3000 grey bats (*Myotis grisescens*)—9000 g/year or 3 g/bat/year. A few caves have summer roosts of bats numbering in the millions, such as Bracken Cave in Texas. The 20 million Mexican free-tailed bats (*Tadarida brasiliensis*) in Bracken Cave are estimated to deposit 50 000 kg of guano annually (Barbour and Davis 1969). In caves with large bat colonies, there is often sufficient guano accumulation for there to be an invertebrate community specialized on this resource (Deharveng and Bedos 2000). Such communities are especially common in tropical caves. At least in southeast Asian caves, a major characteristic of guano communities is their tripartite organization into subwebs. One subweb consists of snakes in the genus *Elaphe* which prey on the bats. Another subweb consists of the giant arthropod community which has one very common primary consumer, a large cricket occurring in high densities, which is preyed upon by four or five predators such as giant centipedes and arachnids. Many of these species retain eyes and pigment and a few leave the cave at night but most are restricted to guano deposits in these caves. The third subweb, the meso- and micro-invertebrate community, includes a diverse assemblage of Diptera, staphylinid beetles, millipedes, cockroaches, moths, and woodlice that are the primary consumers. These species, being much smaller, rarely leave the guano piles. Many of the species retain eyes and pigment, but there are also many species with reduced eyes and pigment.

Although guano piles are the quintessential high energy habitat, their inhabitants are often not included in the classification systems of cave animals—troglobionts, troglophiles, and so on. Some biologists have coined a separate term for cave guano specialists—guanobites (Decu 1981). However they are classified, they are specialists in this habitat and most guano species of southeast Asia are specialized to exploit definite kinds of high-energy resources under special microclimate constraints (Deharveng and Bedos 2000).

When guano is not concentrated but scattered in small amounts, the animals utilizing guano are not guano specialists but rather part of the terrestrial fauna generally adapted to take advantage of almost any source of organic carbon in food-poor habitats in caves.

Bats are not the only source of guano. For example, in Mammoth Cave, Poulson (2012) distinguished five major types of guano, each with its own patterns of spatial and temporal variability, size, and digestibility (Table 2.5). The available guano ranges from the highly unpredictable raccoon guano to highly predictable cave cricket and cave beetle guano. Several of these guano types have distinct, specialized communities.

Table 2.5 Differences in characteristics of guano found in Mammoth Cave, Kentucky, USA.

	Raccoon	Pack rats (*Neotoma* sp.)	Bats (*Myotis* sp.)	Cave crickets (*Hadenoecus* sp.)	Cave beetles (*Neaphaenops* sp.)
Caloric density	High	Moderate	Low	Very low	High
% easily digestible	Moderate	Low	Low	High	High
Fecal pellet size (mg)	13,700	60	20	5	0.1
Spatial heterogeneity	High	High	Low	Low	Moderate
Seasonal heterogeneity	Moderate	Low	High	Low	Low
Number of species specialized to fecal type	Low	Low	High	High	High

Source: adapted from Poulson (2012), with permission of Elsevier Ltd.

A particularly interesting spatial subsidy of carbon is that resulting from the movement of crickets out of the cave for night-time feeding. In many caves there are large cricket populations that feed in surface habitats at night and stay in the cave to avoid predators and temperature extremes, and to lay eggs (Lavoie *et al.* 2007). An individual cricket lays between 150 and 200 eggs per year, and the population of crickets in even a small cave can be more than 5000 (Hobbs and Lawyer 2002). As is discussed in detail in Chapter 5, there are beetles that are specialized cricket egg predators. In addition to egg-laying, cricket guano is also an important food source for many cave species.

2.2.6 Roots

Tree roots are not uncommon in shallow limestone caves and lava tubes. In some cases, most notably in the lava tubes of the Hawaiian Islands and Canary Islands, a specialized terrestrial fauna has evolved to take advantage of this resource. The roots in lava tubes may extend several metres into the cave passage (Fig. 2.6). They are best developed when the water table is too far beneath the surface for roots to reach and relative humidity in cave passages is high. In many lava tubes in Hawaii, tree roots are the primary source of organic carbon. There are no regular cave visitors such as bats and there are rarely active streams. In fact, the lava tubes of Hawaii were thought to be highly depauperate until tree roots were examined (Howarth 1972). The terrestrial fauna of tree roots is quite exotic, including planthoppers and moths (Fig. 2.7). Cave planthoppers that are specialized on tree roots in lava tubes have also been described from the Cape Verde Islands (Hoch *et al.* 1999) and the Canary Islands (Oromí and Martín 1992). On many lava flows, lava tubes may present a more hospitable environment, with more food and less temperature fluctuation, than surface environments. This situation led Howarth (1980) to propose the adaptive shift model of cave colonization (see Chapter 7). Briefly, under this hypothesis, populations actively invade caves because they represent a superior, unexploited habitat compared to surface habitats in lava fields.

Fig. 2.6   Roots of *Metrosideros polymorpha* coming through the ceiling of Lanikai Cave, Hawai'i. Photo by H. Hoch, with permission. See Plate 8.

Aquatic tree roots, or more precisely, tree root mats—compact structures with many finely branching rootlets—also occur in some shallow caves. They are best developed in arid areas with little soil moisture or humidity in the caves themselves, making cave streams one of the only sources of water (Jasinska and Knott 2000). A diverse aquatic fauna is known from aquatic root mats in caves in western Australia. One of the most important large primary consumers is the amphipod *Austrochiltonia subtenuis*. A wide variety of organisms, including amphipods, isopods, leeches, and even fish depend on the energy provided by root mats.

In Chapter 4, we will look at the separate but related question of whether subterranean ecosystems are carbon limited, or whether they are limited by other factors such as phosphorus or nitrogen (see Schneider *et al.* 2010). Energy limitation is also considered to be a major selective agent in adaptation to subterranean environments, and we consider this question in Chapter 6.

2.3 Summary

Although subterranean habitats in general and caves in particular are often held to be extremely energy-poor (oligotrophic) environments, not all are. In fact they range from extremely oligotrophic deep aquifers to extremely

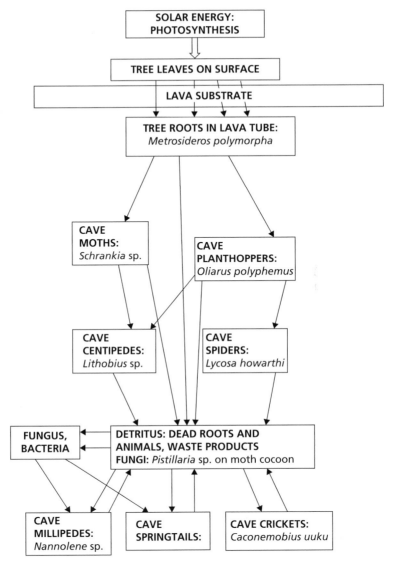

Fig. 2.7   Food web of tree root communities in lava tubes on the island of Hawai'i. Arrows indicate the direction of energy flow. Modified from Stone *et al.* (2012). Used with permission of Elsevier Ltd.

eutrophic caves harbouring immense bat colonies. Compared to surface habitats, subterranean habitats are nutrient-poor, especially because there is no photoautotrophic production and chemoautotrophy appears to be uncommon. On the other hand, these differences are not always pronounced. For example, the quantities of carbon fluxes in cave streams are in the range of those reported from surface streams.

In some subterranean systems, chemoautotrophy is the main source of energy, but more typically subterranean communities depend on allochthonous sources of organic carbon. The major source of carbon in interstitial habitats is DOM from surface waters. The major sources of carbon for cave communities are (1) water percolating from the surface, (2) sinking streams that enter caves, and (3) activities of animals moving in and out of caves. All of these three major sources of organic carbon show considerable variation, both spatially and temporally. Even guano, which might seem to be a very homogeneous source of carbon, varies in quality and quantity.

3 Survey of Subterranean Life

3.1 Introduction

A wide variety of organisms are found in subterranean habitats and they occur there for a variety of reasons, ranging from trivial and temporary to profound and permanent. We first consider some of the temporary residents of caves—roosting bats, hibernating bears, and the like—as well as the residents of cave entrances. We next try to simplify and make sense of what can be best described as a terminological jungle of terms used for ecological and evolutionary classifications of subterranean organisms. Because caves can be visited directly (whereas one cannot take a trip into the interstitial, for example), cave organisms loom large in all of these classifications. Perhaps the main distinction among organisms (although nothing in the area of terminology is universally accepted) is between permanent and temporary residents in subterranean habitats. The obligate, permanent residents of subterranean habitats are from a wide variety of taxonomic groups and we briefly review these. We consider the successes and failures in maintaining and breeding cave animals in the laboratory. Finally, the specialized equipment and techniques that have been developed to collect the denizens of the underground are reviewed.

3.2 Temporary subterranean visitors

Animals are in subterranean habitats for a variety of reasons. Some may be there by chance or accident. Chapman (1993) describes the odd behaviour of *Heleomyza* flies in Otter Hole Cave in Wales. They arrive in the autumn and remain in a semi-torpid state for months, many eventually dying from starvation or pathogenic fungi. Chapman suggests that the flies are victims of seasonally reversing air currents. They enter the cave by following inward-flowing air streams and then mistakenly try to exit by flying into the wind, the problem being that the direction of air flow has reversed. Of course, a variety of animals blunder into caves, perhaps escaping summer heat, and

The Biology of Caves and Other Subterranean Habitats. Second Edition. David C. Culver and Tanja Pipan. Published 2019 by Oxford University Press. © David C. Culver and Tanja Pipan 2019. DOI: 10.1093/oso/9780198820765.001.0001

the list of mammals that do so is quite large, including domesticated animals such as sheep and wild animals such as foxes, wolves, and even elephants in African caves. Some of these blunderers end up as a nutrient not as a resident. Aquatic animals also get swept or washed into a variety of subterranean habitats, where they also end up as sources of nutrients. Chapman (1993) provides examples of fish such as the brown trout, *Salmo trutta*, being washed into caves where they lose pigment and eventually die, in the process looking like 'cave fish' to the unpractised eye.

Some animals occupy caves for part of the day, the best known example being bats that utilize caves in the summer as a day roost. For example, the Brazilian free-tailed bat, *Tadarida brasiliensis*, is present in many caves in the southwestern United States and Mexico in the summer months (Barbour and Davis 1969). It uses caves as shelters and places to avoid predators and summer heat during the day while it sleeps and then exits the cave at night to forage for insects. Many other mammals also use caves as their shelter. The European dormouse, *Glis glis*, utilizes caves in this way, regularly exiting the cave at night to forage (Tvrtković 2012). The Allegheny wood rat, *Neotoma magister*, uses eastern North American caves in a similar way. Invertebrates, including cave crickets in many North American caves (see Chapter 5), also enter and exit caves on a regular, often daily basis (Lavoie *et al.* 2007). In all of these cases, feeding occurs outside the cave.

Many bat species utilize caves in one way or another. These include day roosts, courtship and mating sites, maternity roosts, and hibernacula (Kunz *et al.* 2012). Nearly half of all genera of bats use caves; the main genera are listed in Table 3.1. Why do bats use caves so much? Kunz *et al.* point out several benefits. First and foremost, caves provide a structurally and climatically stable appropriate microclimate. What this is depends on the species and how it utilizes the cave. One important microclimate parameter is high humidity. Since bats have high surface area to volume ratios, they risk extensive water loss, which can be minimized in high humidity environments in caves. Caves also provide protection against adverse weather and protection against predators. Since bats utilize caves primarily for a set of physical conditions, it is not surprising that mines sometimes provide appropriate habitats as well (Barbour and Davis 1969).

Balanced against the benefits of cave use by bats are some costs. Large maternity or hibernating colonies may be susceptible to disease and there may be high 'commuting' costs. Environments in the immediate vicinity of the cave are unlikely to provide all the food resources needed; therefore, bats may need to fly several kilometres to find food (tens of kilometres in the case of the large Brazilian free-tailed bat colonies). There is also predation by snakes and birds of bats emerging from caves and occasional flooding of roosts, but these factors are probably of small quantitative effect.

Many temperate zone bats in the families Vespertilionidae, Rhinolophidae, and Molossidae form large hibernating colonies. Among the largest of these

Table 3.1 Bat genera that frequently occupy caves.

Suborder	Family	Genera found in caves
Megachiroptera	Pteropodidae	*Balionycteris*
		Eonycteris
		Penthetor
		Rousettus
Microchiroptera	Rhinopomatidae	*Rhinopoma*
	Emballonuridae	*Taphozous*
	Craseonycteridae	*Craseonycteris*
	Rhinolophidae	*Rhinolophus*
	Hipposideridae	*Hipposiderus*
	Noctilionidae	*Noctilio*
	Mormoopidae	*Pteronotus*
	Phyllostomidae	*Artibeus*
		Brachyphylla
		Erophylla
		Macrotus
		Phyllonycteris
	Miniopteridae	*Miniopterus*
	Vespertilionidae	*Barbastella*
		Eptesicus
		Myotis
		Nycticeius
		Pipistrellus
		Plecotus
	Molossidae	*Cheiromeles*
		Sauromys
		Tadarida
	Nycteridae	*Nycteris*

Source: From Woloszyn (1998). Used with permission of International Society of Subterranean Biology.

are the hibernacula of the endangered North American grey bat *Myotis grisescens*. Tens of thousands of bats congregate in a few deep, vertical caves with cold air traps (Barbour and Davis 1969). In addition to day roosts and hibernacula, bats also use caves for large maternity roosts, something *M. grisescens* also does. *M. grisescens* is one of the most cave-dependent bats in the world, using caves throughout the year, although different caves in different seasons. It forms large maternity roosts in caves other than the hibernating caves. Bats employ several strategies to minimize energy costs for selection of hibernation sites (Kunz *et al.* 2012). Some seek out an ambient temperature within their thermal neutral zone (they are homeotherms) and others seek out a low temperature where they go into torpor. Other species modify the environment by forming large colonies in chambers with little air flow and raise the ambient temperature with their metabolic activity, and yet others increase temperature locally by dense clustering.

Perhaps the most spectacular case of mammals using caves to hibernate was the extinct European cave bear *Ursus spelaeus* (Kurtén 1968). Many skeletons of the cave bear are preserved in the Slovenian cave, Križna jama and in the Romanian cave, Peştera Urşilor. At least in desert environments, hyenas (*Hyaena hyaena*) enter caves to avoid temperature extremes and to eat large scavenged prey (Kempe *et al.* 2006). Frogs and many invertebrates also overwinter in caves. For example, the moths *Scoliopteryx libatrix* and *Triphosa* species overwinter in caves throughout Europe and North America (Graham 1968; Chapman 1993). Many other arthropods, including Diptera and Opiliones, also overwinter but this phenomenon has been little studied.

A very different kind of seasonal use of subterranean habitats is quite common in some interstitial habitats. Some stonefly species in genera such as *Isocapnia*, *Paraperla*, and *Kathroperla* live in the total darkness of the hyporheic zone of streams and rivers for one or more years before returning to the stream to emerge as terrestrial adults to mate (Stanford and Gaufin 1974, Gibert *et al.* 1994). These Plecoptera spend nearly all their entire life in subterranean waters, but require both subterranean and surface waters. Apparently, the stonefly *Protonemura gevi* spends its entire life cycle in the stream in Cueva del Naciemento del Arroyo de San Blas in Spain (López-Rodrígues and Tierno de Figueroa 2012).

3.3 Residents of cave entrances

Many organisms, not just animals but plants and algae as well, are typically or even exclusively found around cave entrances. These are not organisms of aphotic, true subterranean environments, but rather inhabitants of dimly lit twilight zones. Several species of birds nest near entrances or in shallow aphotic zones in caves. One of the most interesting is the oilbird, *Steatornis caripensis*, found in caves in Trinidad and northwestern South America, where colonies can be as large as 5000 individuals. A fruit-eater, this large bird, with a wingspan of 1 m, nests in caves and cave entrances, and has a major impact on seed dispersal in the surrounding forest (Thomas *et al.* 1993; Thomas 1997). Similar to bats, oilbirds, along with some swiftlets, have the ability to echolocate, navigating through the cave by using clicking noises. It is called the oilbird because of the high oil content of their bodies, especially in the young (Day and Mueller 2005). In tropical caves, swiftlets and swallows are common around cave entrances. One of the world's great vertical caves, with a depth of 512 m, Sotano de las Golondrinas, Mexico, is named for the swallows that are abundant in the vertical shaft. In temperate zone caves, swifts, swallows, phoebes, owls, and even the occasional vulture are found nesting in cave entrances. Presumably the cave entrance provides some protection from predators and perhaps also more favourable environmental conditions.

A variety of other species can be found in the entrance zone. Some are there to escape summer heat and some are there to avoid predators, but some are entrance-zone specialists, often predators of the animals coming into the entrance. The North American cave salamander, *Eurycea lucifuga*, feeds mostly on Diptera (Peck 1974). It is really misnamed since it is more common around entrances than deep in the cave (Camp and Jensen 2007). Invertebrate predators commonly found in cave entrances in Europe and North America are web-building spiders in the genus *Meta*. The sticky webs of the rather large European *M. menardi* and American *M. ovalis* are almost always within the limits of daylight penetration and in the path of flying insects. Ants are often important predators in twilight zones throughout the world (Pape 2016).

In the twilight zone of caves many species of algae, mosses, and ferns occur. The algae and cyanobacteria form an interesting community specialized to low light conditions (Mulec *et al.* 2008). In the most intensive study to date of this community in several Slovenian caves, they found that species such as the cyanobacterium *Chroococcus minutus* and green algae in the genus *Chlorella* increased production of accessory pigments, such as carotenoids, under low photon flux. In situations where speleothems (stalactites and stalagmites) were in light, which occurs in caves with very large entrances, some of the algae promote growth of stalactites and stalagmites (Mulec *et al.* 2007).

Less is known about mosses, lichens, and ferns in caves. It is a common observation that the rock around cave entrances often has a rich moss and fern flora. For example, Dobat (1998) reported that over half of the cave entrances in the Jura Mountains of France had the moss *Thamnobryum alopecurum*. Cave entrances are often known for their rich growth of ferns. Descriptions of tropical caves often mention the profusion of ferns at the entrance, and several caves in the United States and Great Britain are known as Fern Cave. Most of the ferns and mosses around cave entrances are classified as calcicoles, or calcium-loving, and it is unlikely that they are specialized for low light. However, in some situations cave entrances are more than just a slab of wet limestone from the point of view of ferns. The American hart's tongue fern, *Asplenium scolopendrium* var. *americanum* is rather common in Canada, but rare in the United States where it is listed as a threatened species under the US Endangered Species Act. It occasionally occurs in northern Michigan and New York on carbonate rock. The only other known sites are nearly 1000 km to the south around the entrances to three vertical pits in Tennessee and Alabama (Evans 1982; Elliott 2000). It seems very likely that the summertime moist, cool temperatures around the pit allow this northern species to survive in southern locations.

3.4 Ecological and evolutionary classifications

If the biota of caves consisted entirely of the temporary visitors and entrance zone residents described above, it is unlikely that any special categories would have been constructed (although a special name is given to the hyporheic insects—amphibionts (Gibert *et al.* 1994)—that spend most of their life in subterranean waters). But of course much of the biota of caves consists of populations that spend their entire life cycles there.

The real interest in the biota of caves came with the discovery of species that were eyeless, depigmented, and with elongated appendages. The fact that subterranean inhabitants from many disparate groups shared this morphology, a phenomenon we now attribute to convergent evolution (Culver *et al.* 1995; see Chapter 6), led Schiner (1854) to propose an ecological classification based on habitat, which included a category for species that were limited to caves. In addition to being obligate inhabitants of subterranean habitats, most also shared the morphological syndrome of lack of pigment, eyelessness, and elongated appendages. Although this classification system, usually called the Schiner–Racoviță system (Sket 2008), is based on the distribution of animals, it is the morphology that is the obvious attribute of many subterranean animals (but not all as we will see in the following sections). Indeed a major impetus for the inclusion of non-cave subterranean habitats within the purview of speleobiologists and cave biologists was the presence of organisms with this same morphological syndrome in these habitats. Racoviță ([1907] 2006) was perhaps the first to comment on the extension of 'cave' environments, and the appearance of 'cave' animals in interstitial and shallow subterranean habitats gave a unity to the study of the subterranean environment. The category of *troglobiont* (sometimes 'troglobite') is used for terrestrial species (and sometimes for both terrestrial and aquatic species) and the parallel category *stygobiont* (used for aquatic species, sometimes only non-cave subterranean species; Trajano and de Carvalho 2017) include all those species that are only found in subterranean habitats.

Recognizing that stygobionts and troglobionts were ecologically defined, Christiansen (1962) proposed a term to reflect the convergent morphology of obligate subterranean-dwelling animals—*troglomorphy*.[1] This convergent morphology, which is the subject of Chapter 6, is characterized by both losses (such as reduction of eyes, pigment, and cuticular thinning in arthropods), and gains (such as increases in extra-optic sensory structures, appendage elongation, and increased egg volume). The convergence is truly remarkable and makes troglomorphy easy to identify (Fig. 3.1). What is simultaneously interesting and confusing is that not all stygobionts and troglobionts are

[1] One could use the term stygomorphy for aquatic species but the use of a single term, troglomorphy, emphasizes the convergence of both aquatic and terrestrial species.

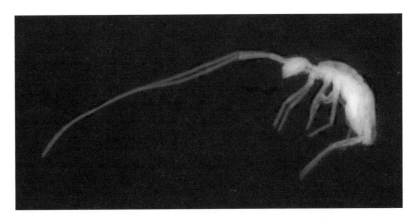

Fig. 3.1   Photograph of an undescribed Collembola (Entomobryoidea) in a cave in Laos, the most troglomorphic Collembola known. In addition to the long antennae, other appendages are lengthened, and it has no eyes or pigment. The species is a giant in the family at 2 mm, double the size of most species in the family. It is also an exception to the rule that troglomorphy is more pronounced in temperate zones. Photo by L. Deharveng, with permission.

necessarily troglomorphic. Species recently isolated in subterranean habitats may not have had time to evolve troglomorphic features or they may be in subterranean environments where selective pressures have not led to troglomorphy (Culver 1982). For example, species living on guano piles in caves are often not troglomorphic. Trajano (2012) and Trajano and de Carvalho (2017) suggest that troglomorphy is really a mosaic of modified and unmodified features, suggesting troglomorphy is not a fixed category. The converse is also not universally true—not all troglomorphic species are stygobionts or troglobionts. For example, there are a few species of the entirely troglomorphic amphipod genus *Niphargus* found in surface streams (Ginet and David 1963), most probably the result of invasion of surface habitats from subterranean habitats.

The concepts of troglobionts and troglomorphy can also be extended from species to individual populations. This has proved to be a very useful extension because many of the best studied cases of adaptation and evolution of subterranean species are ones where there are troglomorphic populations of non-troglobiotic and non-stygobiotic species. These include the Mexican cave fish *Astyanax mexicanus*, the European isopod *Asellus aquaticus*, and the North American amphipod *Gammarus minus*.

The clarity and usefulness of ecological classification systems breaks down when we move beyond stygobionts and troglobionts. There are several other kinds of species that live in subterranean habitats. There are species for which there are both surface-dwelling and subterranean-dwelling populations that are permanently in one habitat or the other. For example the springtail (Collembola) *Pseudosinella violenta* is widespread in caves in

Texas and in surface habitats throughout much of the United States (Christiansen and Culver 1969). These species are called troglophiles, or sometimes eutroglophiles (Sket 2008; Trajano and de Carvalho 2017). Other species, such as stoneflies found in hyporheic habitats, cave-crickets, and cave-roosting bats, are bound to the surface for some biological function, such as feeding or reproduction. These species are usually called trogloxenes or sometimes subtroglophiles (Sket 2008; Trajano and de Carvalho 2017). Species unable to sustain a population in caves but which occasionally occur there are called *accidentals*. Unfortunately, different authors have defined the terms in different ways (Table 3.2), creating a terminological mess. Trajano and de Carvalho (2017) provide a clear guide through this terminological jungle, but we will use these terms sparingly except for troglobiont, stygobiont, and troglomorphy. Table 3.2 is designed to serve as a guide for different authors' terminologies.

In practice, it is often very difficult to determine the ecological status of a species. Typically, very little information is available except its distribution, in particular, whether it occurs outside of subterranean habitats, and its morphology. Species like *Rana iberica* were thought to be either accidentals or occasional visitors, but Rosa and Penado (2013) report that it reproduces in the cave Serra da Estrela, Portugal, making it a troglophile. Many species living in soil, epikarst, and interstitial habitats are eyeless and depigmented, and may be confused with cave dwellers even though they usually have shortened appendages and small size (Culver and Pipan 2014).

3.5 Taxonomic review of obligate subterranean species

Summarizing the subterranean fauna is a formidable task. Relatively few taxonomic generalities are possible for residents of subterranean habitats that hold true on a global basis. For example, there are many troglobiotic species of the hexapod order Collembola in French caves, but few in Slovenian caves. The crustacean order Syncarida is an important constituent of interstitial faunas in Europe, southern Australia, and New Zealand, but not in eastern North America. Most species have highly restricted ranges, often known from only one or two sites, and so new species are being discovered and described at a high rate, making generalizations difficult. Finally, the subterranean fauna is relatively species-rich. On the basis of an educated guess and some 'back of an envelope' calculations, Culver and Holsinger (1992) estimate that there are approximately 50 000 species of stygobionts and troglobionts worldwide, if all species were known and described. The number of described species is much less but still large. Deharveng *et al.* (2009) report 930 stygobiotic species from six European countries (Belgium, France, Italy, Portugal, Slovenia, and Spain) and Zagmajster

Table 3.2 A comparison of definitions of the Schiner–Racoviţă categories (except for troglobionts) by different authors.

	Schiner 1854	Racoviţă 2006 (1907)	Thinès and Tercafs 1972	Holsinger and Culver 1988, also Barr 1968	Sket 2008 (based on Ruffo 1957)	Trajano 2012
Trogloxenes (Schiner's hôtes occasionels = occasional visitors)	Animals found in caves, but also at the surface, everywhere 'when one find those constraints typical of their life style'.	Animals lost or occasional visitors of caves attracted by humidity or food, but that do not live continuously or reproduce in caves.	Organisms that live in the surface, but, due to very precise reasons, they colonize temporarily the subterranean environment.	Species habitually found in caves or similar cool, dark habitats outside caves, but they must return periodically to the surface or least to the entrance zone of a cave for food.	*Subtroglophiles* Need to utilize the surface environment for a least one vital function (e.g., reproduction or feeding).	Source populations in epigean habitats, with individuals using subterranean resources.
Troglophiles	Animals inhabiting regions where daylight still penetrates, which may exceptionally be found at the surface or that only have photophilous forms.	Permanently inhabiting the subterranean domain, but preferably in superficial regions, they frequently reproduce there, but may also be found outside.	Organisms that live in the subterranean environment as well as in the surface.	Species able to complete their life cycles within a cave but may also occur in ecologically suitable habitats outside caves.	*Eutroglophiles* Essentially epigean species able to establish more or less permanent subterranean populations.	Some populations both in epigean and hypogean habitats, with individuals regularly commuting between these habitats, promoting the introgression of genes selected under epigean regimes into subterranean populations (and vice-versa).
Accidentals				Species that wander, fall, or are washed into caves and generally exist there temporarily.	*Trogloxenes*	Organisms introduced into caves by mishap or entering in search of a mild climate; may survive temporarily, but the inability to orient themselves and to find food leads to their eventual demise. Not evolutionary units responding to subterranean selective regimes.

Troglobionts are defined as obligate subterranean dwellers by all authors. From Trajano and de Carvalho (2017).

et al. (2008) report 282 troglobiotic beetle species in two families in the Dinaric karst of Italy, Slovenia, Croatia, Bosnia & Herzegovina, Serbia, Montenegro, and Albania. We take up the geography of subterranean biodiversity in Chapter 9.

After a brief discussion of the taxonomy and functional groups of microbes, we consider the 21 invertebrate orders with at least 50 stygobionts or troglobionts (Table 3.3) and the two vertebrate groups with stygobionts— salamanders and fishes. More detail on systematics and biogeography can be found in the 2294-page, three-volume *Encylopaedia Biospeologica*, edited and largely written by Juberthie and Decu (1994–2001). Also discussed are a few exceptional relic groups with fewer species. For more than a century, subterranean biologists have recognized that some exceptionally old and relic phyletic lineages are found in subterranean habitats. The struggle to explain these relics is one of the recurring themes in subterranean biology (see Chapter 6). Here we give a face to some of these relics.

Table 3.3 Invertebrate orders with more than 50 stygobionts and troglobionts.

Phylum	Class (or major group)	Order, Subclass, or family (idea)
Platyhelminthes	Acentrosomata	Tricladida
Annelida	Clitellata	Lumbriculidae
		Naididae
Mollusca	Gastropoda	Caenogastropoda (subclass)
Arthropoda	Maxillopoda	Cyclopoida
		Harpacticoida
	Ostracoda	Podocopida
	Malacostraca	Bathynellacea
		Amphipoda
		Isopoda
		Decapoda
	Entognatha	Collembola
		Diplura
	Insecta	Coleoptera
		Hemiptera
	Chilopoda	Lithobiomorpha
	Diplopoda	Chordeumatida (=Craspedosomida)
		Julida (=Iulida)
	Chelicerata	Araneae
		Opiliones
		Pseudoscorpiones

Data from various sources.

In some regions, the troglobiotic and stygobiotic fauna is not even the majority component of subterranean communities, and troglophiles and stygophiles predominate. This is generally the case in the tropics and in north temperate regions that were glaciated. As one example, the number of stygophilic, troglophilic, stygoxenic, trogloxenic, and accidental species recorded from California caves is more than ten times that of stygobiotic and troglobiotic species (1224 compared to 114) (Elliott *et al.* 2017). The difficulty is that information on troglophiles and stygophiles is generally not available, and troglophiles and stygophiles are rarely distinguished from trogloxenes, stygoxenes, and accidentals. Hence, we limit coverage to stygobionts and troglobionts.

3.5.1 Microbes

Bacteria, Archaea, Fungi, and Protista are nearly ubiquitous in subterranean habitats. They range from common surface-dwelling bacteria to some of the most metabolically exotic organisms known. They play a critical role in (1) the processing and breakdown of organic matter (Barton 2015), (2) the production of organic matter (chemoautotrophy, see Chapters 2 and 4), and (3) dissolution of rock and deposition of minerals (Engel *et al.* 2004; Paterson and Engel 2015), nitrogen fixation (Barton 2015), and others.

The early study of microbes in caves was largely limited to culturing microorganisms collected in well-oxygenated subterranean environments. These microbes are for the most part identical to surface strains. One of the most visible groups of bacteria seen in caves are colonies of actinomycete bacteria, evident as reflective white dots on moist rock (Northup and Lavoie 2004). Although Caumartin (1963) called these bacteria from well-oxygenated environments contaminants (he was looking for chemoautotrophs), in fact they are natural and critical parts of nearly all subterranean habitats. Their critical role is in the processing and breakdown of organic matter. In a study of microbes in Old Mill Cave, Virginia, USA, and the surrounding forest floor, Dickson and Kirk (1976) found that the density of microbes in the caves was less than the surrounding forest soils, not surprising given that organic carbon levels were lower as well. Caumartin (1963) suggested that, as organic matter is broken down, chemoautotrophs would come to predominate. This does not seem to be the case, largely because there are continuing inputs of organic carbon. Where chemoautotrophs have been found is in redox environments (ones where both oxidation and reduction are occurring, usually under reduced oxygen conditions) in a few caves (Table 2.2) and deep groundwater habitats. Chemoautotrophic processes identified in these environments include iron oxidation, sulfur oxidation, iron reduction, ammonia oxidation, methanogenesis, and methanotrophy. Among chemoautotrophic Bacteria and Archaea, carbon fixation pathways are distinctly related to the phylogenetic position of the respective organisms. Engel (2012a,b) and

Northup and Lavoie (2001) provide details on metabolic functions and phylogenetic position of the various subterranean chemoautotrophic microbes.

Overall, the microbial community is highly diverse, much more so than the multicellular organisms, and its metabolic functions are only partly understood. Furthermore, the overall composition of microbial communities is not just a random sample of nearby surface bacterial communities (Fig. 3.2), supporting the often cited dictum, attributed to Baas Becking, that 'all microorganisms are everywhere, the environment selects.'

3.5.2 Invertebrates

Planarians are known from a variety of aquatic subterranean habitats in Europe, North Africa, and North America, and can be quite common in cave streams and pools. They are characterized by loss of eyespots, pigment, lower metabolic rate relative to surface-dwelling relatives, and strangely, an increase in chromosome number (Gourbault 1994). They feed on living or dead animals. Planarians engulf food particles, extruding a long and muscular pharynx through their mouth, and sucking up the food into their gastrovascular cavity. More than 35% of the 400 described triclad species are stygobionts (Dumnicka 2012). They can reproduce both sexually and asexually. A related order, the Temnocephala, consists of external parasites of Crustacea, feeding mainly on secretions and body fluids of the host. About 15 stygobiotic species are known from Europe but they have not been looked for elsewhere (Matjašič 1994).

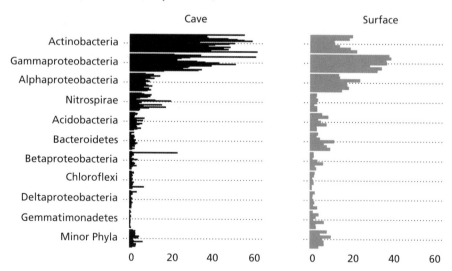

Fig. 3.2   Staggered bar chart of relative abundance of microbial OTUs in and outside of lava tubes in Lava Tube National Monument, California, USA. Major and minor phyla by all cave samples and all surface samples. Cave samples have a greater relative abundance of Actinobacteria and Nitrospirae. From Lavoie *et al.* (2017).

Oligochaetes in the orders Lumbriculida and Tubificida are also common in subterranean waters throughout the world, and are known from all continents except Antarctica (Dumnicka 2012). For example, they were found in over 60% of all subterranean sites in a recent comprehensive sampling of both interstitial and cave habitats in the Jura Mountains of France (Creuzé des Châtelliers et al. 2009). In studies of the fauna of epikarst drips in Slovenian caves, oligochaetes were the third most abundant group in drips, exceeded only by copepods and nematodes in a group of six caves (Pipan 2005) and the most common in a more isolated cave (Pipan et al. 2008). They are little studied outside of Europe, and pose special problems for taxonomists interested in subterranean species. Special collection and preservation techniques are usually required, and so they are often not included in faunal studies. Species richness of stygobiotic species rivals that of surface-dwelling species in areas where both have been studied.

They show little morphological adaptation for subterranean life. Nearly all oligochaete species, whether in surface or subterranean waters, have little in the way of body or visual pigments and so there are no obvious troglomorphic features. As a group they are pre-adapted to subterranean life since they are not visually oriented, typically can endure low oxygen levels often found in interstitial waters, and feed on a variety of microorganisms (Creuzé des Châtelliers et al. 2009). Some species also have the ability to form cysts to allow them to survive periods of desiccation that may occur in epikarst and cave pools. Indeed the distinction between surface and subterranean habitats is not clear-cut for groups like Lumbriculida and Tubificida. Few, if any, species occur in well-lit habitats, and most are found in a similar kind of physical environment of gravels and finer sediments. One family, the Parvidrilidae, is exclusively stygobiotic, with species from North America and Europe (Martínez-Ansemil et al. 2012).

A variety of snails live in both terrestrial and aquatic subterranean habitats, but it is only in the family Hydrobiidae, that relatively large numbers of species are found—more than 350 (Bole and Velkovrh 1986). Morphological characteristics of stygobiotic hydrobiids and other snails include a thin, often translucent shell that is usually white, a depigmented body, and reduced, depigmented eyes. They are known from a variety of aquatic subterranean habitats in both north and south temperate regions, but there are no undoubted cases of stygobionts occurring in the tropics. The most obvious feature of hydrobiid snails is their miniaturization—they are typically the size of a large sand grain, 2 mm or so. Features associated with the miniaturization of subterranean snails include complex coiling of the intestine, loss or reduction of gills, simplification of gonadal morphology, and loss of sperm sacs. Although these characteristics are an interesting case of convergent evolution (see Chapter 6), they also pose a difficult taxonomic challenge because of the simplifications of body structure. Much of the taxonomy is based on shell morphology, but mtDNA sequencing of the COI gene in the terrestrial Carychiidae clarified the taxonomy of the family, which has two genera with troglobionts—*Carychium* in North America and *Zospeum* in Europe (Weigand *et al.* 2011). Little information is available about ecology of subterranean hydrobiids but it is likely that they are detritus feeders (Culver 2012a). The global hotspot of hydrobiid diversity is the Dinaric Mountains of central Europe. In the Slovenian part of the Dinaric Mountains, there are 37 species of aquatic (mostly hydrobiid) and 11 species of terrestrial gastropods. In contrast, in all North American mountains as a whole there are only 23 species of aquatic and 5 species of terrestrial subterranean snails known.

There is only one undoubted species of a stygobiotic clam, but it is an exceptionally interesting and well-studied one. The Dinaric cave clam, *Congeria kusceri*, is a relic species and 'living fossil', the last survivor of a lineage that flourished in late Miocene Seas about 5 million years ago (Morton *et al.* 1998). The genus *Congeria* was common in the western Balkans, Hungary, and Romania during that time. Its near-extinction occurred when the ancestral Mediterranean Sea dried up (the Messinian Salinity Crisis). *C. kusceri* escaped extinction by colonizing the subterranean waters that exited into the ancestral Mediterranean. Its current distribution includes caves in Croatia and Bosnia & Herzegovina.

The crustacean class Remipedia has only 24 living species, all stygobionts, and it hardly qualifies as numerically important. What makes it noteworthy are its habitat, its morphology, and its distribution. Almost everything about remipedes is unusual. First discovered by Yager (1981), superficially a remipede looks similar to a polychaete worm (Fig. 3.3). It has a long (9–45 mm) body and uses its multiple paddle-like appendages to swim. Many aspects of the appendage morphology, which is less differentiated compared with other crustaceans, suggest that it is a very old group (Yager 1994). Remipedes also have a poison gland that indicates a predaceous habit. They are found in anchialine caves, marine caves with a freshwater lens (see Chapter 1), and are typically found just below the lens in haline, poorly oxygenated water, swimming into the freshwater portion for prey—a highly specialized habitat and style of feeding for what is apparently a very primitive group. While they are only found in anchialine caves, remipedes have a wide geographical range, occurring in the Yucatan Peninsula of Mexico, Cuba, Bahamas, Turks and Caicos, Canary Islands, and Australia (Hobbs 2012). This highly disjunct distribution suggests either that it is a very ancient group, perhaps predating the break-up of the northern and southern supercontinents about 150 million years ago, or that present-day populations may be connected by hypothetical deep conduits in the ocean bottom. All in all the distribution is enigmatic. Its phylogenetic position is also in doubt (Fanenbruck *et al.* 2004). Some morphological and molecular sequence studies support a position basal to the insects and their allies and the Malacostraca Crustacea (crabs, shrimp, etc.); other studies support a basal position relative to all Crustacea; and yet other studies support an intermediate phylogenetic position but basal to groups

Fig. 3.3   Photo of the remipede *Lasionectes entrichoma* showing male and female reproductive systems, ventral view. Photo by D. Williams and J. Yager, with permission. See Plate 9.

such as Copepoda. Sequencing of nuclear and mitochondrial DNA resulted in a complete realignment of families (Hoenemann *et al.* 2013).

Copepoda, comprising a subclass of the crustacean class Maxillopoda, are a major component of the fauna of nearly every freshwater habitat, both surface and subterranean. Two copepod orders (Cyclopoida and Harpacticoida) are the most important copepods in groundwater habitats (Fig. 3.4). These tiny crustaceans, ranging in size from 0.3 to 2.0 mm, have been found wherever they have been looked for in all subterranean habitats on all continents (Rouch 1994). In some habitats, especially the hyporheic and epikarst, they are dominant both numerically and in terms of species richness. Rouch (1991) found more than 15 harpacticoid copepod species in a 10 m reach of small stream in the Pyrenees, and Pipan and Culver (2007a) report 15 copepod species from epikarst drips of a single cave system—Postojna–Planina Cave System in Slovenia. Globally, there are 330 stygobiotic species of Cyclopoida and 640 stygobiotic species of Harpacticoida, approximately one-third of the total freshwater copepod fauna (Galassi *et al.* 2009). In Europe, where subterranean copepods are best studied, approximately half of the freshwater species of cyclopoids and harpacticoids are stygobionts. Morphological modifications for subterranean life include reduced or absent eyes and pigment and larger eggs (Rouch 1968). For interstitial species, miniaturization and reduction of segmentation, and even number of appendages, is a common theme. Presumably, these morphological changes aid in the ability of copepods to move and wriggle in the interstices of the habitat. Similar to oligochaetes, copepods, particularly harpacticoids with their worm-like body structure (Fig. 3.4), are pre-adapted to interstitial life. Harpacticoids for the most part are grazers of biofilms, while cyclopoids are usually predators, often of harpacticoids.

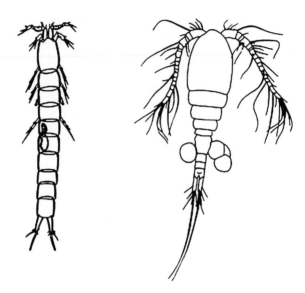

Fig. 3.4   Diagrams of representatives of the two main orders of subterranean copepods. Left: Harpacticoida: *Parastenocaris dianae*; Right: Cyclopoida: *Kieferiella delamarei*. Modified from Rouch (1994). Used with permission of the International Society of Subterranean Biology.

Harpacticoids tend to be more common in interstitial habitats while cyclopoids tend to be more common in caves (Rouch 1994), perhaps because harpacticoids tend to be smaller and may be better able to move about in the spaces between sands and gravels. The ratio of the two orders varies not only between interstitial waters and cave waters, but also geographically and among different kinds of interstitial waters and different kinds of cave waters. For example, Pipan and Brancelj (2001) found that harpacticoids were more common in samples taken directly from epikarst as opposed to samples obtained from drip pools, a more open habitat.

Ostracods are widespread in marine and freshwater habitats throughout the world, including subterranean habitats. Ostracods are bivalved, with a shell that completely engulfs the body, and are typically less than 1 mm long. Because of the shell, ostracods have a rich fossil history, dating back to the Ordovician, 450 million years ago. Even under a magnifying glass, they look like seeds with a few appendages sometimes sticking out. Over 300 species, mostly in the order Podocopida, are stygobionts. They are more common in interstitial habitats than in caves, but they are also common in anchialine caves (Martens 2004). Stygobiotic species are typically eyeless and pigmentless, with reduced setae on the limbs, and often reduced in size. The overall morphological differences between stygobiotic and other species of ostracods are large enough that Danielopol (1981) was able to use the ostracod fauna to determine whether some wells in Greece were contaminated by surface water. Many ostracod groups have relict distributions. For example, the genus *Namibcypris* is known only from karstic springs in Namibia (Martens 2004). This has led some specialists such as Martens (2004) to suggest an ancient age, up to 150 million years, for many subterranean species, while others, such as Danielopol *et al.* (1994) suggest that they are much more recent, actively invading subterranean habitats, rather than being stranded in them. The discussion of dispersal and colonization is elaborated in Chapter 7.

The crustacean order Bathynellacea is exclusively stygobiotic, and almost all of the approximately 200 described species are found in the interstitial, only occasionally being found in caves or superficial subterranean habitats (Coineau 1998). They are eyeless and the body is thin and cylindrical, ranging in length from 0.4 to 3.5 mm, with a very distinctive appearance (Fig. 3.5). All adult bathynellans are paedomorphic (retaining larval traits in the adult), resulting in reduced appendage complexity and miniaturization. In spite of their size, they are slow-growing and long-lived—up to 2.5 years (Coineau and Camacho 2004). They lay a single egg with abundant yolk. When hatched, the young resemble the adults in structure. All in all they have a life history typical of subterranean species living in energy-poor environments (see Chapter 6). The group has an old evolutionary history, with members of the superorder Syncarida, of which Bathynellacea are the major living order, already well established by the Carboniferous, 300 million years ago. The present-day distribution of species also indicates an ancient history. The Parabathynellidae show a Tethyan and Gondwanaland distribution (Fig. 3.6). The Tethys Sea formed in the late Cretaceous about 100 million years ago, and separated the ancestral northern (Laurasia) and southern (Gondwanaland) supercontinents. The remnants of the Tethys Sea formed the present-day Mediterranean. Some parabathynellid genera are found only on the southern continents, the supercontinent Gondwanaland, which lasted until the late Jurassic, 180 million years ago. The distribution of Parabathynellidae can most easily be explained if the group originated when the southern continents were together. The distribution also suggests a marine origin for the group. Other crustacean groups share features of the distribution of Parabathynellidae. For example, the order Spelaeogriphacea has only four species, two of which are known from a single site, and all sites are caves or wells in karst aquifers. Its distribution is in southern Africa, Brazil, and northwestern Australia (Hobbs 2012).

Crustaceans in the order Amphipoda are, in most aquatic subterranean habitats in many regions, the predominant macroscopic invertebrate. Amphipods are also common in surface waters, both freshwater and marine. The morphology varies but superficially many amphipods resemble small shrimps. More than 750 species in 159 genera in 29 families are stygobionts (Hobbs 2012). There are even a few terrestrial troglobiotic amphipods from lava tubes in Hawaii (Holsinger 1994, 2004). Amphipods have colonized

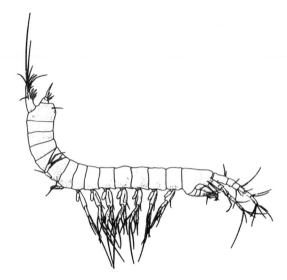

Fig. 3.5   Lateral view of an adult male (length 0.91 mm) of the bathynellid *Agnatobathynella ecclesi*. From Coineau (1998). Used with permission of the International Society of Subterranean Biology.

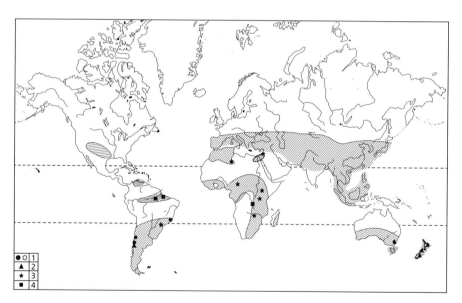

Fig. 3.6   Distribution of Parabathynellidae. Circles represent *Parvulobathynella* and *Acanthobathynella*; triangles represent *Chilibathynella* and *Atopobathynella*; stars are *Cteniobathynella*; and squares represent *Thermobathynella*. Symbols refer to different genera. From Coineau (1998). Used with permission of the International Society of Subterranean Biology.

subterranean waters numerous times and places. In Europe, 140 of the 297 species of amphipods are stygobionts (Sket 1999). In Slovenia, a hotspot of stygobiotic richness, 40 of 52 freshwater species are stygobionts. There are five families of amphipods with more than 50 species of stygobionts—Bogidiellidae, Crangonyctidae, Hadziidae, Melitidae, and Niphargidae (Hobbs 2012), and we briefly consider each.

All of the 110 bogidiellid species from 23 genera are stygobionts. They are found in aquatic subterranean habitats with a nearly worldwide distribution except for continental Africa south of the equator and continental Australia. The family is most probably of marine origin. The crangonyctids have a Holarctic distribution, with the majority of species found in North America. A remarkable pair of species, one in the genus *Crangonyx,* and the other in a unique family closely related to *Crangonyx* (Kornobis and Pálsson 2013), were found in Iceland, where they survived the Pleistocene under the ice sheet (Kristjánsson and Svavarsson 2007). Most of the species in the family are subterranean but they are also common in streams and lakes. Stygobiotic species have no eyes, reduced pigment, and elongated appendages. The largest genus, *Stygobromus*, has nearly 140 described species (Holsinger, 2009). They occur in a variety of subterranean habitats, with species nearly equally divided between interstitial, cave, and superficial subterranean habitats (Fig. 3.7). The Hadziidae, with 78 species in 26 genera, are exclusively subterranean. They are found in caves in Australia, Europe, Mexico, and many islands, including Bahamas, Caribbean, Fiji, and Hawaii (Hobbs 2012). This distribution indicates multiple origins from marine ancestors. The Melitidae, another exclusively stygobiotic family, is known from 54 species in 23 genera. Occurring in a wide variety of subterranean habitats, it is especially common in anchialine caves in islands throughout the world, including the Balearic, Canary, Caribbean, Galapagos, Hawaiian, Philippine, and other Pacific Islands. The final amphipod family with more than 50 species is the Niphargidae with 8 genera. The genus *Niphargus* has more species than any other stygobiotic or troglobiotic genus, with more than 300 species and subspecies, with many more to be described (Fišer *et al.* 2006; Fišer 2012). Niphargids are common throughout Europe and Asia Minor. Although predominately subterranean, a few species have invaded surface habitats, retaining their troglomorphic features (Ginet and David 1963), although some of these habitats may actually be subterranean habitats (see Chapter 1). McInerney *et al.* (2014), based on molecular evidence, claim that the ancestral *Niphargus* was in Great Britain tens of millions of years ago, but this is disputed by Esmaeili-Rineh *et al.* (2015), based on additional molecular evidence from Middle Eastern species. A final amphipod example comes from yet another family—the Ingolfiellidae. While not especially speciose, this exclusively subterranean family is found in both caves and interstitial habitats, and this makes possible comparisons of adaptive morphologies in the two habitats. While all species are troglomorphic, there are important

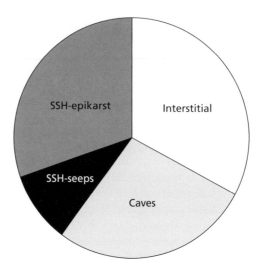

Fig. 3.7   Subterranean habitats of species of the subterranean amphipod genus *Stygobromus* in North America. Note the relatively high frequency of occurrence in shallow subterranean habitats (SSHs).

differences between troglomorphs in caves and troglomorphs in interstitial habitats (Fig. 3.8). Coineau (2000) showed that species from interstitial habitats were much smaller (more than 10 times or so) and had relatively shorter appendages.

Subterranean isopods in many ways parallel the situation with amphipods. A general morphological difference is that isopods are dorsoventrally flattened while amphipods are laterally flattened. Similar to amphipods, a large number of orders and families of isopods have stygobiotic and troglobiotic representatives. They occur in all subterranean habitats throughout the world. Nearly 1000 species are described from 35 families (Coineau *et al.* 1994, Hobbs 2012). Of the 168 described freshwater isopod species in Europe, 105 are stygobionts (Sket 1999). There are five families of aquatic isopods—Cirolanidae, Sphaeromatidae, Asellidae, Stenasellidae, and Microcerberidae—and one family of terrestrial isopods—Trichoniscidae—that have more than 50 species each.

Cirolanids (93 species in total) (Hobbs 2012) are found both in cave and interstitial habitats, mostly in the Caribbean, Mediterranean, and Africa. Many cirolanid species live in marine waters, and the subterranean groups probably had a direct origin from marine waters rather than surface freshwater habitats (see Chapter 7). An enigmatic species is *Antrolana lira*, found in caves in the Shenandoah Valley of Virginia and West Virginia, USA, hundreds of kilometres from any marine shore in the past 100 million years (Holsinger *et al.* 1994; Hutchins *et al.* 2010).

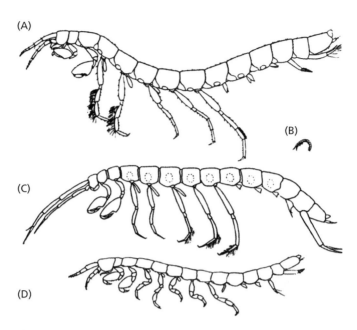

Fig. 3.8   Diagrams of stygobiotic ingolfiellid amphipods. (A) *Trogloleleupia leleupi* (12–20 mm) from a cave; (B) *Ingolfiella* sp. (<1 mm) from interstitial habitats, shown at same scale as (A); (C) *Trogloleleupia opisthodorus* (24–28 mm) from a cave; and (D) *Ingolfiella petkovskii* (1 mm) from an interstitial habitat. From Coineau (2000). Used with permission of Elsevier Ltd.

Asellids are exceptionally diverse, with more than 260 species. They have a mostly northern distribution (Holarctic) in a variety of subterranean habitats. Many species occur in surface freshwater habitats, and the subterranean species are probably derived from them. The stenasellids, with 71 species, are in some ways the tropical equivalent of the temperate asellids. The Microcerberidae, with 63 species, are exclusively interstitial, and are all miniaturized. The family Sphaeromatidae, also of marine origin, has 51 stygobiotic species in caves and interstitial habitats in southern Europe.

The lone terrestrial isopod family with more than 50 species is the Trichoniscidae, with more than 200 species in 54 genera. It is found both in caves and in a shallow subterranean habitat—the milieu souterrain superficiel (MSS, see Chapter 1). Taiti (2004) highlights the differences between cave and MSS species—species in the MSS 'creep' on short appendages and cave species 'run' on long appendages. Most of the species are found in Mediterranean Europe and in North America. A few trichoniscids are amphibious. *Titanethes albus*, a common inhabitant along the banks of streams in Planinska jama in Slovenia, appears to move with impunity in and out of the stream.

The final order of crustaceans that has more than 50 stygobiotic species is the Decapoda—crabs, shrimps, and crayfishes. These large crustaceans are usually the largest invertebrate in caves, and often the largest animal since fish and salamanders are by no means universally present. Typically they are also at the top of the food chain, and as such, are most likely to experience the effects of nutrient scarcity. Two of the groups, shrimps and crabs, are found in caves with relatively high-energy fluxes—tropical and anchialine caves. There are 111 species of carideans (shrimps) found in caves, usually anchialine caves (Hobbs 2012). They are usually found in the tropics (Holthuis 1986) but *Troglocaris* is rather common in caves in the Balkans. There are 41 species of crabs (Anomura) known from tropical caves, especially in the southwestern Pacific. Crayfishes have a quite different distribution pattern—they are neither tropical nor are they in caves with especially high-energy fluxes. There are 41 species of stygobiotic crayfishes, all of them from Cuba, Mexico, and the eastern United States (Hobbs 2012). Being highly omnivorous, they are less predaceous than the crabs and shrimps. The life history of stygobiotic crayfishes is highly adapted to low-energy fluxes. They live for more than 20 years (but not as long as 100 years; Culver 1982) and produce relatively few large eggs (Venarsky *et al.* 2012a).

Collembola, or springtails, are numerically the most abundant arthropod in most terrestrial habitats (especially leaf litter and the soil), including subterranean habitats (Thibaud and Deharveng 1994). They are minute (1–2 mm) six-legged arthropods characterized by a furcula or 'springtail'. They are common in caves and the MSS throughout the world, often found near organic matter or standing water. The Collembola fauna of tropical caves in the southwest Pacific, although largely undescribed, is very diverse (Deharveng and Bedos 2000, 2012). Deharveng and Thibaud (1989) tallied 240 troglobionts among the 1500 species described from Europe. They also reported nearly as many non-troglobiotic Collembola species in caves as there were troglobionts, as did Christiansen and Culver (1987) for the US cave fauna. It is a frequent observation in many caves and MSS habitats that the abundance of non-troglobionts rivals that of troglobionts. In tropical caves many troglobiotic Collembola are not troglomorphic, and Deharveng and Bedos (2012) report the frequency of troglomorphy among troglobiotic Collembola decreasing with increasing mean annual temperature and decreasing latitude. Tropical caves also harbour a number of Collembola that are specialized on bat guano—they are guano specialists rather than cave specialists. The trophic position of Collembola is near the base of the food web, feeding on particulate organic matter, bacteria, and microfungi, both on terrestrial substrates and on the surface of standing water. Many European and US troglobionts are from genera that are also widespread in the leaf litter and soil, such as *Pseudosinella*. A few are from genera known only from caves, such as *Bessoniella*.

In one of the most thorough studies to date of adaptation (see Chapter 6) in subterranean organisms, Christiansen (1961, 1965) demonstrated that morphological and behavioural changes occurred that allowed troglomorphic

Pseudosinella and *Sinella* to walk across water surfaces, where they feed on organic matter on the surface (Fig. 3.9). It is worth noting that this adaptation is not one made in response to low food per se but rather for the ability to move across water and wet surfaces both to avoid being trapped and to find food. Troglomorphic Collembola generally have lower fecundity, longer development, and higher resistance to starvation, all traits associated with an energy-poor environment.

Diplurans, similar to Collembola, are wingless arthropods with six legs. There are about 100 species of troglobionts described (Bareth and Pages 1994), mostly from Europe and primarily in the family Campodeidae. There are approximately 300 described species overall, so the frequency of troglobiotic species in the order is very high, matched only by some minor orders of Arachnida (see later in this section). In spite of extensive collections from North American caves, nearly all species are undescribed. In surface habitats they are common in leaf litter and soil, although often overlooked because of their small size (<5 mm) and fragility. Even surface-dwelling campodeids look like subterranean animals—no pigment, no eyes, and relatively elongated appendages, and their surface habitats are in fact subterranean-like, with dim light and high humidity. Troglobionts have larger body size and even longer appendages and antennae—the antennae often exceed the body length. Their ecological role in subterranean habitats is not understood. On the basis of surface-dwelling species, campodeids are likely to be omnivores, and may be important predators of mites and Collembola.

There are more troglobiotic species of beetles (Coleoptera) than any other order of invertebrates. There are more than 2100 described species from 15

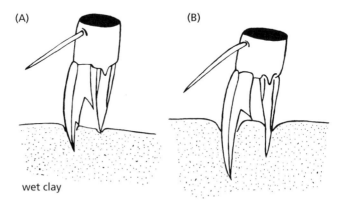

(A)   (B)

wet clay

Fig. 3.9   Convergent changes in structure of collembolan claw as a result of adaptation of walking on wet substrates. (A) Claw of non-troglomorphic species; (B) claw of troglomorphic species. Modified from Culver (1982). Reprinted by permission of the publisher from *Cave Life: Evolution and Ecology* by David C. Culver, p 33. Cambridge, Mass.: Harvard University Press. Copyright ©1982 by the President and Fellows of Harvard College.

families (Decu and Juberthie 2004), and they have diverse ecological roles, including predators (Carabidae and Pselaphidae), detritus feeders (Cholevidae), and guano feeders (Tenebrionidae). Two families—Carbadidae and Cholevidae—account for more than 80% of the total. Several genera have more than 100 species. Among the Carabidae, they include *Trechiama* from Japan, *Duvalius* from the European Mediterranean Coast and the Caucasus, and *Pseudanophthalmus* from the eastern United States; among the Cholevidae, they include *Speonomus* and *Bathysciola*, both from Mediterranean Europe (Decu and Juberthie 1998).

These and other species-rich genera are likely examples of speciation (see Chapter 7), where a series of populations, sharing a common surface ancestor, are independently isolated in caves (and the MSS), and become reproductively isolated from each other. The result is a series of very similar species each with a very narrow geographical distribution, often restricted to one cave. The distribution of *Pseudanophthalmus* beetles in caves in Virginia, USA, corresponds to this model (Holsinger and Culver 1988), often with one species of *Pseudanophthalmus* per cave (see Chapter 7). However, speciation without subsurface dispersal and high endemism cannot explain all the patterns, especially diversity hotspots where several species of closely related beetles co-occur. Subsurface dispersal followed by additional speciation is needed to explain hotspots of endemism and species richness (Christman *et al.* 2005) (see Chapter 7).

Although troglobiotic Coleoptera are known from nearly all cave regions, almost 70% of the species are known from Europe and the Mediterranean region. While, as with other groups, subterranean habitats in Europe are better studied, this concentration is likely real, at least with respect to other north temperate regions. For example, the cave beetle fauna of North America has been extensively studied by Barr (2004), but diversity, especially at the family and generic level, is much lower than Europe (Deharveng *et al.* 2012). Only Carabidae show extensive speciation in North American caves. Adaptation to subterranean life has been extensively studied in cave beetles, especially in the Cholevidae (Deleurance-Glaçon 1963). There are different degrees of troglomorphy in these lineages, which are often interpreted as levels of adaptation and perhaps different ages of isolation in caves. An equally plausible explanation is that the species are more or less equally adapted to different conditions in the cave (see Chapter 6).

The Homoptera, a large order of phytophagous insects, would appear to be unlikely troglobionts. However, nearly 60 species of troglobiotic and troglomorphic homopterans are known (Hoch 1994). They are all specialized feeders on the root exudates of plants that penetrate the ceiling of caves, especially lava tubes (see Fig. 2.7). More than 10 species are known from lava tubes in Hawaii and from lava tubes in the Canary Islands, with a scattering of species found in the rest of the world (Australia, New Zealand, Galapagos,

Mallorca, Samoa, and southern France), usually, but not always, in lava tubes. In Hawaii, surface ancestors are still living on the surface of lava flows, leading Howarth (1987) to formulate his adaptive shift hypothesis. Howarth argues that extinction of surface relatives is not necessary for the evolution of troglomorphy and that species such as *Oliarus* in lava tubes actively invaded the habitat (see Chapter 7). Hoch (2000) showed that the planthoppers in lava tubes communicate with each other via low-frequency signals which are especially important for mate recognition. They also appear to have some morphological modifications to the claws, which enables them to move more easily over the hard surfaces in the lava tubes, an adaptation analogous to that demonstrated by Christiansen for Collembola.

Chilopoda (centipedes) are forest litter-dwelling predators that have colonized caves and the MSS, primarily in temperate Europe (Culver and Shear 2012). Most of the troglobionts are in the order Lithobiomorpha (Table 3.3). Some troglobiotic species, such as *Lithobius matulicii* from Bosnia & Herzegovina, have more than 100 segments in their antennae (Fig. 3.10). Caves in the Pyrenees Mountains of Spain and France are a hotspot of centipede species richness, with 20 species. The rest of Europe has another 20 species. In contrast, only one centipede species is known from North American caves.

Diplopoda (millipedes) are forest litter-dwelling detritivores that have colonized caves in both temperate and tropical areas (Harvey *et al.* 2000). They are adapted for cutting and chewing hard material, such as wood or dead leaves. Millipedes are often numerically dominant in temperate and

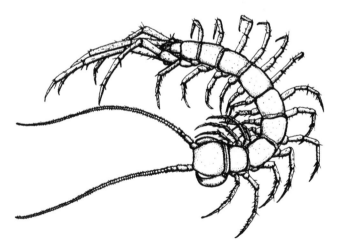

Fig. 3.10   Illustration of the cave centipede *Lithobius matulicii* from the cave Vjetrenica in Bosnia & Herzegovina, one of the most diverse caves in the world. Drawing by S. Polak, with permission.

tropical caves, frequently found in large numbers around food sources such as dead animals. It is not unusual to see dozens of millipedes, both troglobiotic and non-troglobiotic, in a small area in a cave. In North America, the Chordeumatida are dominant, with nearly 50 troglobiotic species. In Europe, both the orders Chordeumatida and Julida are common in caves and the MSS. Several species of Julida have mouth parts modified to comb bacteria from surfaces and to filter particulate organic matter from water. As a result, they are amphibious, and often collected in pools and streams. Troglobiotic millipedes tend to have elongated legs and antennae. In some species the organic fractions of the cuticle may be greatly reduced, resulting in a very brittle cuticle (Harvey *et al.* 2000). Compared with surface species, troglobiotic millipedes tend to have fewer body segments and shorter bodies.

Arachnids (spiders and related groups) are often common in caves throughout the world, but especially in arid regions, where arachnids predominate in the surface fauna as well. In many caves they are, together with carabid beetles, the major predators. Although not rich in species, the orders Amblypygi, Ricinulei, Palpigradi, and Schizomida deserve special mention. Between 20% and 40% of all known species in these orders are troglobionts, a percentage matched only by the Diplura (see earlier in this section). All of these orders are primarily tropical and subtropical in distribution.

There are more than 1000 species of troglobiotic spiders (out of a total of 50 000) known from 30 of the 60 families of Araneae (Reddell 2012). At least an equal number of species are found in caves, but not limited to them. Spiders are found in caves and the MSS throughout the world, but tropical species are less frequently troglomorphic than temperate zone species. Troglomorphic spiders show the typical morphological changes associated with subterranean life—reduced eyes and pigment and lengthening of appendages (Fig. 3.11). Many of the species, such as those in the family

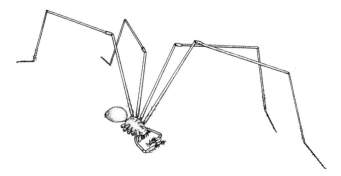

Fig. 3.11   Diagram of the highly troglomorphic spider *Ochyrocera peruana* from a Peruvian cave. From Ribera and Juberthie (1994). Used with permission of the International Society of Subterranean Biology.

Linyphiidae are tiny, with bodies less than 2 mm. Many species are free-ranging predators but some build simple webs. Most troglobiotic spiders belong to genera or at least families that have surface-dwelling species in the same geographical area, suggesting that most cave spiders are not from phylogenetically old groups. An exception is *Telema tenella*, the only species in an otherwise tropical family found in the temperate zone (Juberthie 1985).

There are 80 troglobiotic Opiliones (harvestmen) known worldwide out of a total of more than 5500 species (Reddell 2012), scattered in caves and MSS habitats throughout the world. They are predators, especially of Collembola. They are important components of cave communities, not only in the rich cave regions of Europe and North America, but also in the western United States, Australia, and New Zealand (Rambla and Juberthie 1994).

The final invertebrate order with more than 50 species is the Pseudoscorpionida, minute predators of microarthropods. There are about 400 troglobionts from 15 families out of a total of about 3000 species (Reddell 2012). They are found both in tropical and temperate areas and some families (Chernetidae) are associated with bat guano. Caves in the United States rival those of Europe in species richness, unlike the case in other groups, such as Coleoptera, where European cave faunas are more diverse. A total of 136 species are known from US caves, and are a major component of the hotspot of species richness in northeast Alabama (Culver *et al.* 2006b).

The above march through the subterranean invertebrates was based on the number of stygobionts and troglobionts in different groups, and only those groups with at least 50 species were considered (Table 3.3). It was a look at invertebrates from an ecological point of view, which is one of common-ness of different groups in subterranean environments, with species number serving as a surrogate for abundance. Another way is a more evo-lutionary one—one that assesses how frequent stygobionts and troglobi-onts are within a taxonomic group. This is a measure of the propensity of different taxonomic groups to be found in subterranean habitats. A very different picture emerges (Fig. 3.12). For troglobionts, some orders with many species (Araneae and Coleoptera) actually have a low frequency in subterranean habitats, relative to the total number of spider and beetle species on the surface. On the other hand, relative to other taxonomic groups, they are dominant in the subterranean environment (the ecological point of view). It is the minor arachnid orders plus Diplura and Collembola that are really 'subterranean' orders, from the evolutionary point of view. The reasons for this are complicated but the 'subterranean' orders tend to be ancient groups, ones that are basal in phylogenetic position. The pattern is similar to stygobionts with some of the undoubted ancient groups—Bathynellacea and Remipedia—limited to subterranean sites. Overall, none of the largely subterranean orders has a large number of species. For both stygobiotic and troglobiotic 'subterranean' orders, most

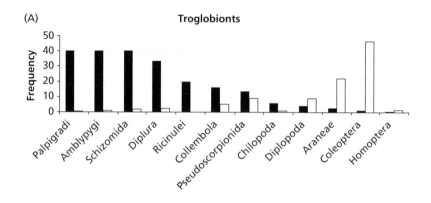

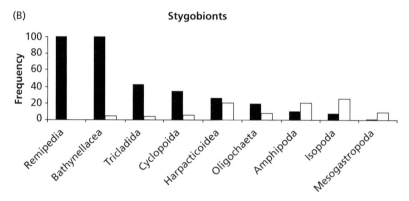

Fig. 3.12   Comparison of frequency of (A) troglobionts (solid bars) and (B) stygobionts (solid bars) for different subterranean orders relative to number of species in the order (open bars). For Collembola, numbers are based on the European fauna only; otherwise global estimates are used. Data are from various sources.

surface-dwelling lineages have gone extinct. This is why some biologists emphasize the relictual nature of the subterranean fauna.

3.5.3 Vertebrates

Salamanders are known from caves and deep cavities in phreatic water that is accessible only by wells. Stygobiotic salamanders are limited to two regions—the southern United States, especially Texas, with a total of 8–10 species, depending on the taxonomic authority, and the Dinaric Mountains of south-central Europe, with one (Weber 2004). Non-stygobiotic species can play important roles as predators in some caves (Peck 1974) and stygobiotic species are important predators in the sites where they are found. The North American species are all plethodontids—lungless salamanders. Some US species—*Eurycea tridentifera*, *Eurycea rathbuni*, *Eurycea robusta*, and *Haideotriton wallacei*, —are known primarily from wells. They live in deep cavities only occasionally intersected by caves accessible to humans. *Gyrinophilus* species live in cave streams, as does the Ozark cave salamander, *Eurycea spelaea*. *T. spelaeus* has a very interesting life cycle—larvae typically live in springs and surface habitats, the adults metamorphose, colonize caves, and show considerable eye and pigment reduction. Although often listed as stygobionts because of their troglomorphy, *T. spelaeus* is clearly not a stygobiont. Their diet is also interesting. Fenolio *et al.* (2006) showed that this apparent predator is actually coprophagous, eating bat guano. Except for *Gyrinophilus subterraneus*, all the stygobiotic salamanders, including the European *Proteus anguinus* are neotenic, with the aquatic larvae reaching sexual maturity. The loss of the adult, terrestrial stage may be adaptive if the energy costs of transformation are high or food is less abundant in the terrestrial habitat (Weber 2004).

Probably the most famous cave animal, and the first described one (by Laurenti in 1768), is *P. anguinus*, the 'olm' or 'human fish' (Fig. 3.13). It has a large range for a stygobiotic species, known from 300 subterranean sites, mostly caves, extending more than 60 000 km^2 along the Adriatic Coast (Goričky *et al.* 2017). Nearly all the populations have a very similar morphology with reduced pigmentation and vestigial eyes. They are long-lived, reaching sexual maturity at 16 years. Non-optic sensory structures, such as the inner ear, lateral line system, and magnetic field detectors, are well developed (Bulog 2004). One population, described as a separate subspecies *P. anguinus parkelj*, has retained eyes and pigment (Sket and Arntzen 1994). Molecular analysis (Goričky and Trontelj 2006; Trontelj *et al.* 2009; Goričky *et al.* 2017) has indicated that the species is actually made up of a series of clades (a group of populations or species that have a common ancestor) that colonized subterranean habitats between 1 and 8 million years ago, depending on the clade. *P. anguinus* has no close surface ancestors and is in its own family—Proteidae.

Fig. 3.13   Photograph of *Proteus anguinus*. Photo by G. Aljančič, with permission. See Plate 10.

There are over 200 stygobiotic species of fish (out of a total 24 000 described species) in 19 families and 10 orders (Proudlove 2010), many in the families Cyprinidae and Balitoridae. With few exceptions, the ancestors of the subterranean fishes were surface-dwelling freshwater species rather than marine species. Most of the species live in streams in caves but a few live in deep phreatic cavities that have only been sampled through wells. The wonderfully named *Satan eurystomus* is known only from a series of wells, all over 100 m deep, in the Edwards Aquifer, Texas, USA. An even more unusual habitat is that of *Ituglanis epikarsticus*, known only from drip pools in São Mateus Cave in Brazil (Bichuette and Trajano 2004). Many Chinese cavefish in the genus *Sinocyclocheilus* have a large head-horn of adipose tissue, apparently an adaptation for resistance to starvation (Ma and Zhao 2012). Compared to their anatomy and physiology, the ecology of fishes is relatively little studied. In part this is because some of the best known cave fishes, such as *Astyanax mexicanus*, live in habitats that seasonally flood and are inaccessible during that time (Espinasa and Espinasa 2016). From an ecological point of view, the best studied fishes are the North American amblyopsid fishes, which are important predators in many cave streams in central United States (Poulson 1969; Noltie and Wicks 2001; Neimiller *et al.* 2012a).

The distribution of subterranean fishes is in sharp contrast to the distribution of most groups of invertebrate stygobionts, which are concentrated in temperate zones (see Chapter 8). There is only one stygobiotic cave fish from Europe (Behrmann-Godel *et al.* 2017) and a relatively modest number of six species described from the United States, two of which are only found in deep phreatic sites. The sole European species, a yet undescribed loach in the genus *Barbatula*, is likely a post-Pleistocene invader of the large karst basin resurging at Aach spring in Germany. It is also the most northern stygobiotic cave fish known. In contrast, there are 46 species known from China, 15 from Brazil, and 11 from Mexico. The concentration of species is between 16°N and 25°N, where nearly half all species are found. That is, fishes tend to be found closer to the equator than do most stygobiotic invertebrates. It may be that the relatively higher surface productivity (and ultimately the higher secondary productivity in subterranean habitats, see Chapter 4) of these more tropical latitudes makes the successful invasion of fishes in subterranean habitats possible. Although fishes are larger than most other stygobionts and tend to be at the top of subterranean food webs, they are especially sensitive to levels of productivity. One of the most interesting groups of fishes from a biogeographical point of view is the genus *Sinocyclocheilus* in China. There are at least 49 species living in caves (Ma and Zhao 2012). On the basis of mitochondrial DNA sequences, each stygobiotic species separately invaded caves (Xiao *et al.* 2005).

The morphology of cave fish, especially *A. mexicanus*, has been extensively studied (Fig. 3.14). It is a particularly interesting species to investigate because there are both cave and surface stream populations that can be interbred in the laboratory. Much of the focus of the anatomical work was on eye degeneration rather than any elaborated features, such as taste buds and cranial neuromast size. Eye degeneration was obvious and some of the most important students of this fish, such as Thinès (1969) and Wilkens and Strecker (2017), held that natural selection was relatively unimportant in the evolution of cave fish. More recently, Jeffery (2005a,b) has listed both elaborated and reduced characters in the species (Table 3.4). We return to the question of adaptation and regressive evolution in Chapter 6.

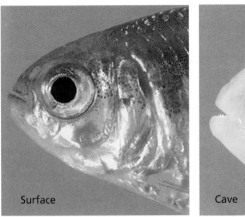
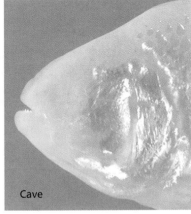

Fig. 3.14   Photograph of individuals from river and cave populations of *Astyanax mexicanus*. Photo by W. Jeffery, with permission.

Table 3.4 Some constructive and regressive changes in *Astyanax* cavefish.

Constructive Changes	Regressive Changes
Jaw size	Eyes
Taste bud number	Optic tecta
Tooth number	Pigment
Cranial neuromast size and number	Schooling behaviour
Nasal chamber size	
Forebrain size	
Feeding behaviour	
Fat content	

Source: From Jeffery (2005b). Used with permission of the International Society of Subterranean Biology.

3.6   Subterranean organisms in the laboratory

The laboratory study of subterranean animals has its origins in the nineteenth century when residents living near Postojnska jama sold live specimens of the European cave salamander, *P. anguinus*, to visitors (Shaw 1999). All in all some 4000 specimens were probably sold, many of which were taken to various laboratories and museums throughout Europe. Purchasers were given instructions to change the water daily and at least a few individuals made it to their destination alive. As far as Shaw could determine, very little scientific study was actually done on these specimens.

Those that survived were treated as curiosities. For example, several were on display at the London Zoo. In fact, it is difficult to keep *Proteus* alive, let alone establish a breeding colony. There have been only two successful efforts to maintain *Proteus* in the laboratory, and both of these involved using a cave as the site of the population. In Kranj, Slovenia, the biologist Marko Aljančič established populations of *Proteus* in 1960 in the cave Tular in a series of large artificial pools (Aljančič 2008). It took three years to even maintain adults in Tular. Captive breeding was finally successful in 1991, a testament to the longevity of *Proteus* and the dedication of Aljančič. Among the problems he encountered was egg cannibalism (Aljančič *et al.* 1993). A population was also established in the cave laboratory associated with the Laboratoire Souterrain in Moulis, France. Beginning in the 1930s, plans were made to construct an in-cave laboratory in Postojnska jama, Slovenia, primarily for the captive breeding of *Proteus* and other troglobionts and stygobionts (Dudich 1932–1933). World War II intervened and the project was never completed. In France, the Laboratoire Souterrain was established in 1948, and the cave laboratory was the cornerstone of this institute (Vandel 1964). There have been several attempts to establish captive breeding programs for Texas stygobiotic salamanders, with some success (Najvar *et al.* 2007; Gorički *et al.* 2012), but it takes at least several years for the salamanders to reach sexual maturity.

The reason that so much effort was expended on trying to establish captive breeding programmes for *Proteus* and other stygobionts and troglobionts is that there is much that can only be learned through the study of living, captive populations. This includes information on development, physiology, and of course genetics, especially if hybridization of populations is possible. The difficulties in maintaining subterranean populations remain a major impediment in contemporary studies, especially in developmental and evolutionary studies.

One of the reasons that the Mexican cave fish, *A. mexicanus*, has figured so prominently in the study of subterranean cave biology is the relative ease with which breeding populations can be maintained in a laboratory. It is widely available commercially, and wild-caught populations are relatively easy to establish in the laboratory (W. Jeffery, personal communication). The list of other stygobionts and troglobionts that can be maintained or bred in the laboratory is a short one. Proudlove (2006) lists all cases of captive populations of cave fish, involving 26 species. Aside from *A. mexicanus*, only six species have been bred successfully in the laboratory. The record for invertebrates is even more abysmal. Among crustaceans, the only case of captive breeding known to us is that of Fong (1989) with the amphipod *G. minus*. Among troglobionts, Juberthie-Jupeau (1988) successfully maintained breeding populations of leptodirid beetles in the cave laboratory at Moulis. Christiansen was able to maintain breeding populations of a number of troglobiotic Collembola in a laboratory in Iowa, USA. All of these efforts required a great deal of husbandry, and except for *Astyanax* populations, most workers do not sustain these efforts over the decades.

3.7   Collecting stygobionts and troglobionts

Collecting in caves in general is decidedly low-tech. The absence of vegetation makes visual searching easier, but the rarity of animals makes it harder. Pitfall traps, both baited and unbaited, have been extensively used for terrestrial cave fauna, especially in the past. They are often quite effective, perhaps too much so. In some cases, hundreds of troglobionts such as millipedes, Collembola, and beetles can be captured in a single baited trap after only a few days, raising concerns about over-collecting (Elliott 2000). Aquatic fauna is often collected with the aid of nets, especially fine mesh ones such as plankton nets. Baiting is less common in aquatic cave habitats, but crayfish traps and the like have been frequently used not only for crayfish, but for macro-crustaceans such as amphipods.

One of the main impediments to the study of shallow subterranean habitats and deep wells has been the absence of effective sampling devices. It is not an overstatement to say that the invention of the Bou–Rouch pump (Bou and Rouch 1967) revolutionized the study of hyporheic and other shallow interstitial habitats. The pump (Fig. 3.15) allows the sampling of relatively shallow interstitial water, especially in the beds of streams and rivers, and along their banks. The device consists of a mobile stainless steel pipe that is driven to various depths in the bed sediment with a pump fixed on top. The principle of the method is to create a disturbance and maintain an interstitial flow around the pipe that is sufficient to dislodge interstitial organisms. Because of its high discharge rate, the pump probably samples both swimming organisms and species intimately linked to sand particles (Malard 2003). The pumped water is filtered through a net typically with a mesh size of 150 μm. Practical experience with the pump indicates that most of the species are collected in the first five litres of water pumped. Since the publication of

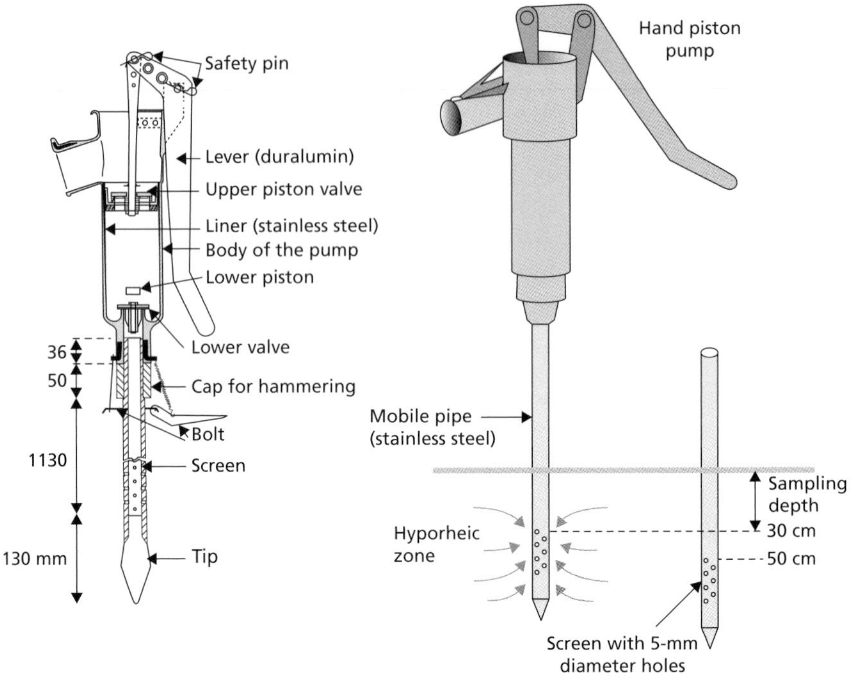

Fig. 3.15 Diagram of Bou–Rouch pump used to sample interstitial and hyporheic invertebrates. From Bou (1974). Used with permission of CNRS Editions.

the original paper by Bou and Rouch, there have been over 250 papers citing it, an indication of its impact.

The sampling of deeper aquifers is almost always accomplished through the use of existing wells. Unfortunately, because the standard centrifugal pumps used in many wells damage or destroy macroorganisms, special pumps are needed (Malard *et al.* 1997). For larger diameter wells (>50 cm), a special phreatobiological net (Cvetkov 1968) can be used if the well is not too deep. The lower end of the net consists of a container closed with a valve that prevents the animals from escaping once they are caught. Successive downward and upward movement of the net captures animals that swim in the well water.

Pipan (2005) achieved a breakthrough in sampling epikarst. Rarely if ever is it possible to sample this habitat directly, so Pipan concentrated on a technique for indirectly but quantitatively sampling water exiting the epikarst through dripping water. In her sampling device, water from a ceiling drip is directed via a funnel into a 500 mL rectangular filtering bottle fitted on two sides with plankton netting of 60 μm mesh size, and the filtering bottle is placed within a sampling container (Fig. 3.16). Each sampling container has a drain 3 cm from its base such that collected animals and a small amount of water remain in the filtering bottle while most of the water passes through

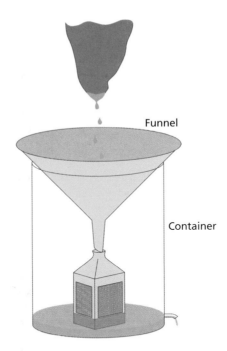

Funnel

Container

Fig. 3.16   Diagram of filtering device to continuously collect invertebrates from drips. From Pipan (2005). Used with permission of ZRC SAZU, Založba ZRC.

the filtration unit. Samples are collected at weekly or monthly intervals to reduce mortality of the fauna collected. Pipan and Culver (2007a) give guidelines for how many such filters need to be used, and for how long, to sample completely the fauna. This sampling device can be combined with devices that continuously measure discharge and water chemistry to provide a more quantitative analysis.

Two problems make collecting in terrestrial SSHs especially difficult. The first is that almost any trapping design requires the destruction and recreation of the habitat. In order to sample the aphotic zone of terrestrial SSHs, rocks and soil must be removed and then replaced. Second, because of low densities of animals, traps that can be left in place weeks or months are needed. A trap designed especially for the MSS by López and Oromí (2010) solves these problems.

They developed a permanent MSS pitfall trap which produces minimal disturbance once it has stabilized. The main component of the trap is PVC pipe with many small holes drilled along its surface (Fig. 3.17). A plastic tray is fitted with a nylon cord as a handle, with a bait container at the centre of the tray. The third component of the trap is a silicone cap closing the top of the pipe.

The most exciting possibility for collecting subterranean animals, particularly aquatic ones, involves not direct collection, but rather collection of

Fig. 3.17   Basic features of an MSS trap. A. Silicone cap sealing the pipe, with lateral cuts for easy removal. B. PVC pipe with abundant holes. From López and Oromí (2010).

environmental DNA (eDNA) that was shed by the animals into the water column (Ficetola *et al. 2008*). This is a non-invasive sampling technique that should minimize the problem of false negatives in sampling. Pioneering the use of eDNA for species detection was the work of Gorički *et al.* (2017) on both dark and light forms of the European salamander, *Proteus anguinus* (Fig. 3.13). Niemiller *et al.* (2017) were also successful in detecting two stygobiotic amphipod species (*Stygobromus tenuis potomacus* and *S. hayi*) in hypotelminorheic habitats in the vicinity of Washington, D.C. This study was especially significant because these species are hard to locate and *S. hayi* is on the U.S. Endangered Species List.

3.8 Summary

A wide variety of organisms are found in subterranean habitats and they have varying degrees of dependence and permanence in these habitats. Some species, stygobionts and troglobionts, have an obligate dependence on subterranean habitats, and are found nowhere else. Other species have an obligate dependence on caves and other subterranean habitats, such as bats and aquatic insects, but only spend part of their life cycle in caves. There are 21 invertebrate orders that have over 50 stygobiotic and troglobiotic species. Among vertebrates, salamanders and especially fishes are also well represented in caves and deep aquifers. Although the information obtained is very informative, very few subterranean species have been maintained successfully in the laboratory. There are a few specialized collecting techniques that are very useful, especially in non-cave subterranean habitats.

4 Ecosystem Function

4.1 Introduction

In its simplest version, an ecosystem is a black box with inputs, which are altered by the biological activities inside the black box to produce the output (Odum 1953). Although subterranean ecosystems are literal black boxes to match the conceptual black boxes of ecosystem studies, subterranean biologists in general have been slow to take up the conceptual framework of ecosystem studies. The problem was that the appropriate size of the black box and what the inputs and outputs were was not clear.

Interestingly, one of the very first detailed ecosystem studies was of the surface component of a karst spring (Odum 1957)—Silver Springs in Florida. Silver Springs, similar to most springs in Florida, is essentially a window dissolved through the bedrock into the underlying shallow groundwater aquifer (the Floridan Aquifer). However, for Odum the ecosystem began with boils of water emerging in the springs. From the point of view of subterranean ecosystems, the object of study is immediately upstream of the emerging spring, with the input of Odum's study becoming the output of a groundwater ecosystem study. It was not until the 1970s when stream ecologists became aware of the need to incorporate groundwater into stream studies (Hynes 1983) that conceptualizations for both karst and interstitial ecosystems were developed. Rouch (1977) provided an appropriate conceptual model for subterranean karst systems, while Stanford and Gaufin (1974) did the same for fluvial aquifers.

In this chapter, we discuss the scale and extent of subterranean ecosystems, paying special attention to organic carbon availability and fluxes because subterranean ecosystems are likely to be carbon limited. We also consider the special situation where chemoautotrophic sulfur-oxidizing bacteria are 'ecosystem engineers', modifying their environment by dissolving the limestone around them.

The Biology of Caves and Other Subterranean Habitats. Second Edition. David C. Culver and Tanja Pipan. Published 2019 by Oxford University Press. © David C. Culver and Tanja Pipan 2019. DOI: 10.1093/oso/9780198820765.001.0001

4.2 Scale and extent of subterranean ecosystems

An ecosystem is usually simply defined as the plants, animals, and micro-organisms plus their environment. Ecosystem science re-parameterized ecology, moving the focus away from the numbers of individual organisms and species to the amounts and fluxes of energy and nutrients, that is, carbon, nitrogen, phosphorus, and so on. The scale is left undefined and can be of different sizes for different purposes but there is an element of 'natural-ness' so that inputs and outputs can be defined and measured.

The critical conceptual question related to understanding subterranean energy and nutrient fluxes is what are the components of the ecosystem, particularly what are the inputs from the surface. The size of the ecosystem black box is less critical, or at least can have many answers. One of the simpler systems to consider inputs and outputs is a fluvial aquifer underneath and lateral to a stream. The hyporheic zone and the groundwater zone have inputs from the surface at the downwellings of the surface stream (Fig. 1.16). In other interstitial aquifers, especially deeper ones, there is a recharge zone of inputs where water slowly percolates through the soil and rock into the aquifer, the input in this case being more distant from the aquifer itself than in the case of fluvial aquifers. Some aquifers at depths of 1 km or more may not have any surface inputs at all. The water present is connate, water trapped in a sedimentary rock during its deposition. These aquifers often have chemoautotrophic activity with Bacteria and Archaea present (Frederickson *et al.* 1989; see Chapter 2).

Inputs in subterranean karst systems are a bit more complicated. In his studies of the subterranean fauna of the area around Moulis, France, Rouch began, as do most subterranean biologists, with a study of the fauna of caves (Rouch 1968). In his sampling of copepods in Grotte de Sainte- Catherine, Rouch could only find an occasional copepod, and he concluded that he was basically looking in the wrong place. The right place turned out to be the spring that drained the cave and the surrounding area—the Baget basin. He found that, instead of a handful of copepods, hundreds of thousands of copepods were exiting the spring.

Using a net 0.4 m in diameter, he collected more than 18 000 copepods in 19 months of continuous filtering. He estimated that 500 000 copepods were washed out of the system annually and that more than 10 million copepods were washed into the subterranean system from surface systems, especially in times of flood (White *et al.* 1995). Taking advantage of the morphological difference between subterranean and surface harpacticoids (troglomorphy, see Chapter 3), Rouch was able to trace connections between inputs and outputs. He demonstrated that the major inputs in the system were not sinking streams but rather percolating water, especially in the epikarst zone. In a series of 20 papers on Baget (Rouch 1986), a small 11.4 km^2 karst basin, Rouch defined inputs and outputs, described temporal variability, and pointed out the intimate connection between surface and subsurface. What he did not do was to parameterize the system into units of organic carbon, nitrogen, and so on. The inputs were sinking streams, percolating water, and deep groundwater, and the outputs were springs. Not all of these inputs that occur in Baget are present in every karst basin. In some karst basins there may be no input from groundwater (see Fig. 2.4) and in some there may be no sinking streams.

A second conceptual question, but one that can have many answers, is what the appropriate scale of analysis is. Analysis can be at different size scales depending on the questions being asked and the system being studied. In subterranean ecosystems, the different scales of study range from a stream reach, to a cave, to a karst basin.

The stream reach is the standard unit of analysis in stream ecology, and this has been adopted for groundwater ecosystem studies, especially in fluvial aquifers, especially the hyporheic zone. There are two ways that the size scale of the stream reach has been used in subterranean ecosystem studies. The first approach is to conceptually turn the stream upside down. Instead of focusing on the stream, it is the underside of the stream—the hyporheic and the underlying groundwater—that is the focus, and the stream becomes the input. An example of this approach is Marmonier *et al.* (2000) who studied the fluvial aquifers of the Rhône River near Lyon, France.

A second approach is to analyse a reach of stream inside a cave. In this case, the cave stream is conceptually similar to a surface stream with a roof, but a leaky roof. The leaks from the roof are water percolating from the epikarst. This is the approach exploited by Simon and colleagues (Simon and Benfield 2001, 2002; Simon *et al.* 2003) for stream reaches inside Organ Cave, West Virginia, USA.

In karst systems, the scale of analysis can be the cave, a scale already encountered in Chapter 2, for the movement of carbon in Organ Cave, West Virginia, and Postojna–Planina Cave System (PPCS), Slovenia. In this case, the inputs are sinking streams and percolating water, and the output is a spring (Simon *et al.* 2007a). A very different kind of cave—Lower Kane Cave in Wyoming, USA—was the scale of analysis of a study of chemoautotrophy by Engel and colleagues (Engel *et al.* 2003, 2004; Engel 2007). Here the input is hydrogen sulfide-rich groundwater and there are no surface inputs.

Finally, there have been two basin-wide studies. Gibert (1986) analysed a karst basin—Dorvan–Cleyzieu—in the Jura Mountains in France. Because of the details of the physical setting, inputs were separated into different caves making measurement of carbon and so on from epikarst, sinking streams, and groundwater possible. Hutchins *et al.* (2013, 2016) analysed the sources and movement of organic carbon in the very large and economically important Edward Aquifer of Texas, USA. A schematic diagram of the different scales of analysis is shown in Fig. 4.1.

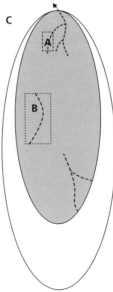

Fig. 4.1   Schematic representation of scale and extent of subterranean ecosystem models. The large ellipse represents the karst drainage basin, the shaded ellipse the extent of karst within the basin, the heavy dashed lines subterranean stream passages, and the arrow the exit of the water from the spring. The rectangles (A, B, and C) are possible scales of analysis.

4.3 Stream reaches

Simon and colleagues (Simon and Benfield 2001, 2002; Simon *et al.* 2003) studied several stream reaches in Organ Cave, West Virginia. Organ Cave is a multientrance, multilevel cave with streams fed entirely by percolating water and streams largely fed by streams sinking from the surface (Culver *et al.* 1994), allowing for spatial replication of measurements. Simon and colleagues began by asking the question of how these streams differed from surface streams. Several features were in fact quite similar. Leaf breakdown rates, measured by measuring loss rates from leaf packs, were similar to those of surface streams. Cave streams fed by sinking surface streams had higher breakdown rates, in part because they had populations of invertebrates, especially the amphipod *Gammarus minus*, that were able to directly break apart the leaves (shredders). They also found that the pattern of microbial colonization of leaves in cave streams was similar to that of surface streams. So what was different? On the basis of a variety of evidence, they concluded that these streams were not limited by nitrogen or phosphorus, but by carbon. Nitrate levels in the streams were relatively high, and the addition of ammonium had relatively little impact on the stream ecosystem (Simon and Benfield 2002). Nitrogen, and other nutrients, 'spiral' between abiotic and biotic components, and the shorter the spiral is, the more limiting the nutrient is likely to be. Spiral lengths in streams in Organ Cave were very long relative to surface streams and the low amount of benthic organic carbon (BOC) rather than nitrogen is the most probable limiting factor. Cave streams obviously differ from surface streams because cave streams have a roof which prevents logs, branches, and even leaves from falling directly into the stream, so that most cave streams have little coarse particulate organic matter (CPOM). Leaves and wood entering the cave only travels a short distance (usually less than 50 m) before they are transformed into fine particulate organic matter (FPOM), and are not important in the metabolism of the cave stream community. Community-wide metabolism was 32.9 g $C/m^2/$ year. The surface streams most similar in this respect to Organ Cave streams were the Kuparak River and Monument Creek in the tundra of Alaska's North Slope. The energy source for the macroscopic invertebrate community (snails, amphipods, and isopods) is the biofilm that forms in the stream. Even though there is more FPOM than dissolved organic matter (DOM), it is DOM that is the base of the food web in Organ Cave streams (Simon *et al.* 2003).

Venarsky *et al.* (2012b, 2014, 2017) studied several stream reaches in caves in Alabama and Tennessee, USA, with the intent of testing the hypothesis that these donor-controlled detrital ecosystems were carbon limited. Using experimental leaf packs of corn (*Zea mays*) and red maple (*Acer rubrum*), they found that invertebrate biomass and litter breakdown rates were not correlated with the ambient abundance of organic matter (Venarsky *et al.* 2012b), although Tony Sinks Cave had the highest organic matter and consumer biomass of the four caves studied. Much stronger evidence that cave stream reaches are carbon limited came from the demonstration (Venarsky *et al.* 2014) that almost all available detritus was required to support the invertebrate population with little left for fish and salamander predators (Table 4.1). A very interesting finding was that the numerical response of invertebrate populations to increased organic matter came almost entirely from the stygophilic and stygoxenic populations (see Table 3.2 for terminology), and the resident stygobiotic species (primarily isopods and crayfish) did not increase in numbers (Venarsky *et al.* 2017). The stygobiotic and non-stygobiotic components are trophically connected because the majority of the diet of the stygobiotic salamander *Gyrinophilus palleucus* is non-stygobionts (Huntsman *et al.* 2011).

Souza-Silva *et al.* (2012) looked at the riparian community in the 450 m-long cave Lapa da Fazienda Extrema I in Goiás, Brazil. They found evidence that this community was also carbon limited because of the correlation of biomass transport and stream richness and abundance over time (Fig. 4.2).

There are many studies of surface streams that include the hyporheic, but the number of studies that focus on the groundwater and associated hyporheic

Table 4.1 Estimates of mean detritus supply rates, macroinvertebrate production, *Orconectes australis* demand, and detritus and macroinvertebrate surplus in Hering, Limrock, and Tony Sinks Caves, Alabama, USA.

	Hering Cave	Limrock Cave	Tony Sinks
Detritus supply rate	13 (7–19)	48 (5–91)	79 (27–131)
Macroinvertebrate production	0.13 (0.05–0.21)	0.26 (0.24–0.28)	1.61 (0.90–2.32)
Macroinvertebrate detritus demand	4.03 (1.65–6.41)	8.03 (2.10–13.96)	48.7 (27.18–70.22)
Crayfish (*O. australis*) detritus demand	0.33 (0.20–0.46)	0.27 (0.18–0.36)	4.32 (1.77–6.87)
Crayfish (*O. australis*) macroinvertebrate demand	0.01 (0.00–0.02)	0.01 (0.006–0.014)	0.09 (0–0.18)
Detritus surplus	9 (2–46)	40 (–4–84)	26 (–30–82)
Macroinvertebrate surplus	0.12 (0.04–0.20)	0.26 (0.06–0.46)	1.51 (0.79–2.23)

All values are g ash-free dry mass m^{-2} year^{-1}. Numbers in parentheses are 95% confidence intervals.

Source: From Venarsky *et al.* (2014). Used with permission of Springer Nature.

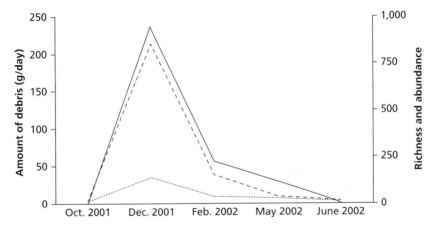

Fig. 4.2   Rate of organic matter (OM) import per month (grams/day, solid line), richness of invertebrates (dotted line) and abundance of invertebrates (dashed line) in a Brazilian cave. From Souza-Silva *et al.* (2012). With permission of the Karst Research Institute at ZRC SAZU.

is relatively small. Among the areas studied from this perspective are Lobau wetlands of the Danube River near Vienna, Austria (Danielopol *et al.* 2000), South Platte River of Colorado, USA (Ward and Voelz 1994), Flathead River of Montana, USA (Stanford *et al.* 1994), Lachein Creek in France (Rouch *et al.* 1989), and the Rhône River in France (Marmonier *et al.* 2000). The emergent theme of all these studies is the intimate connection between surface and groundwater, and the concept of an ecotone (Gibert *et al.* 1990), a zone of tension and transition between two communities, has proved to be very useful in understanding interstitial ecosystems (Gibert 1991). If the physical analogy of an ecotone in karst is of a leaky roof with many small holes (epikarst) and a few large holes (sinking streams), the ecotone of fluvial aquifers can be likened to a highly permeable sheet. Exchange occurs more or less everywhere, although it is hardly homogeneous.

The long-term, large-scale study of the Rhône fluvial aquifer is an exemplar of these studies (Dole-Oliver et al. 1994; Marmonier et al. 2000). In the upper reaches of the Isère River in the French Alps, a part of the Rhône drainage, they investigated three habitats—the benthic layer of the stream, the interstitial habitat 40 cm deep in the stream channel, and the interstitial habitat of gravel bars 40 cm below the surface. The fauna of the stream and the interstitial was very different, with benthic habitats dominated by insects and interstitial habitats dominated by microcrustaceans and nematodes. In spite of these faunal differences, the distribution of organic carbon was quite similar in the two habitats. Organic content of the sediment was 2.2 mg organic matter/g of dry sediment. Dissolved organic carbon (DOC), which varied between 1 mg/L and 2 mg/L, increased slightly from surface to groundwater, perhaps as the result of interstitial biofilms acting to increase organic content of interstitial sediments. Oxygen content decreased significantly from surface to groundwater, as a result of community respiration. Oxygen is not replenished in groundwater systems because there is no photosynthetic activity.

In addition to these habitat differences, there are also differences along the longitudinal reach of the stream. Even at the scale of a few metres, the scale of upwellings and downwellings (see Fig. 1.16), there were significant faunal differences. Dole-Olivier and Marmonier (1992) investigated this heterogeneity in the Canal de Miribel, a regulated channel of the Rhône River. Surface-dwelling crustaceans (ostracods and amphipods) and insect larvae were dominant in the shallow (50 cm) interstitial habitat of the well-oxygenated, organic carbon-rich downwelling zones. Widespread stygobionts, especially amphipods, were more abundant at a depth of 100 cm. In upwelling zones, stygobionts with narrow tolerance ranges dominated in the lower, poorly oxygenated, carbon-poor deeper sections of the habitat.

Finally, there are differences in the transverse direction. In yet another Rhône site, the Grand Gravier section further downstream along the Rhône (Marmonier et al. 2000), differences were found between interstitial habitats below the river, at the shoreline, and 1.5 m up the bank of the river (Fig. 4.3). Sampling was carried out at various depths (20–100 cm) throughout the year. Total organic matter, DOC, bacterial abundance, and hydrolytic enzymatic activity (measured on biofilm) all were highest in the sites in the river channel and lowest at the sites on the bank. Further analysis of DOC showed an interesting pattern—organic matter was removed from infiltrating water in the downwelling zone by physical processes (adsorption on fine particles) and by bacterial activity. When microbial activity is low, some of this DOC migrates downwards and physical processes of filtration dominate. When microbial activity is high, DOC is rapidly assimilated into the bacterial population except for the refractory component, which is adsorbed only.

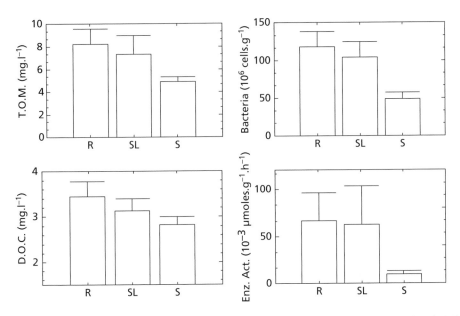

Fig. 4.3   Total Organic Matter (TOM) content of the sediments, Dissolved Organic Carbon (DOC) content of the interstitial water, bacterial abundance, and hydrolytic enzymatic activity of biofilm fixed on fine sediment of the Rhône River in the Grand Gravier sector. Three sampling points were studied: at 1.5 m in the river (R), at the shoreline (SL), and 1.5 m up the bank (S). Mean values ($\pm$ SE, n = 12) were calculated for three depths (20, 50, and 100 cm in the sediments) and four dates (January, May, August, and November 1992). Differences between sampling points were significant for bacterial abundances and enzymatic activity (ANOVA, $p<0.05$), differences between TOM were marginally significant ($p<0.10$). From Marmonier *et al.* (2000). Used with permission of Elsevier Ltd.

4.4 Caves

Two very different kinds of caves have been studied from an ecosystem point of view. Simon and colleagues expanded the stream reach perspective used in Organ Cave to include the entire cave, especially inputs and outputs (Simon *et al.* 2007a, b). They also used this cave-wide approach for Postojna–Planina Cave System (PPCS). These caves have the 'typical' inputs of sinking streams and percolating water. Schneider *et al.* (2010, 2011) studied very small open air pits (less than 20 m in depth and 25 m in length) and did whole-cave resource removal experiments. The other kind of cave that has been studied from an ecosystem perspective is sulfidic caves with no inputs from the surface and only hydrogen sulfide-rich groundwater as an energy input (Hose *et al.* 2000; Engel 2007; Kumaresan *et al.* 2014).

4.4.1 Caves with surface input

Simon *et al.* (2007a) measured the standing crop of DOC for inputs, outputs, and some interior points for Organ Cave and PPCS. The conceptual model of carbon flux is shown in Fig. 2.4. In both caves, there are two types of inputs—percolating water and sinking streams. In contrast to interstitial aquifers where differences in organic carbon concentrations in different components were not particularly large, organic carbon concentrations in the various cave components are nearly an order of magnitude different (Table 4.2). Sinking streams in both caves and the cave stream in PPCS had the highest DOC values and epikarst drips in both caves and the resurgence of Organ Cave had the lowest DOC values.

There were interesting differences between the two caves. Organ Cave is in a largely agricultural area and PPCS is in a forested area, and so it is not surprising that organic carbon levels in sinking streams and epikarst drips in Organ Cave are higher than in PPCS. There is also a difference in the in-cave stream organic carbon concentrations, which in both caves derive from a mixture of water from sinking streams and percolation water, but mostly water from sinking streams (Simon et al. 2007a). In Organ Cave, cave stream DOC values are less than 20 per cent of the sinking stream values, suggesting considerable in-stream processing. This hypothesis is also supported by Simon and Benfield's (2002) measurement of whole metabolism, which was in the range of surface streams; these also have considerable in-stream processing of organic matter. The situation in PPCS is somewhat complicated because the DOC values at the resurgence are 50 per cent lower than the in-cave stream values. This is because a second cave stream (Rak River) joins the measured one (Pivka River) and likely has lower values of DOC. The streams in PPCS are much larger than in Organ Cave, so it is not surprising that the reduction in DOC for PPCS in the output is less than for Organ Cave.

As with interstitial systems, there is considerable heterogeneity within the components of the systems. Flow rates and organic carbon levels of cave streams obviously vary with the season, but there is also considerable spatial heterogeneity as well (Simon et al. 2007b). In PPCS, the mean values of DOC for epikarst drips showed variation at several scales. PPCS consists of a series of connected caves, each named for the entrance that provides access. Drips in Črna jama, Pivka jama, and Postojnska jama were measured (Postojnska jama is about 2 km from the other two, which are about 100 m apart), and median DOC values varied from 0.42 mg/L in Postojnska jama

Table 4.2 Estimates of dissolved organic carbon in mg/L from Organ Cave and Postojna–Planina Cave System.

	Organ Cave	Postojna–Planina Cave System
Input: sinking streams	7.67±1.03	4.36±0.46
Input: percolation water	1.10±0.15	0.70±0.04
In cave: streams	1.08±0.32	4.75±1.57
Output: resurgence	0.90±0.17	2.67±0.80

Source: From Simon et al. (2007a). Used with permission of the National Speleological Society (www.caves.org).

Table 4.3 Comparisons of percent C, %P, and molar C:P among detritus, troglobiotic, and facultative cave invertebrates.

	%C	%P	C:P
Detritus (removed from caves)	34.77	0.22	913.11
Detritivores:			
*Pseudotremia fulgida** (millipede)	28.84 (7)	1.39 (27)	60.51
Pseudotremia hobbsi (millilpede)	32.31 (5)	1.50 (19)	62.82
Collembola	47.52 (6)	1.36 (18)	101.91
Predators:			
*Pseudanophthalmus fuscus** (beetle)	44.54 (3)	1.21 (2)	107.28
*Pseudanophthalmus grandis** (beetle)	50.83 (13)	0.77 (17)	192.51

Troglobionts are marked with an asterisk. Numbers in parentheses are sample sizes.
Source: Data from Schneider *et al.* (2010).

to 0.39 mg/L in Črna jama to 0.79 mg/L in Pivka jama. Within each cave, there was also spatial variation. In Pivka jama, individual drips ranged from 0.53 mg/L to 2.05 mg/L. This variation in some ways is analogous to the variation observed along stream reaches with upwellings and downwellings, but in the case of epikarst we do not know what the equivalents of upwellings and downwellings are. It may be that percolating water with more DOC has been in the subsurface longer, in analogy with downwellings in fluvial aquifers.

Schneider *et al.* (2010, 2011) provide information about possible carbon and phosphorus limitation in small pit caves. They removed all visible detritus from these small caves, and followed colonization of both leaf packs and dead rats, and found that species composition depended on which carbon source was added, suggesting that at least species composition was sensitive to carbon quality (Schneider *et al.* 2011). In a study of possible phosphorus limitation using C:P stoichiometric ratios, Schneider *et al.* (2010) found that millipedes (both troglophiles and troglobionts) were possibly phosphorus limited because of the high P content of their bodies and the low P content of their detrital food source (Table 4.3).

4.4.2 Sulfidic cave systems

The idea of a cave ecosystem that does not even indirectly derive its energy from photosynthesis has intrigued speleobiologists for decades. Although the possible chemical basis for chemoautotrophy in caves has been known for a long time, it was widely held to be insufficient to support ecosystem-level processes (e.g., Poulson and White 1969). The discovery of chemoautotrophic deep sea hydrothermal vents in the late 1970s, coupled with the discovery of Peștera Movile in Romania in 1986, toppled the dogma that chemoautotrophy was insufficient to sustain ecosystems.

Most of the known cases of chemoautotrophy in caves involve sulfidic environments, although other electron acceptors such as iron and manganese oxides may also be important (Peck 1986; Carmichael and Bräuer 2015). Chemoautotrophy in deep aquifers may involve terminal electron acceptors other than SO_4^{2-} (Engel 2012), but these ecosystems are poorly understood and are chemically very complex (Frederickson et al. 1989). The percentage of caves world-wide that are formed by sulfuric acid speleogenesis is estimated to be 5 per cent (Palmer 2013). The number of caves world-wide is in the hundreds of thousands, so the number of sulfidic caves is itself in the thousands, including a number of show caves (see Klimchouk et al. 2017). Not all of these caves have chemoautotrophic ecosystems, but many do. Sulfidic caves are often disagreeable and dangerous places to work. The distinct smell of H_2S emanating from a cave in Virginia, USA, led discoverers to name it Cesspool Cave, even though it is at least 500 m from the nearest dwelling. More importantly, H_2S is very toxic and its presence is often accompanied by reduced oxygen levels.

The most thoroughly studied chemoautotrophic cave from a geomicrobiological perspective is Lower Kane Cave, Wyoming, USA. Hydrogen sulfide enters the cave in several springs. It is oxidized to sulfuric acid in the following reaction:

$$H_2S + 2O_2 \leftrightarrow H_2SO_4$$

with a Gibbs free energy, ΔG, of -798 kJ/mol (Engel et al. 2004). Sulfur-oxidizing bacteria promote and obtain energy from this reaction. A variety of sulfur-oxidizing bacteria are involved in this process, but it is Epsilonproteobacteria that dominate and these are found in all sulfidic caves (Engel et al. 2003). Epsilonproteobacteria are also found in other sulfur-rich habitats, especially in marine sediments and hydrothermal vents. Lower Kane Cave is the first non-marine system known to be driven by the activity of Epsilonproteobacteria, which typically form microbial mats with other sulfur oxidizers that either attach to substrates or float (Engel 2007). Other reactions with H_2S are also biologically important. Microbes such as *Beggiatoa* produce ammonium, an important nitrogen source for the ecosystem, from the oxidation of sulfur dioxide and nitrate (Engel 2007):

$$4H_2S + NO_3^{2-} + 2H^+ \leftrightarrow 4S + NH_4^+ + 3H_2O$$

This reaction, linking the sulfur and nitrogen cycles, may also account for the deposits of elemental sulfur found in some caves such as Lechuguilla Cave in New Mexico, USA. Nitrogen fixation is common in chemoautotrophic systems (Desai et al. 2013) although its functional role needs to be further elucidated (Barton 2015).

The final key reaction is the one that reduces sulfate to hydrogen sulfide, thus completing the cycle:

$$4H_2 + SO_4^{2-} + H^+ \leftrightarrow 4H_2O + HS^-$$

HS^- is then further reduced to H_2S. This reaction is promoted by the δ-proteobacteria, and occurs in anoxic (lacking oxygen) conditions.

Estimates of overall chemoautotrophic primary productivity and heterotrophic activity from four caves (Grotta di Frasassi in Italy, Peştera Movile in Romania, Lower Kane Cave in Wyoming, and Cesspool Cave in Virginia) show considerable variation but in all four systems, chemoautotrophic productivity greatly exceeds heterotrophic productivity (Table 4.4). The macroscopic invertebrate community that uses the chemoautotrophic microbial mats as the food base varies from 19 stygobionts and 27 troglobionts in Peştera Movile, one of the most biologically diverse caves in the world (Culver and Sket 2000), to no stygobionts or troglobionts in Cesspool Cave (Engel *et al.* 2001). The differences among the caves can be accounted for by differences in geological and hydrological settings, size, age, and extent of isolation. Peştera Movile is old (probably several million years), very isolated, and part of a relatively large aquifer. Cesspool Cave is small (<20 m), less than 10 000 years old, and in a small aquifer. Sulfidic caves, provided they are old enough, large enough, and isolated enough from the surface, are likely to harbour diverse communities, given that non-chemoautotrophic caves are likely to be generally carbon limited (see section 4.3).

The feature that makes sulfidic caves really interesting is that the chemoautotrophic processes also contribute to cave formation (Fig. 4.4). When hydrogen sulfide-rich water reaches the cave, it is oxidized to sulfuric acid, and this reaction involves microbes, especially Epsilonproteobacteria. Epsilonproteobacteria are 'ecosystem engineers' in the sense that these microbes significantly alter their environment and the environment of succeeding generations (Jones *et al.* 1994). They modify their environment by creating the cave itself, dissolving calcium carbonate:

$$H_2SO_4 + CaCO_3 + H_2O \leftrightarrow CaSO_4\ 2H_2O + CO_2$$

The gypsum produced readily dissolves in water, resulting in cave enlargement. Thermodynamically, this process does not require microbial involvement but microbes seem to be universally present and can greatly increase

Table 4.4 Summary of biomass production (μg C/mg dw/h) for chemolithoautotrophic (^{14}C-Bicarbonate) and heterotrophic (^{14}C-Leucine) metabolisms from sulfidic karst environments.

	^{14}C-Bicarbonate	^{14}C-Leucine	Bicarbonate/Leucine
Cesspool Cave, Virginia, USA	0.03 ± 0.01	$1.7 \times 10^{-4} \pm 2.5 \times 10^{-5}$	176.5
Lower Kane Cave, Wyoming, USA	0.10 ± 0.01	not measured	not measured
Grotte di Frasassi, Italy	1.01 ± 0.47	$1.5 \times 10^{-2} \pm 4.1 \times 10^{-3}$	67.3
Peştera Movile, Romania	4.72 ± 0.78	$5.5 \times 10^{-5} \pm 5.3 \times 10^{-5}$	85 454.50

Source: From Porter *et al.* (2009).

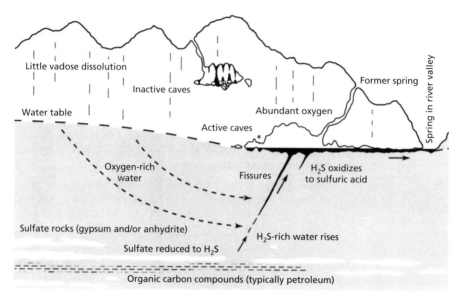

Fig. 4.4   A typical setting for the origin of sulfuric acid caves. Deep flow of oxygen-rich water may be scant or absent so that nearly all dissolution occurs at or near the water table. Enlargement by sulfuric acid also occurs in air-filled parts of the cave where H₂S and oxygen are absorbed by moisture on bedrock surfaces. From Palmer (2007). Used with permission of Cave Books.

the rate of cave dissolution and enlargement. The process of carbonate dissolution also alters the chemistry of the stream in Lower Kane Cave. Stream pH is buffered to near neutrality (7) by ongoing $CaCO_3$ dissolution. All in all, Epsilonproteobacteria are remarkable ecosystem engineers, certainly rivalling and even exceeding the impact that beavers, the classic example of ecosystem engineers, have on their environment. Microbial involvement in cave enlargement by sulfuric acid is exemplified by Cueva de Villa Luz in Tabasco, Mexico (Hose *et al.* 2000, Hose and Rosales-Lagarde 2017). The cave is approximately 250 m long with several shafts that open to the surface, and at least 26 groundwater inlets. Some of these groundwater inlets have a pH of around 0. H_2S concentrations reach 300 ppm in some seeps, and concentrations in the air can reach the dangerously high level of 200 ppm. In many places in the cave, sulfur-oxidizing bacteria form flexible, rubbery speleothems with the consistency of mucus. The cave is enlarging by several processes. One, similar to that in Lower Kane Cave in Wyoming, is the conversion of limestone to gypsum by sulfuric acid in the cave stream. Another method of cave dissolution is by drops of acidic water from the bacterial speleothems falling on the cave floor. Both of these, especially the second, are bacterially mediated.

Recent work has made it clear that the nitrogen cycle, as well as the carbon and sulfur cycles, are involved in chemoautotrophic caves (Kumaresan *et al.* 2014; Barton 2015). The different cycles are shown in Fig. 4.5.

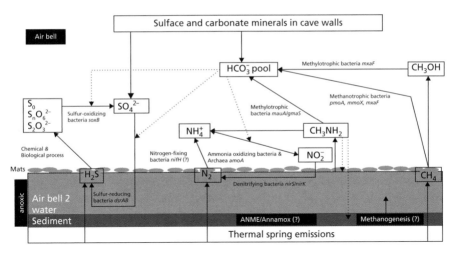

Fig. 4.5   Schematic representation of microbial sulfur, carbon, and nitrogen cycling in Peştera Movile, Romania. Evidence for metabolic pathways comes from functional gene analyses. From Kumaresan *et al.* (2014). Used with permission of Walter de Gruyter GmBH. See Plate 11.

4.5 Karst basins

Gibert (1986) studied the Dorvan–Cleyzieu basin in east-central France. This basin drains the water of an area of 10.5 km^2 in the foothills of the Jura Mountains. Water infiltrating the karst exits at the spring at Grotte du Pissoir and several other nearby springs. Higher up in the basin is a cave (Grotte du Cormoran) that is a convenient site for the collection of epikarst water, which makes it possible to distinguish between percolating water exiting the epikarst (Grotte du Cormoran) and base level flow (Grotte du Pissoir).

The hydrological budget (Table 4.5) divides the precipitation into components of evapotranspiration, surface runoff, infiltration, and extraction. Relative to surface runoff, infiltration was more than twice that of surface runoff. These percentages correspond approximately to proportions of the basin covered by soluble rock and by insoluble rock. Such hydrological budgets have been performed for other karst basins (see Jones 1997); the proportion of different components varies, of course, but the main categories (evapotranspiration, runoff, and infiltration) remain the same.

Gibert estimated both the standing crop of organic carbon in different components as well as their fluxes. Estimates of density of organic carbon (Table 4.6) show that there is a substantial reservoir of carbon in sediments. Similar to the results of Simon *et al.* (2007a) (see Fig. 2.4), most of the carbon entering through epikarst is dissolved rather than particulate.

Most investigators, when beginning the study of organic carbon in subterranean environments, are mindful of the large numbers of animals entering

Table 4.5 Water budget for the Dorvan–Cleyzieu basin in east-central France.

Component	Percent of precipitation	mm of precipitation
Evapotranspiration	41	663
Surface runoff	15	237
Extraction[1]	5	242
Infiltration water exiting at springs	24	389
Infiltration water entering water table directly	15	81

Components are expressed as percentages of total precipitation and the equivalent mm of precipitation, which totaled 1612 mm.
Source: Data from Gibert (1986).
[1]Used by the villages of Cleyzieu and Mont de Lange

Table 4.6 Standing crop of organic matter in various components of the Dorvan–Cleyzieu basin.

Major components	Categories	Concentration
Sediments	Surface	74 g/kg
	Grotte du Cormoran (epikarst)	16 g/kg
	Grotte du Pissoir (groundwater)	25 g/kg
Terrestrial biomass (in water)		
	epikarst drips in Grotte du Cormoran	40 µg/m³
	epikarst stream in Grotte du Cormoran	4 µg/m³
Total organic carbon in water		
	Grotte du Cormoran (epikarst)	0.92 mg/L
	Grotte du Pissoir (groundwater)	2.11 mg/L

Carbon concentrations are approximately half that of organic matter.
Source: Data from Gibert (1986).

the system from the surface, as shown by Rouch (1991), and the large number of animals entering the system through epikarst drips, as shown by Pipan (2005), but the reality is that DOC is much more plentiful. Gibert (1986) paid special attention to the particulate component of the drift of terrestrial animals. Carbon concentrations due to terrestrial drift were negligible compared with DOC but what little there was appeared to have been utilized, because concentrations were an order of magnitude lower in the small stream fed by epikarst than in the drips themselves (Table 4.6). Total organic carbon in epikarst drips was similar to that found by Simon *et al.* (2007a) in caves in Slovenia and the United States. Gibert also estimated the standing crop of organic matter in the living component in the water at 33 µg/m³ (Fig. 4.6).

Fluxes of organic matter are of course much harder to measure because they are rates, but they also provide more insight into the ecosystem. Gibert was able to make estimates of the yearly flux for several components. At the spring exit of the ecosystem at Grotte du Pissoir, approximately 3000 kg of DOC and 5 kg of particulate organic carbon exited per year. In the portion of the basin that drains into Grotte du Cormoran through epikarst drips, she estimated that 100 kg of DOC and 400 g of POC entered the system per year.

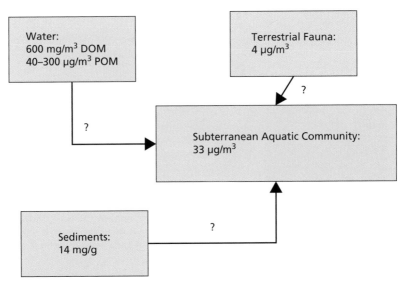

Fig. 4.6   Schematic diagram of the carbon reservoirs that contribute to the carbon in the aquatic community in the cave. Fluxes are unknown. Data from Gibert (1986).

Studies of subterranean ecosystems in karst at the scale of drainage basins have great potential, largely untapped. There have been no follow-ups to Gibert's pioneering study. Different karst basins represent different challenges in terms of measurement of fluxes and standing crops of carbon and nutrients. There are features of some karst basins that should make such an undertaking easier. They include a well-defined boundary, an exit (spring) that can be monitored, and sites to separately measure percolating water and water from sinking streams. In this regard, isolated karst basins with caves at different levels provide a good starting place. An example is the Paka karst, an isolated area in northeast Slovenia, with a clearly defined drainage, caves at different levels, and an exit spring (Fig. 4.7).

The Edwards Aquifer in Texas, USA, is a very different kind of basin, an artesian aquifer with deep cavities of hypogene origin. Both because of the presence of a freshwater–saltwater interface (FWSWI) at its southern boundary, with petroleum in the saltwater layer, and because it is a highly diverse ecosystem (Longley 1981, 2004), chemoautotrophy has long been suspected. Hutchins *et al.* (2016), using the isotopic signatures of organic carbon in the system, were able to demonstrate that between 25 and 69 per cent of the organic matter utilized by consumers was from chemolithoautotrophic sources. For one species, the cave shrimp *Palaemonetes antrorum*, 88 per cent of its diet came from chemolithoautotrophic sources. Species richness of the macroscopic invertebrate communities was highest near the FWSWI (Fig. 4.8), a clear indication that chemoautotrophy allows higher species diversity (see Chapter 8).

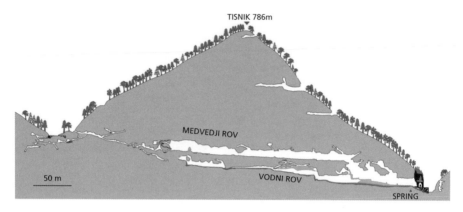

Fig. 4.7   Cross-section of the Paka drainage basin in northeast Slovenia, showing caves at different levels, and the base stream. The entire isolated karst basin is under Tisnik hill. From
Pipan *et al.* (2008). Used with permission of Karst Research Institute at ZRC SAZU.

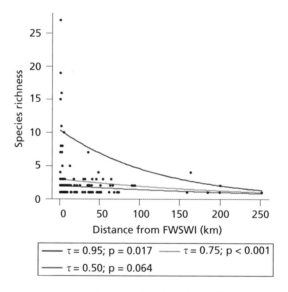

Fig. 4.8   Quantile regression of species richness in the Edwards Aquifer, Texas, as a function of
distance from the freshwater saline water interface (FWSWI). Trendlines are shown for
significant quantiles (τ). From Hutchins *et al.* (2016).

4.6 Summary

An important aspect of nearly all aquatic subterranean ecosystems is the
nature and connectivity of surface inputs, and it was not until the recognition of this by Rouch (1977) for karst systems and Stanford and Gaufin (1974)
for fluvial aquifers that ecosystem studies became possible. Different size
scales have been analysed. Stream reaches in both karst and interstitial sites

have been studied. A theme common to both is the remarkable heterogeneity of inputs and physicochemical conditions that exist at even the smallest scale. In interstitial aquifers, there is heterogeneity of carbon and carbon fluxes in all three dimensions. At least in cave streams, carbon appears to be limiting but phosphorus and nitrogen have been little studied. There is some evidence of phosphorus limitation in terrestrial systems. Studies at the scale of entire caves are of two very different kinds. For caves with surface inputs, inputs from percolation water are quantitatively less important than inputs from sinking streams, but are qualitatively more important because they occur throughout the cave and form the basis for the biofilm. Sulfur-oxidizing bacteria are the trophic base for most chemoautotrophic cave communities. In some cases highly diverse communities are developed in these caves. Sulfur oxidizers are also important in cave formation, making the bacteria ecosystem engineers. Only two ecosystem studies of an entire karst basin have been carried out. For the Dorvan basin in France, most carbon entering the ecosystem is DOC, and there is considerable storage of organic carbon in sediments. In the artesian Edwards Aquifer of Texas, chemolithoautotrophy allows for greater species richness, and contributes to all the components, in a gradient from the freshwater–saltwater interface zone.

5 Biotic Interactions and Community Structure

5.1 Introduction

Subterranean communities are usually simpler, if simplicity is measured by the number of species, than most surface-dwelling communities. On the other hand, interactions among species, such as predation, are no less frequent in subterranean communities than in surface communities, and in fact they are often more obvious and easier to study because the communities themselves are less complex. The best studied systems, which we review, are cave beetles that prey on cricket eggs in many caves in the central United States, an intricate system of interactions between amphipods and isopods in cave streams in the eastern United States, size and shape relationships among *Niphargus* amphipods in the Dinaric karst, and size ratios of aquatic beetles in calcrete aquifers in Australia.

The alternate approach to the study of individual pairwise interactions is to associate each species in a community with a particular set of environmental conditions; that is, to determine its ecological niche. Niche differences are most probably due to, at least in part, past competition and the subsequent evolution of niche separation. It is then possible to analyse associations and interactions in this way. With the ever-increasing computational power available, there has been an explosion in the use of multivariate statistical methods in the analysis of subterranean communities, such as canonical correspondence analysis (CCA) (ter Braak and Verdonschot 1995; Pipan *et al.* 2006a) and outlying mean index (OMI) (Doledec *et al.* 2000; Dole-Olivier *et al.* 2009b).

5.2 Species interactions—generalities

Individuals of different species can interact in a variety of ways, ranging from the trivial to the deadly. For instance, two species of amphipod may encounter

The Biology of Caves and Other Subterranean Habitats. Second Edition. David C. Culver and Tanja Pipan. Published 2019 by Oxford University Press. © David C. Culver and Tanja Pipan 2019. DOI: 10.1093/oso/9780198820765.001.0001

Demographic Impact of
Species B on Species A

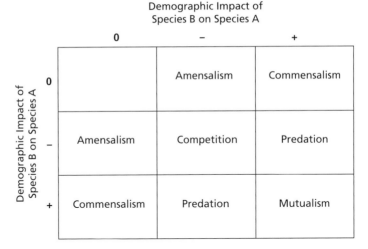

Fig. 5.1   Classification of pairwise interspecific interactions. When both species have a negative effect on each other this is competition; when both species have a positive effect on each other this is mutualism; when one has a positive effect and the other a negative effect this is predation. Less familiar are the highly asymmetric interactions of commensalism and amensalism.

each other in a pool and simply avoid each other after initial contact. However, depending on the species, this encounter may turn into a predator–prey interaction with one amphipod eating the other (Culver *et al.* 1991).

A useful way to categorize and compare species interactions is by their demographic impact (Fig. 5.1). When two species interact in an ecologically significant manner, the interaction affects population size and growth. If species A decreases the number of individuals of species B and species B decreases the number of individuals of species A, then A and B are competitors. If species A increases population size of species B but species B decreases population size of species A, then B is a predator of A. In this demographic sense, parasites also fall into the category of predators. All interactions can be classified according to the effect that an encounter of an individual of species A with species B has (Fig. 5.1). Two-way interactions are quite familiar—competition, mutualism, and predation. One-way interactions are less so—amensalism and commensalism. Ecologically, amensalism is a very one-sided case of competition, and commensalism is a very one-sided case of mutualism.

It is likely that many animals initially enter the subterranean realm to escape predators (including parasites) and competitors. Visually oriented predators cannot effectively find their prey in the darkness of a cave. If a species has a competitor that is more visually oriented rather than, for example, olfactory oriented, the visually oriented competitor will be at a disadvantage in subterranean habitats. Parasites may also find it difficult to make the transition to subterranean environments perhaps because it is more difficult to locate a host in the absence of light.

In subterranean environments, predation and competition are relatively well known. The other two-way interaction—mutualism—has not been studied in subterranean environments. The closest approach to this is a study by Hobbs (1975) of ostracods living on the exoskeletons of cave crayfish. Ostracods gain an advantage from the interaction because they feed on microorganisms and detritus that accumulate on the host exoskeleton. Crayfish may gain some advantage from the ostracods directly from exoskeleton cleaning. Hobbs found that the interaction seemed to decrease in intensity with increasing cave adaptation. In Pless Cave, Indiana, USA, stygobiotic *Orconectes inermis inermis* had an average of 13.5 ostracods per individual. *Cambarus tenebrosus*, a stygophile also common in surface streams, had an average of 31 ostracods in Pless Cave, more than twice as many. Strictly surface-dwelling species, such as *Cambarus bartonii*, have even more ostracod ectocommensals— sometimes more than 100 (Culver 1982).

Relative to free-living predators, there are few studies of parasitic interactions in subterranean environments. There are a few parasites that have special-ized on subterranean species, in some ways the *ne plus ultra* of extreme specialization. One of the most spectacular examples is Temnocephalida, parasitic flatworms which parasitize cave shrimp living in Balkan caves, probably feeding on the haemolymph of the shrimp that they access through thin parts of the exoskeleton (Matjašič 1994). Matjašič (1958) reported that seven species and several genera of Temnocephalida are found only on the cave shrimp *Troglocaris schmidti*, with each species specializing on a par-ticular region of the body. For example, *Subtelsonia perianalis* is only found around the anus of *T. schmidti*. There are remarkably few parasitic species known to be limited to stygobionts or troglobionts in spite of the fact that parasites often speciate in parallel with their hosts. Protozoans may also prove to be important parasites of subterranean animals as one of the few studies of this phenomenon indicates (McAllister and Bursey 2004).

5.3 Predator–prey interactions—beetles and cricket eggs in North American caves

Aside from bats, the most obvious organisms in many caves are cave crickets in the family Rhaphidophoridae. Rhaphidophorids are widespread in caves in the United States, Mexico, southern Europe, China, the Indo-Pacific, southern Australia, and the tips of Africa and South America. Around dusk, it is common to see hundreds and sometimes thousands of cave crickets massed near cave entrances (Fig. 5.2). Morphologically, they show signs of

Fig. 5.2 Concentration of cave-crickets, *Ceuthophilus stygius*, on the ceiling of Dogwood Cave, Hart Co., Kentucky, USA. Photo by H. H. Hobbs, with permission. See Plate 12.

both subterranean and surface life—they have eyes (sometimes reduced) but appendages are elongated relative to surface species. Many individuals, especially those near the entrance, leave the cave at night to forage for food during the warmer times of year (Di Russo and Sbordoni 1998). During the day in the summer, and during day and night in winter months, they stay in the caves, where they typically lay their eggs. In addition to elongated antennae, they show other adaptations to cave life. They have a thinned cuticle that allows them to survive in the moisture-saturated atmosphere of caves, are resistant to starvation, and have lower metabolic rates compared with surface-inhabiting relatives (Lavoie *et al.* 2007).

Cave crickets, and their fates and those of their eggs, are especially noticeable and well studied in Mammoth Cave and nearby caves in central Kentucky, USA. The most common cricket is *Hadenoecus subterraneus*. It is more probable that species such as *H. subterraneus* became associated with caves because caves were a place to avoid predation. Crickets and other relatively large-bodied orthopterans have many enemies—many birds during the day and small rodents at night. Individuals in caves should be relatively safe from predation from vertebrates, although protection is relative. Viele and Studier (1990) show that the white-footed mouse, *Peromyscus leucopus*, tends to concentrate its foraging near cave entrances where it eats *H. subterraneus*. The main defence of *H. subterraneus* and other orthopterans against such predators is their prodigious jumping ability (Helf 2003). Capture–recapture studies on cave crickets from Texas caves indicate that cave crickets often forage up to 100 m from the entrance (Taylor *et al.* 2005). *H. subterraneus* only leaves the cave to forage when it is dark and conditions on the surface are similar to those in the cave (nearly 100 per cent humidity and, in the case of Mammoth Cave, 15°C). Foraging *H. subterraneus* are omnivores, eating a wide variety of foods, including mushrooms, dead insects, animal droppings, berries, and flowers (Lavoie *et al.* 2007). Inside the cave, *H. subterraneus* contributes to energy flow in two important ways. First, as is the case with any regular visitor to caves, it leaves faeces, which is an important food source for many troglobionts (Poulson 2017a). Second, they lay eggs in passages with sandy substrates. Mammoth Cave is especially rich in passages with sandy substrates because it underlies a sandstone caprock that both is a source of sand and preserves high, upper-level passages that disappear during erosional cycles in most caves (Palmer 2017). The population of *H. subterraneus* in Mammoth Cave and elsewhere tends to be subdivided into a subpopulation within 20 m of the entrance with individuals of all sizes and ages, and a deep cave subpopulation with large adults and very young crickets that have recently hatched from eggs (Lavoie *et al.* 2007). The entrance subpopulation consists of individuals that are likely to be foraging on the surface within a day or two, and the deeper subpopulation consists of ovipositing females, their eggs, and recently hatched young.

Several species of North American beetles, including *Rhadine subterranea* (Mitchell 1968), *Darlingtonea kentuckensis* (Marsh 1969), and *Neaphaenops tellkampfi* (Kane *et al.* 1975), are cricket egg predators. The size of these beetles is larger than most other troglobiotic carabids in North American caves. *N. tellkampfi* and other cricket egg predators are usually about 7–8 mm long, and most *Pseudanophthalmus*, the dominant troglobiotic beetle genus in North American caves, are less than 5 mm. The size of a cave cricket egg relative to that of *N. tellkampfi* (other cricket egg predators are of similar size) indicates

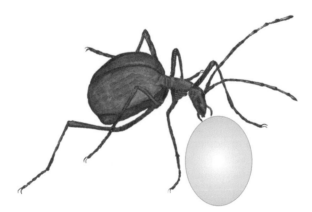

Fig. 5.3   Sketch of a beetle eating a cricket egg. Modified from a drawing by S. Polak, with permission.

the potential energetic importance of cave cricket eggs (Fig. 5.3). Studier (1996) found that the dry weight of a *H. subterraneus* egg (2.26 g ± 0.03) is nearly three-quarters that of the dry weight of an adult *N. tellkampfi* (3.02 g ± 0.07). A single egg is sufficient food for a beetle for weeks and it takes approximately 50 days for the beetle to return to its pre-feeding weight (Griffith and Poulson 1993).

This interaction between beetles and crickets is a model system for the study of predator–prey interactions not just in caves, but in general, because of the near-exclusivity of the interaction—*H. subterraneus* eggs are nearly the only prey for *N. tellkampfi* during the season when eggs are available and *N. tellkampfi* is nearly the only predator of the eggs of *H. subterraneus*. Therefore, the patterns of foraging by *N. tellkampfi* and predation avoidance by *H. subterraneus* are most probably the results of this relatively simple interaction.

The intensity of egg predation is high. In sandy areas in Mammoth Cave, beetles consume over 90 per cent of the cricket eggs laid throughout the year, and more than half are found and eaten by beetles within 15 days of being deposited (Kane and Poulson 1976; Griffith and Poulson 1993). Unpredated eggs hatch in about 12 weeks (Lavoie *et al.* 2007). Beetles dig in the substrate to find eggs, and when they are successful, they remove the egg, pierce it with their mandible, and the egg contents are pumped into the gut. The impact of this resource on beetle population dynamics is striking, and reproduction in beetles closely follows the time of maximum rate of egg depositions by crickets (Fig. 5.4). The quantitative impact of egg predation on *H. subterraneus* populations is unknown but it must be considerable.

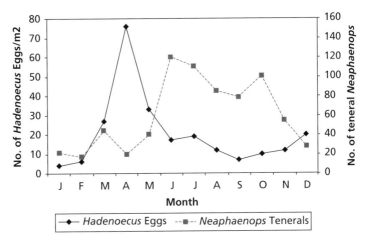

Fig. 5.4 Seasonal change in number of eggs per m² of *Hadenoecus subterraneus* and a visual census of newly emerging (teneral) adult *Neaphaenops tellkampfi* in Edwards Avenue in Mammoth Cave, Kentucky, USA. From Culver (1982). Reprinted by permission of the publisher from *Cave life: evolution and ecology* by David C. Culver, p. 103. Cambridge, Mass.: Harvard University Press. Copyright © 1982 by the President and Fellows of Harvard College.

Eggs are deposited at a depth of about 10 mm in the sand. Hubbell and Norton (1978) showed that the ovipositors of *H. subterraneus* females in populations that are subject to egg predation by *N. tellkampfi* are 1 mm longer than those in unpredated populations. This 10 per cent increase in depth is effective in reducing predation because beetles often dig holes that are less than the depth of the eggs (Lavoie *et al.* 2007). Extensive intraspecific competition occurs among beetles digging for cricket eggs, and as a result many holes are dug that are taken over by other beetles (Griffith and Poulson 1993). In the autumn, egg laying declines and eggs become scarce. During this season, some beetles continue to dig for cricket eggs (and find most of them) but many *N. tellkampfi* move away from sandy passages with cricket eggs and closer to the cave entrance where they likely act as generalized predators (Kane and Poulson 1976; Kane and Ryan 1983).

5.4 Competition and other interactions in Appalachian cave streams

The beetle–cricket egg interaction is of course an interaction between trophic levels. A thoroughly studied case of interaction within a trophic level is that of the amphipods and isopods that occupy many stony-bottomed streams in Appalachian caves from Maryland to Alabama (USA). In these streams, as in any stony-bottomed stream, there is an alternation between deeps (pools) and shallows (riffles). The amphipods and isopods are highly concentrated in riffles as a result of the concentration of food (especially leaf detritus), increased oxygen levels, and the absence of predators, especially salamanders, which live in pools. Nearly all of the species feed on the biofilm on the leaves. Most species do not skeletonize leaves although at least one species—the amphipod *Gammarus minus*—does. There are differences in habitat preference but they seem to be of a very obvious kind—larger individuals prefer the underside of larger rocks more than do smaller individuals (Culver and Ehlinger 1982). Both the amphipods and isopods typically crawl along the substrate, rather than swim in the water column.

In riffles, there are three obvious kinds of interactions. First, species may compete for energy sources and nutrients in leaves and biofilm. Second, smaller individuals can become prey for larger individuals. Third, species may compete for space (the underside of rocks). The underside of rocks serves as a place to avoid the brunt of the current. All three kinds of interactions (competition for food, competition for space, and predation) can and do occur in particular situations but the most universal (and easiest to analyse) is competition for space on the underside of rocks in riffles.

There is a general risk associated with life on the underside of rocks—the risk of washing out into the current. Individuals that wash out into pools run the risk of being eaten by salamanders, the primary predators living in pools (Culver 1975). Individuals also run the risk of damage from buffeting by the current. As with many subterranean species, the amphipods and isopods have long, thin appendages (see Chapter 6) that are easily broken. Amphipods and isopods with broken appendages are not only easy prey for salamanders but are also attacked and eaten by other amphipods and isopods. Amphipods may be at more risk than isopods in currents. The cave stream amphipod species are not good swimmers and their laterally compressed body shape, compared to the dorsoventrally flattened body shape of isopods, is not hydrodynamically efficient in moving water. It is very easy to observe the behavioural response to most encounters even in a small dish in the laboratory—one or both individuals rapidly move away. More realistic laboratory experiments can be performed in a small artificial riffle, where the washout rate of individuals put in the riffle in various combinations can be measured (Culver 1973; Culver *et al.* 1991).

An example is shown in Fig. 5.5, involving two pairs of species from Organ Cave. When the amphipod *Stygobromus spinatus* is in an artificial riffle with the isopod *Caecidotea holsingeri*, *S. spinatus* has nearly double the washout rate that it has when it is alone, indicating that *C. holsingeri* is its competitor (Fig. 5.5A). However, the converse is not true. *S. spinatus* seems to have no effect on *C. holsingeri*. Thus the overall interaction is one of amensalism (Fig. 5.1), a highly one-sided competition in this case. This asymmetry was not the result of size differences because these two species are of roughly the same size, about 5 mm. The isopod is the superior competitor perhaps because of its more hydrodynamic body shape.

Another pair of species, the amphipod *G. minus* and the isopod *C. holsingeri*, displayed competitive behaviour, and both species had a higher washout rate in the presence of the other (Fig. 5.5B). This would seem to be a simple interaction, but it proved to be more complex than first thought. The complication was that some *C. holsingeri* 'disappeared' during the day-long washout experiments, and where they disappeared to was the gut of *G. minus*. The interaction between these two species had elements of predation (hence the 'disappearing' isopods) and of competition (hence the mutual avoidance by both species when in a riffle).

In Thompson Cedar Cave and other nearby caves in Lee County, Virginia, USA the amphipod *Crangonyx antennatus* and the isopods *Caecidotea recurvata* and *Lirceus usdagalun* are the only three amphipods and isopods, and they illustrate the importance of competition in determining distribution. The standard equations of competition are:

$$dN_1/dt = r_1N_1(K_1 - N_1 - \alpha_{12}N_2) \text{ and}$$

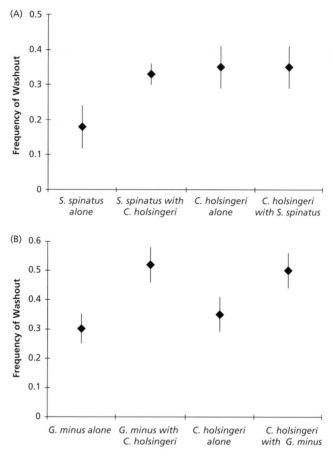

Fig. 5.5   Mean frequency, together with standard errors, of washout rates from riffles of *Gammarus minus*, *Caecidotea holsingeri*, and *Stygobromus spinatus* in various combinations in laboratory stream experiments. Data from Culver *et al.* (1991).

$$dN_2 / dt = r_2 N_2 (K_2 - N_2 - \alpha_{21} N_1).$$

The subscripts refer to the different species, r's are the intrinsic rates of increase of species 1 and 2, N's their population size, K's their carrying capacity, and α's the effect of species j on species i (see Fig. 5.1). The larger the α's, the stronger the interspecific competition is. With artificial riffle experiments, estimates of the competition coefficients among *C. antennatus*, *C. recurvata*, and *L. usdagalun* were possible (Table 5.1). The strongest competition in the laboratory riffle was between *C. antennatus* and *L. usdagalun* and in nature these species barely coexist in the same cave stream and never in the same riffle. The weakest competition was between *C. antennatus* and *C. recurvata*, and they routinely coexist in the same riffle, utilizing different sizes of rocks. The intensities of competition measured in the laboratory were such that, if they accurately reflect the situation in the field, only one of the three pairs (*C. antennatus* and *C. recurvata*) should be able to persist in the same riffle. Pairs involving *L. usdagalun* should not coexist in the same riffle because competition was predicted to be too strong. In fact, competition is evidently strong enough to determine the pattern of distribution of species within a cave and even to determine the overall species composition of the fauna of a cave stream (Table 5.1) (Culver 1976).

The impact of competition is epitomized by the distribution of these three species in Thompson Cedar Cave (Fig. 5.6). In various parts of this small cave stream, all three species occurred together, *L. usdagalun* occurred by itself, and *C. antennatus* and *C. recurvata* occurred together, but in no place did the pair *L. usdagalun* and *C. recurvata* or the pair *L. usdagalun* and *C. antennatus* occur. These two pairs of course are the stronger competing pairs (Table 5.1). In addition, all three species occur together, a result that was also predicted from the laboratory stream results. This seemingly paradoxical

Table 5.1 Intensity of competition in laboratory riffles and distribution in cave streams.

Species pair	Competition coefficient	Distribution pattern
Crangonyx antennatus—Caecidotea recurvata	0.30	Different microhabitats within a riffle
Caecidotea recurvata—Lirceus usdagalun	0.65	Different riffle
Crangonyx antennatus—Lirceus usdagalun	1.50	Different streams or marginally in same stream

Source: From Culver (1973, 1976).

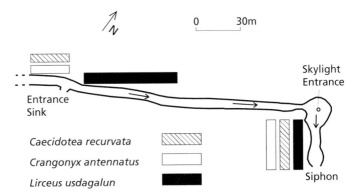

Fig. 5.6 Map of distribution of *Caecidotea recurvata*, *Crangonyx antennatus*, and *Lirceus usdagalun* in Thompson Cedar Cave, Virginia, USA. From Culver (2012b). Used with permission of Elsevier Ltd.

result that strongly competing pairs (ones involving *L. usdagalun*) can be stabilized by a third competitor is predictable from the standard competition equations expanded to include three species (Culver 1994). It is really a consequence of the old adage that 'an enemy of an enemy is a friend'.

The third example of species interactions that has been intensively studied is the isopod community in Alpena Cave, West Virginia, USA (Culver and Ehlinger 1982; Culver 1994); one that demonstrates that competition is not universal in cave stream communities. Two species occur in the same stream—the isopods *Caecidotea cannula* and *C. holsingeri*. Superficially, they would seem to be competitors. Neither laboratory stream studies nor field perturbation experiments detected any evidence of competition between these two species. They both occur in the same riffles throughout the cave stream. It is possible that competition between the two species did exist in the past. Typically, *C. cannula* is larger than *C. holsingeri* and this difference is enhanced in Alpena Cave when the two species occur together, thus reducing competition. However, the size of the isopods is strongly correlated with the size of the rocks in the streams and it turns out that Alpena Cave has a bimodal distribution of gravel sizes. Therefore, we can only say that at present competition is not occurring.

Luštrik *et al.* (2011) investigated a case of two co-occurring amphipods— *Gammarus fossarum* and *Niphargus timavi*—that has some similarities to and many of the complexities of the cases of the Appalachian stream amphipods and isopods. In this case, the two occur in a small groundwater-fed surface stream. Only *N. timavi* occurs at the headwaters spring and spring-run that sinks after 250 m. The spring re-emerges after 150 m, and both species are present for a 1 km reach of the stream (Fišer *et al.* 2007). *G. fossarum* has few modifications for subterranean life while *N. timavi* is strongly troglomorphic. Both species show evidence of size-dependent intra- and interspecific competition. Both species are also cannibalistic and predaceous, but the troglomorphic *N. timavi* is more so, in contrast to the situation in Appalachian cave streams where *Gammarus* was more predaceous than *Stygobromus*, a rough ecological equivalent of *Niphargus*. Juvenile *N. timavi* were especially vulnerable to cannabilization and predation by both species. In spite of the strong interactions between the two species, it is not clear what factors both allow coexistence in the lower reaches and do not allow coexistence in the upper reach.

5.5 Morphological consequences of competition

If interspecific competition is a selective pressure in subterranean communities, then we would expect divergence in morphology when species occur

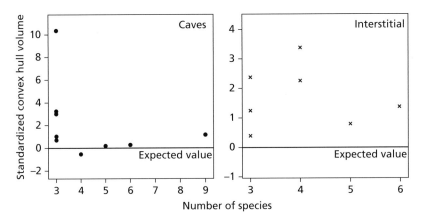

Fig. 5.7   Convex hull volumes (CHV) as measures of functional diversity of subterranean amphipod communities in caves and interstitial groundwater. Expected value is set to zero, each community is standardized (community CHV—median value of simulated CHV)/(1st quartile of the simulated CHV—3rd quartile of the simulated CHV). Positive values indicate higher morphological and hence functional diversity than expected if an equally sized community were assembled at random. Data from Fišer *et al.* (2012).

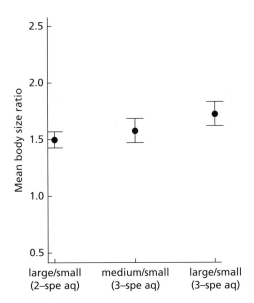

Fig. 5.8   Mean body size ratios (with standard error) of coexisting species of dytiscid beetles in calcrete aquifers in the Yilgarn, Western Australia. Mean body size ratios between coexisting species belonging to consecutive ranks did not differ significantly (F=1.5416, p=0.23). From Vergnon *et al.* (2013). Used with permission of University of Chicago Press, conveyed through Copyright Clearance Center, Inc.

together. Fišer *et al.* (2012) did in fact find divergence among species of the amphipod genus *Niphargus* inhabiting the same cave or the same interstitial site (mostly the underflow of rivers) in the Dinaric karst. Based on measurements of body length, antennal length, and pereopod and coxal plate size, they created multidimensional convex hull volumes for nine cave and seven interstitial communities with three or more *Niphargus* species. They then compared these actual convex hull volumes with those generated by 1000 random communities drawn from a regional pool of 94 *Niphargus* species. The results are shown in Fig. 5.7. All of the seven interstitial sites and six of the nine cave sites clearly have communities whose morphological composition is more diverse than expected. Three caves (Luknja, Podpeška jama, and the Postojna–Planina cave system) had communities with morphological divergence very similar to random expectation. In a companion paper, Trontelj *et al.* (2012) showed that, while there was divergence within a community, there was convergence among communities. They identified several ecomorphs common to either cave or interstitial habitats in the Dinaric karst, including small pore species, cave stream, cave lake, lake giants, and the enigmatic daddy-longlegs.

Classic competition theory predicts that there is a limiting similarity to species, i.e., they must be sufficiently different in order to coexist (MacArthur and Levins 1967). While the theoretical development of this idea is rich, there are relatively few empirical examples. Vergnon *et al.* (2013) provide excellent examples from the subterranean diving beetles inhabiting a series

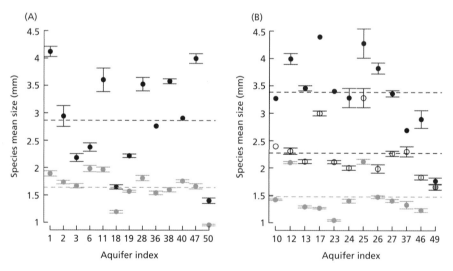

Fig. 5.9   Mean species body sizes in millimetres with standard errors in two-species (A) and three-species (B) aquifers. Different colours correspond to different species size ranks: small (grey), medium (white), and large (black). Vergnon *et al.* (2013). Used with permission of University of Chicago Press, conveyed through Copyright Clearance Center, Inc.

of calcrete aquifers in the Yilgarn, Western Australia. They studied the diving beetles (Dytiscidae) from 34 calcrete aquifers, all with different species. For aquifers with either two or three species, the ratios of body sizes were remarkably constant, averaging approximately 1.6 (Fig. 5.8), indicating that these species strongly compete for resources. This result is all the more remarkable because of the considerable variation in absolute body sizes (Fig. 5.9). For example, the largest beetle in the 13 aquifers with two species ranged from 1.5 to 4.1 mm (Fig. 5.9).

This demonstrates that both interspecific competition among *Niphargus* in both cave and interstitial habitats and competition among aquatic diving beetles in calcrete aquifers contradict the widely held idea that subterranean communities are dominated by convergent evolution. This conjecture dates back to at least Eigenmann (1909) who stated that cave fish were dominated by convergence and epigean fish were dominated by divergence.

5.6 Competition as a result of eutrophication

Species interactions also play a key role in the replacement of a specialized cave fauna in a cave stream subjected to eutrophication. In the late 1950s, the Pivka River in Slovenia (see Fig. 1.9) carried a heavy load of organic pollutants into Postojnska jama. Where the Pivka River entered the cave, oxygen concentrations (a measure of eutrophication) in the river were 10 per cent of saturation. One kilometre into the cave, oxygen levels recovered to nearly 50 per cent of saturation (Sket and Velkovrh 1981). Six kilometres into the cave, oxygen concentration was nearly at saturation and other measures of eutrophication indicated that it had largely disappeared. The very rich obligate cave stream fauna, including the very interesting and intensively studied isopod *Asellus aquaticus cavernicolus*, was largely extirpated in the first several kilometres of stream passage. The fauna was replaced by surface-dwelling aquatic insects, surface-dwelling amphipods in the genus *Gammarus*, and the surface-dwelling isopod species *Asellus aquaticus aquaticus* (Sket 1977). These species were able to invade because of the high energy and nutrient levels in the eutrophic Pivka River. As a result of their invasion, the stygobionts (such as *A. aquaticus cavernicolus*) were pushed further into the cave, not because they could not survive under the higher food conditions, but because they are outcompeted. In extreme cases, such as probably occurred near the entrance to Postojnska jama, low oxygen levels can prevent the survival of stygobionts, but most areas did not have extremely low oxygen values. Especially interesting are those areas of the underground Pivka River that came to be dominated by aquatic insects and *Gammarus*, both of which are not species of highly polluted waters, but rather stygophiles. Stygophiles were probably better competitors of surface-dwelling species than were stygobionts in polluted cave streams. Of course, stygophiles can survive and reproduce in non-polluted surface streams.

It also seems likely that competition with surface-dwelling species and predation by surface-dwelling species is a major factor in preventing the movement of subterranean species on to the surface. Sket (1986) shows that in the absence of competitors and predators, stygobionts may forage in surface environments. Of course, other factors prevent the movement of subterranean species to the surface. Many subterranean species are sensitive to light and also unable to cope with environmental fluctuations.

5.7 Community analysis—generalities

The intensive studies of the beetle–cricket egg interaction and the cave stream invertebrate interactions provide fascinating case studies of interspecific interactions that have proved interesting to ecologists in general. However, the number of such studies is small. This is because the conditions that made the detailed study of these interactions possible are not very common in subterranean habitats. These special circumstances are that the species involved were abundant for collection and study and that the communities themselves had few species when compared to other subterranean communities. For example, most terrestrial habitats in Mammoth Cave have at least a dozen species—it is only sand-floored upper-level passages that are nearly the exclusive province of crickets, cricket eggs, and their beetle predators. Many cave streams, especially in central and southern Europe, have many more species than the three or four found in Appalachian cave streams, making pairwise analyses impractical. For example, for a community with five species the number of possible pairwise interactions is 10, and for six species it is 15. But most importantly, the species studied in detail were common. Many subterranean species are numerically rare, often known from only a handful of specimens. Obviously, some other approach is needed in these situations.

An alternative approach is the correlation of patterns of species diversity and richness with environmental factors. R. Ferreira *et al.* (2007) used multivariate regressions to investigate the structure of bat guano communities in a Brazilian cave. These are very complex communities and they found a total of 85 species inhabiting guano piles in the cave. They found that number of species in a guano pile was correlated with the size of the pile, distance to the entrance, pH, and organic and moisture content of the piles. Not surprisingly these correlations indicate that species richness is strongly tied to resource quantity (pile size and organic content) and to some environmental conditions (distance to entrance, pH, and moisture content). What this study and ones similar to it cannot do is disentangle the differences among species' preferences and niches. For this a multivariate approach is needed, one that utilizes information on the conditions under which each individual species is found. This requires more detailed information but also makes it possible to make some inferences about the details of community structure. Such niche analyses allows investigation of the evolutionary outcomes of competition in the form of niche separation as well as other factors resulting in niche separation, such as phylogenetic inertia.

5.8 Epikarst communities

Epikarst, a superficial subterranean habitat (see Chapter 1), is both an exceptionally diverse and environmentally heterogeneous habitat. The fauna is dominated by copepods and can only be sampled indirectly by catching the copepods in plankton nets as they fall out of drips. Both their minute size, typically under 1 mm, and the inability to directly sample the habitat make the kinds of detailed study performed on cave stream invertebrates described in section 5.7 impossible. What is possible, and very informative, is to determine the physicochemical characteristics of the water where the different copepod species are found in—that is, a niche analysis (Hutchinson 1958). Although it is difficult, at first glance, to imagine advantages to epikarst communities as model systems for such a multivariate niche analysis, there are important ones. Copepods are relatively abundant in epikarst drips, making quantitative analysis easier, and it is possible to sample extensively enough to collect all or nearly all of the species present even though sampling is indirect (Pipan and Culver 2007a).

Pipan *et al.* (2006a) analysed the extensive samples of Pipan (2005) of copepods and environmental parameters from 25 drips in five caves in central Slovenia over a 12-month period. A variety of physical (drip rate, temperature, ceiling thickness, and surface precipitation in the preceding month) and chemical (conductivity, chloride, nitrate, sulfate, sodium, potassium, calcium, and magnesium) parameters were analysed. CCA allows the simultaneous representation of environmental variables and species' preferences (Fig. 5.10). The axes of the resulting two-dimensional graph show that linear combination of environmental variables that explain the greatest possible amount of the variance of all the environmental variables are taken together. Species lying close to one of the lines for the environmental variables are strongly associated with that variable. For example, *Moraria varica* is found in waters with higher concentrations of NO_3^-, and *Bryocamptus balcanicus* is associated with drips with higher flow rates (Fig. 5.10). Each species niche can be similarly defined by its position with respect to the canonical axes. Species with nearby positions have similar niches. Overall differences among the caves with respect to the environmental variables were found, and these differences help explain faunal differences among the caves.

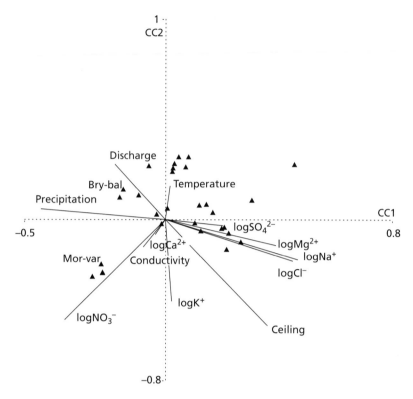

Fig. 5.10   Ordination diagram based on species composition and abundance data in drips in five Slovenian caves. Lines indicate the environmental variables and their orientation on the canonical axes. Triangles indicate different species, and species mentioned in the text are given the following abbreviations: *Moraria varica* Mor-var and *Bryocamptus balcanicus* Bry-bal. From Pipan (2005). Used with permission of ZRC SAZU, Založba ZRC.

The variables that best explained the variation in the 29 copepod species were thickness of the cave ceiling, and concentrations of Na^+, NO_3^-, and K^+. Nitrate is of course a macronutrient, but the reason for a correlation with sodium and potassium is not clear—it may be connected with surface pollution. The negative correlation with ceiling thickness, resulting in the richest fauna from the thinnest ceilings, suggests that the ceiling, the zone of percolation below the epikarst, acts as a filter for fauna falling out of the epikarst.

One of the most common outcomes of a multivariate analysis such as this is the realization that important variables were not measured. Pipan's study is no exception. Other studies of epikarst, including that of Simon *et al.* (2007a) discussed in Chapter 2, suggest that dissolved organic carbon is a key variable. The next study, from a very different habitat, a groundwater aquifer in Lyon, France, focuses on carbon.

5.9 Interstitial groundwater aquifers

Datry *et.* (2005) focused on the role of carbon in a groundwater aquifer in the City of Lyon, France. Many wells are present in the aquifer, including two well clusters that allowed for sampling at depths between 3 and 20 m. Some sites were artificially recharged with storm water, a source of organic carbon and nutrients; other sites were not. Wells at storm water sites that were less than 10 m in depth had approximately twice the amount of dissolved organic carbon (0.8 mg/L) that sites without storm water had. However, at depths greater than 10 m, this difference disappeared, indicating that most of the organic carbon was taken up in the shallow part of the aquifer. This pattern of organic carbon is reflected in both species numbers (richness) and overall numerical abundance (Fig. 5.11). Species abundance and richness is always higher at the shallow water sites and species abundance

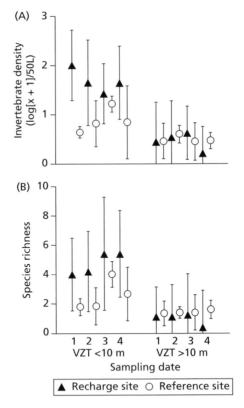

Fig. 5.11   Mean (±1 SD) density (A) and richness (B) of invertebrate assemblages at recharge and reference sites for vadose zone thickness (VZT) < 10 m and VZT >10 m. Sampling dates are 1–15 October 2001, 2–5 April 2002, 3–5 June 2002, and 4–5 October 2002. From Datry *et al.* (2005). Used with permission of the North American Benthological Society.

and richness differs between storm water and reference sites only at shallow sites, where there are differences in dissolved organic carbon. The role of thickness of the aquifer has direct parallels with Pipan's (2005) finding that increased ceiling thickness had a negative effect on copepod richness and abundance in epikarst habitats. Datry *et al.*'s elegant study strongly suggests a carbon-limited system, as was also suggested by Simon and colleagues' study of streams in Organ Cave (see Chapter 4).

A final point of note about this study is that the addition of storm water (and organic carbon) increased both the spatial and temporal heterogeneity of the system, which may also be a necessary prerequisite for increased species richness. Without this heterogeneity, niche separation among species must be reduced and interspecific competition increased. The combination of increased carbon and increased heterogeneity allowed for a diverse community with a total of 26 species, including 10 amphipod species and 12 isopod species (Datry *et al.* 2005).

5.10 Overall subterranean community structure in the Jura Mountains

All of the previous studies covered in this chapter have focused on a single subterranean habitat and a relatively restricted area. The beetle–cricket egg interaction studies only dealt with this species pair in a particular habitat— sandy-bottomed passages in Mammoth Cave and vicinity. The epikarst copepod studies did not consider either copepods in other subterranean habitats or other groups occurring in epikarst. Dole-Olivier *et al.* (2009b) undertook a more ambitious analysis. They did a multivariate analysis of the niches of all stygobionts in a 1200 km^2 area of the Jura Mountains in east-central France. They sampled a total of 192 subterranean sites including both interstitial and karst habitats. Conceptually they employed the same approach as that of Pipan *et al.* (2006a)—exploring the relationship between species' occurrences and environmental variables. However, the range of possible explanatory variables was expanded to include not only physicochemical variables, but also geographical/hydrogeological variables (altitude, hydraulic conductivity of the aquifer, and a qualitative score of pore sizes ranging from caves to spaces between clay particles), land cover variables (e.g., per cent meadows), and historical (distance to the boundary of the Würm glaciation). Similar to Pipan *et al.* (2006a), they did not measure dissolved organic carbon, which unfortunately is rather difficult and time consuming to measure (Emblanch *et al.* 1998). Rather than CCA, they used OMI (Doledec *et al.* 2000).

A regional analysis like this, while complex, has the advantage of analysing substantial numbers of species—62 in Dole-Olivier *et al.*'s study. In contrast,

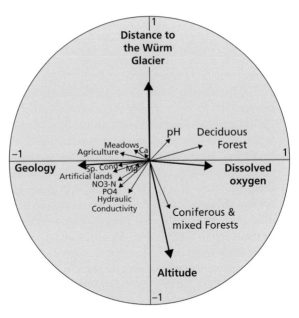

Fig. 5.12   Results of the OMI analysis of the stygobiotic fauna of the Jura Mountains, France. Weights of the 16 environmental variables along axes 1 and 2 are shown. From Dole-Olivier *et al.* (2009b). Used with permission of Blackwell Publishing.

the maximum number of species at a single site was 15, and the average number was only 4. We will return to this pattern of high regional diversity and low local diversity in Chapter 8.

The first OMI axis was largely determined by the amount of dissolved oxygen and the pore size of the aquifer (Fig. 5.12) and accounted for more than 50 per cent of the variability among sites. All of the 62 stygobiotic species except one—the amphipod *Niphargus kochianus*—preferred well-oxygenated aquifers with large pores, either caves or gravel and sand aquifers. Thus there was little niche separation among species, except for *N. kochianus*, which was found exclusively in poorly oxygenated groundwater with small spaces between particles. This is likely a result of competition with other species. The second OMI was largely determined by altitude and distance from the Würm glacier. In this case, species were found at both high and low altitude (although more at low altitude than at high altitude) and at varying distances from the glacier boundary. Species at the boundary or past it (into glaciated areas) are presumably more recent colonists, certainly since the glacier covered the area. What is largely absent as a determinant of species' niches at this scale are physicochemical parameters. Galassi *et al.* (2009) and Martin *et al.* (2009) found similar patterns for groundwater communities in the Lessinian Mountains of Italy and in Belgium, respectively.

5.11 Summary

A general pattern emerges from these and other studies of subterranean communities. At a regional scale, hydrogeological and historical factors exert a controlling influence on many species, and the importance of species interactions in determining them is small. This is the pattern of the Jura Mountain groundwater communities. At a smaller geographical scale and when only a single habitat type is considered, there is little variation in hydrogeological or historical factors. For example, in both the Slovenian epikarst and Lyon aquifer studies (Datry *et al.* 2005, Pipan *et al.* 2006a), there was little if any variation in hydrogeological or historical factors. Species did differ in their occurrence along physicochemical axes, and these differences may well be the result of competition. Of course, they may also be the result of other factors. Finally, some intensively studied communities show high levels of competition and predation, so strong that divergence rather than convergence occurs. There remains a gap between these some-what unusual species combinations (beetles and cricket eggs, Appalachian cave stream invertebrates, Dinaric *Niphargus*, and Australian calcrete diving beetles) and the broader scale community studies.

6 Adaptations to Subterranean Life

6.1 Introduction

The nature of evolution in the subterranean environment is one in which the losses of structures, such as eyes and pigment, seem to dominate. Historically, these losses have not always been viewed as adaptive, and this controversy continues up to the present, with strong adaptationist (e.g., Carlini and Fong 2017) and non-adaptationist viewpoints (e.g., Wilkens and Strecker 2017). Adaptationist viewpoints of eye and pigment losses have a relatively short history, with relatively little mention until the mid-1960s (see Barr 1968 for a review). Non-adaptationist hypotheses have a long history, going back to Lamarck and continuing up to the present in the guise of neutral mutation theory (Wilkens 1988).

Three landmark studies that put losses (regressive evolution) in the context of gains (progressive evolution), that is, a selectionist context, are reviewed in some detail. The first is the study of amblyopsid cave fish, most notably that of Poulson (1963). In the family Amblyopsidae, there are both surface-dwelling species and obligate cave-dwelling species. They were first studied by Eigenmann (1909), one of the last neo-Lamarckians studying cave fauna. In spite of being completely wrong about the nature of inheritance and mutation, he did provide careful descriptions of the eye and extra-optic sensory morphology of surface-, spring-, and cave-dwelling amblyopsids. He also proposed that one could use the degree of modification of these structures, especially eye loss, as a marker of age of cave-dwelling species, an idea that Poulson took up in a neo-Darwinian context. Using a comparative approach, Poulson demonstrated a series of morphological, behavioural, and demographic changes in cave populations that could best be explained by adaptation to the low-resource, aphotic environment of caves. In his analysis, the extent of eye and pigment loss was used as a clock to measure relative time of isolation in caves, an idea originally due to Eigenmann.

The Biology of Caves and Other Subterranean Habitats. Second Edition. David C. Culver and Tanja Pipan. Published 2019 by Oxford University Press. © David C. Culver and Tanja Pipan 2019. DOI: 10.1093/oso/9780198820765.001.0001

The second major study of adaptation is the analysis of cave- and spring-dwelling populations of the amphipod *Gammarus minus* by Culver *et al.* (1995), and Carlini and Fong (Carlini *et al.* 2013, MacAvoy *et al.* 2016, Carlini and Fong 2017). In addition to documenting morphological changes, they showed the genetic component of the traits, using heritability analysis and genetic distances among populations. Their demonstration of natural selection and adaptation followed the requirements outlined by Brandon (1990), a philosopher of science. They did not look to regressive traits as a measure of time of isolation in caves but rather looked for natural selection operating on these traits. More recently they have looked at overall measures of selection from transcriptome analysis as well as selection on the opsin gene.

Finally, Jeffery and his colleagues (Keene *et al.* 2016, Yamamoto *et al.* 2004, Jeffery 2005a, b) studied evolution and development of the eye of cave and surface populations of the Mexican cave fish, *Astyanax mexicanus* (Fig. 3.14), work that extended the earlier analyses of Wilkens and his colleagues (Wilkens 1988, Wilkens and Strecker 2017).[1] Jeffery and colleagues traced the developmental pathways of the genes responsible for eye development. They looked at the adaptiveness of eye loss from a developmental and cellular context.

We also consider more briefly other cave organisms that have been used in the experimental analysis of adaptation, including the Chinese cave fish genus *Sinocyclocheilus* (Meng *et al.* 2013, Chen *et al.* 2016), pigment loss in planthoppers and crustaceans (Bilandžija *et al.* 2012, 2013, 2017), and clock expression in the cave beetle *Ptomaphagus hirtus* (Friedrich *et al.* 2011). The two concluding sections address the question of how long adaptation and regressive evolution takes, and whether any current models are adequate to explain the observed morphological changes.

6.2 History of concepts of adaptation in subterranean environments

For most organisms in most habitats, the immediate observation is one of adaptation. Cheetahs are adapted for running and capturing prey; monarch butterflies are brightly coloured to warn potential predators of their toxicity; and so on. The theme of most nature films is the adaptation of organisms to their environment, and by implication the triumph of evolution by natural selection. Subterranean animals are different in this respect. The most

[1] *A. mexicanus* is one of those species that has suffered through a series of name changes. These include *Anopthichthys jordani, Anopthichthys antrobius, Anopthichthys hubbsi, Astyanax fasciatus*, and *Astyanax jordani* (Proudlove 2006). We follow Jeffery in using *Astyanax mexicanus*.

obvious features of subterranean cave animals are losses, not gains. Consider one of the iconic cave animals, the salamander *Proteus anguinus* (Fig. 3.13). It has long fascinated biologists, dating at least from the time of Lamarck in the late eighteenth and early nineteenth centuries. What makes *Proteus* interesting is in fact what it does not have—eyes and pigment. The recent discovery of a pigmented, eyed population of *Proteus* (Fig. 6.1) allows us to see that much of the bizarre appearance of this species is the result of eye and pigment loss. There are differences in body proportions of the pigmented and unpigmented *Proteus* (Arntzen and Sket 1997), but they are subtle compared to the differences in eyes and pigmentation. *Proteus* was probably better known to nineteenth century naturalists than any other subterranean organism.

It was not clear that the losses shown by *Proteus* were adaptive in any sense. Of course, Lamarck, the great champion of the theory of use and disuse, saw *Proteus* and other subterranean organisms as examples of disuse, and confirming evidence for his theory that morphological change occurred as the

Fig. 6.1   Typical (A) *Proteus anguinus anguinus* and the pigmented, eyed subspecies (B) *Proteus anguinus parkelj*. Photographs by G. Aljančič, with permission. See Plate 13.

result of the direct influence of the environment on the organism and that this change was transmitted to future generations. Writing in 1804, he stated:

> ... it becomes clear that the shrinkage and even disappearance of the organ in question are the results of permanent disuse of that organ. (Lamarck 1984)

Even Darwin saw subterranean animals as examples of eyelessness and loss of structure in general. For him, the explanation was a straightforward Lamarckian one, and one that did not involve adaptation and the struggle for existence.

> It is well known that several animals which inhabit the caves of Carniola [Slovenia] and Kentucky, are blind.... As it is difficult to imagine that eyes, though useless, could be in any way injurious to animals living in darkness, their loss may be attributed to disuse. (Darwin 1859)[2]

Small wonder then that for decades following Darwin, adaptation was not associated with subterranean organisms. Much confusion followed, which we only briefly consider.

At the end of the nineteenth century, one of the leaders of the neo-Lamarckian school of evolution was Packard, who was also the leading American speleobiologist of his time. He was convinced that use and disuse governed the evolution of subterranean animals and gave virtually no role to natural selection and adaptation. He also held that evolution of what we would now call troglomorphy was rapid. Packard knew cave animals well. He had visited several dozen caves in North America and described many species (Packard 1888). Although Lamarckian evolution and the theory of use and disuse are discredited, Romero (2009) points out that the terminology of neo-Lamarckism still pervades speleobiology—for example, the term regressive evolution. Romero (2004) also argues for the importance of phenotypic plasticity in adaptation to caves, and that the evolution of elaborated features is unimportant and largely environmentally determined. This is decidedly a minority viewpoint.

Much closer to a modern view of evolution and adaptation of subterranean cave animals is Racoviţă's *Essai sur les problèmes biospéologiques*, published in 1907. Racoviţă takes Darwin to task for ignoring natural selection in subterranean environments:

> ... he [Darwin] thinks the struggle for life does not exert itself in this environment. It has been seen that this idea is wrong. (Racoviţă 2006)

Enormously influential among European speleobiologists, Racoviţă unfortunately had negligible impact on his American counterparts. Eigenmann (1909), the dominant American speleobiologist in the first decade of the 20th century and a staunch neo-Lamarckian, was apparently unaware of Racoviţă's work.

---

[2] This quotation remained the same through all six editions of the *On the Origin of Species*.

Writing at about the same time as Racoviță, the American biologist Banta (1907) supported what seems to a modern biologist a bizarre theory—that of orthogenesis:

> *Animals do not possess degenerate eyes and lack pigment because they are cave animals.... They are cave animals because their eyes are degenerate and because they lack pigment.... They are isolated in caves and other subterranean abodes because they are unfit for terranean life...* (Banta 1907)

In other words, animals are not blind because they are in caves; they are in caves because they are blind. The major proponent of orthogenesis in subterranean species was Vandel, who wrote the first widely available textbook on subterranean biology, available in both French and English (Vandel 1964, 1965). Vandel also minimized the role of natural selection and adaptation with the following analogy which links the idea of aging individuals to senescent phyletic lines:

> *The idea of adaptation has grown to the point where it has been written that depigmentation and anophthalmy represent 'adaptations to subterranean life'. This is like saying that catarrh [common colds], rheumatism, and presbyopia [far-sightedness] are adaptations to old age.*

The final approach to the evolution of the morphology of cave animals that is not selectionist came from Kosswig. Working in the 1930s, Kosswig was very interested in genetic polymorphism and believed that it held the key to the understanding of regressive evolution. On the basis of his studies of the highly polymorphic isopod *Asellus aquaticus* in the Postojna–Planina Cave System in Slovenia (Kosswig and Kosswig 1940), he believed that mutation was the key to understanding this variability, and he held that the presence of highly polymorphic populations of stygobionts was the result of mutations accumulating that were not subject to selection (Kosswig 1965).

These ideas were greatly elaborated and refined by his student Wilkens (1971, 1988). Wilkens worked on the Mexican cave fish *Astyanax mexicanus*, especially with respect to eye degeneration. In essence Wilkens held that eye and pigment loss was almost entirely the result of the accumulation of morphologically reducing, selectively neutral mutations. He has continued to champion neutral mutation theory as the explanation for regressive evolution (Wilkens and Strecker 2017). The emphasis of Wilkens and his colleagues was strongly oriented toward regressive features and no list of elaborated features in *Astyanax* was published until Jeffery (2001) (Table 3.4), although Schemmel (1974) did some work on the genetics of taste buds and Wilkens (1988) mentioned some elaborated features in his extensive review of regressive features.

Other ideas, often combined with natural selection, have been proposed as being important in explaining the evolution of subterranean organisms (see Barr 1968 for a review). Most prominent of these has been the suggestion of differential responses to light, with smaller-eyed individuals seeking darkness (Lankester 1925, Ludwig 1942).

In spite of this rather long and diverse list of non-selectionist ideas in subterranean biology, the adaptationist paradigm is very strong among contemporary speleobiologists. Nearly all contemporary descriptions of morphology include discussion of how the morphology is adaptive. We have already seen an example of this with the subterranean amphipods in the family Ingolfiellidae where the morphology of interstitial species was compact to match the compact nature of the living space and the cave species had greatly elongated appendages that appear appropriate for the large cavities they live in (Coineau 2000; Fig. 3.8). Furthermore, the loss of structures, regressive evolution, to which Porter and Crandall (2003) apply the term 'evolution in reverse', is universal (Wilkens and Strecker 2017). Structures are lost, just as structures are gained, in all phyletic lineages. It is not a feature unique to subterranean species; it is just more obvious in them (Fong et al. 1995).

The rise in importance of neo-Darwinian thinking can be largely attributed to two speleobiologists working in the 1960s. Christiansen used a comparative approach to study the adaptation of Collembola to darkness and to walking on wet surfaces, including pools (Christiansen 1961, 1965; Fig. 3.9). Christiansen very deliberately set out to establish a neo-Darwinian example from the cave fauna, which he was successful in doing. He used a comparative approach, the preferred method of analysis at the time, and showed consistent morphological differences in claw structure and differences in locomotory behaviour between cave-modified and unmodified species, differences that he found repeated in different lineages of Collembola. Hence he demonstrated that there was convergent evolution (the independent evolution of similar traits) among cave Collembola, and termed characters subject to convergent evolution cave-dependent (troglomorphic) characters.

6.3 Adaptation in amblyopsid cave fish

Working at about the same time as Christiansen, Poulson (1963) studied both demographic and morphological characteristics of fish in the family Amblyopsidae. While Christiansen viewed the selective environment for Collembola as one of locomotion across wet surfaces in darkness, Poulson viewed the selective environment for amblyopsid fish as one of finding scarce energy resources in darkness.

Nine species of Amblyopsidae are known, all from the eastern and central United States (Armbruster et al. 2016). One, *Chologaster cornuta*, is strictly a surface-dwelling species found in freshwater marshes in the Coastal Plain from Virginia to Georgia, a range disjunct from others in the genus (Woods and Inger 1957). *Forbesichthys agassizii* (formerly *Chologaster agassizii*) and *Forbesichthys papilliferus* are known from springs and caves in the central United States (Illinois, Kentucky, Missouri, and Tennessee). These three species have eyes and pigment; the other species in the family have lost their pigment and only vestiges of a non-functional eye remain. *Typhlichthys eigenmanni*, *Typhlichthys subterraneus*, *Amblyopsis spelaea*, *Amblyopsis hoosieri*, *Troglichthys rosae*, and *Speoplatyrhinus poulsoni* are known from caves in the same region (Armbruster *et al.* 2016). Phylogeny within the Amblyopsidae has proved difficult to resolve, in part because the family is an old one (>10 million years; Niemiller *et al.* 2012b, 2013a), morphological convergence is strong (Armbruster *et al.* 2016), and cryptic species are common (Niemiller *et al.* 2012a). The most complete molecular phylogeny (Niemiller *et al.* 2012b) shows multiple, independent loss of eyes (Fig. 6.2). This phylogeny differs considerably from the one implied by Poulson (1963) of a parallel loss of eye structure and length of time isolated in caves. They are all stygobionts but the degree of eye degeneration differs, with *T. subterraneus* showing the least and *S. poulsoni* showing the most. Niemiller *et al.* (2013b) reported that functionality of the *rhodopsin* gene in three independent lineages of amblyopsids was lost, and attributed this to neutral mutation processes, in contrast to cases where the *opsin* genes are conserved

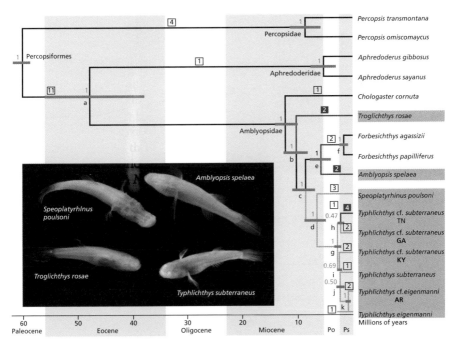

Fig. 6.2   Molecular phylogeny of Amblyopsidae, based on Niemiller *et al.* (2012b). Used with permission of John Wiley and Sons Inc. See Plate 14.

in troglomorphic species of amphipods and crayfish, apparently as a result of pleiotropy (Crandall and Hillis 1997, Carlini *et al.* 2013).

Poulson did not emphasize the reduced (regressive) characters, but rather emphasized those aspects of morphology, behaviour, and life history that indicated adaptation to subterranean life. This switch in emphasis really changed the question from how subterranean animals came to lose their eyes and pigment to how they coped with the harsh subterranean environment.

Neuromast cells in the lateral line system enable fish to detect vibrations in the water. Relative to *Chologaster* and *Forbesichthys*, stygobiotic amblyopsids had more and larger neuromast cells and a larger lateral line system (Fig. 6.3). Among the stygobionts, the neuromast system is least developed in *T. subterraneus* and most developed in *S. poulsoni*.

The brains of amblyopsids show differences related to life in darkness (Fig. 6.4). Most notably, the olfactory lobe is increased and the optic lobe is decreased in stygobionts relative to the other species. In addition, stygobionts have larger heads that displace more water and, therefore, make the detection of obstacles more efficient (Poulson and White 1969). Thus, larger heads are an adaptation to darkness. Of course, there may be other explanations for changes in head size.

Given that amblyopsids are predators, it is likely that they are extremely resource limited. On this basis, Poulson hypothesized that metabolic rates

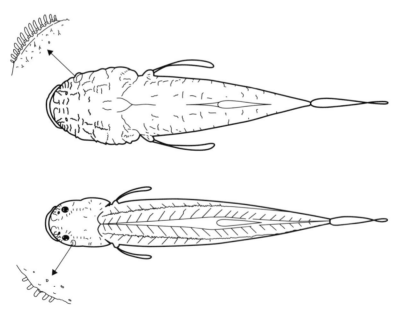

Fig. 6.3   Diagrams comparing the stygobiotic *Amblyopsis spelaea* with its troglophilic relative *Forbesichthys agassizii*. Each of *F. agassizii*'s 'stitches' has fewer neuromasts. Adapted from Eigenmann (1909) and Poulson and White (1969).

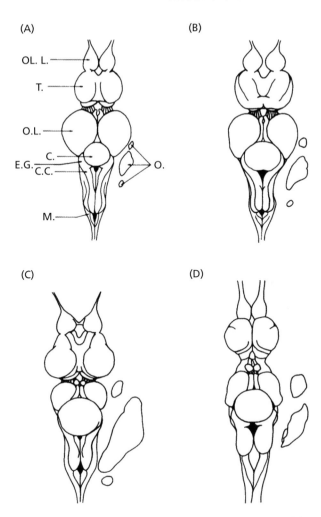

Fig. 6.4   Brain morphology of (A) *Chologaster cornuta*, (B) *Forbesichthys agassizii*, (C) *Amblyopsis rosae*, and (D) *Speoplatyrhinus poulsoni*. Parts labeled are: OL.L., olfactory lobe; T., telencephalon; O.L., optic lobe; C., cerebellum, E.G., eminentia granularis; C.C., cristae cerebelli; M., medulla oblongata; and O, otoliths. From Culver (1982). Reprinted by permission of the publisher from *Cave life: evolution and ecology* by David C. Culver, p. 27. Cambridge, Mass.: Harvard University Press. Copyright © 1982 by the President and Fellows of Harvard College.

should be reduced in stygobionts as a result of adaptation to an energy-poor environment. He measured both standard metabolic rate (metabolic rate with active movement) and routine metabolic rate (metabolic rate at normal activity levels) for *F. agassizii* and all of the stygobionts except *S. poulsoni*. Stygobionts showed at least a one-third reduction of both standard and routine metabolic rates relative to the non-stygobiont *F. agassizii* (Table 6.1).

Table 6.1 Metabolic rates in millilitres of O_2 per gram of fish per hour.

Species	Standard metabolic rate (mL O_2 g^{-1} hr^{-1})	Routine metabolic rate (mL O_2 g^{-1} hr^{-1})
Forbesichthys agassizii	0.0277	0.0415
Typhlichthys subterraneus	0.0157	0.0210
Amblyopsis spelaea	0.0176	0.0276
Amblyopsis rosae	0.0107	0.0114

Source: Data from Poulson (1963),

One of the most interesting aspects of Poulson's study was his evaluation of the life history characteristics of cave fish. The key comparison is between stygobionts and non-stygobionts (Table 6.2). Compared to non-stygobionts, stygobionts have at least a 20 per cent reduction in the number of eggs, at least a 100 per cent increase in the age of first reproduction, at least a 50 per cent reduction in the proportion of the population breeding at any one time, and at least a 40 per cent reduction in growth rate. If the ghost of the great orthogeneticist Vandel were present, he would certainly and correctly point out that these characteristics are a nearly inevitable consequence of starvation, and therefore are not necessarily adaptive. However, Poulson also found demographic characteristics that are much more difficult to dismiss and they make a very strong case for adaptation. These characteristics include a doubling of life span, at least a 40 per cent increase in egg size, and at least a 50 per cent increase in the maximum number of broods as a result of increased longevity (Table 6.2).

Reproductive effort per brood does not show a consistent difference between stygobionts and non-stygobionts, but lifetime reproductive effort does. Lifetime reproductive effort of stygobionts is either similar to or greater than non-stygobionts. This poses a paradox (Turquin and Barthelemy 1985; Culver 2012c). Even though the number of eggs produced in any one brood is small and the possibility of enough resources to allow for reproduction is small, the potential for reproduction of subterranean animals is often quite large. *T. subterraneus* is a good example of this, with a lifetime reproductive effort at least three times that of surface-dwelling species. Turquin and Barthelemy suggest that this paradox is essential to subterranean life. Although there are situations where there is a low but constant flux of organic matter (see Chapter 2), they suggest that organic carbon often comes in pulses or spurts, such as occurs with flooding (Hawes 1939). The combination of life span and ability to expend significant reproductive effort makes life in carbon-poor subterranean environments possible. This allows organisms to 'wait' for organic carbon. It also fits in nicely with the idea that for most of the time, the population growth rate (*r* of the standard growth equations) is

Table 6.2 Life history characteristics of amblyopsid fish.

	Chologaster cornuta	Forbesichthys agassizii	Typhlichthys subterraneus	Amblyopsis spelaea	Amblyopsis rosae
Reproduction					
Age at first reproduction (months)	12	12	24	40	37
Number of eggs	93	150	50	70	23
Egg diameter (mm)	0.9–1.2	1.5–2.0	2.0–2.3	2.0–2.3	1.9–2.2
Reproductive effort per brood (mm3 gm−1 of female)	64	148	452	55	83
Maximum proportion of ovigerous females per year	1.0	1.0	0.5	0.1	0.2
Maximum lifetime number of broods	1	2	3	5	3
Maximum lifetime reproductive effort (mm3 gm−1 of female)	64	297	903	260	249
Growth and longevity					
Longevity (years)	1.3	2.3	4.2	7.0	4.8
Growth rate (mm year−1)	2.4–3.8	1.7–2.2	1.0	1.0	0.9

Source: Poulson (1963)

slightly negative for subterranean populations. At least occasionally the growth rates must be positive, presumably occurring with influxes of organic matter. If not, the population would go extinct.

This adaptationist approach to subterranean organisms has had enormous impact on speleobiology. Numerous studies have duplicated parts of Poulson's analysis with other subterranean organisms. Hüppop (2000) reviewed the adaptations of cave animals to food scarcity and there are many studies that have found that subterranean species have lower metabolic rates and larger and fewer eggs. Poulson's study is important not only because it was the first comprehensive study of adaptation in subterranean species, especially with respect to life history changes, but also because he used phylogenetically appropriate comparisons. Imagine that he had used some surface-dwelling fish such as trout (*Trutta*) rather than *Chologaster* and *Forbesichthys* for comparison with the stygobiotic amblyopsids. In that case, differences might not be due to selection but rather to evolutionary history. For example, higher reproductive rates in the surface-dwelling species might have resulted from the fact that all species in a particular group, irrespective of habitat, tend to have higher reproductive rates. Unfortunately, many comparisons between surface and subterranean species are not phylogenetically appropriate, such as comparisons of distantly related amphipods in the genus *Gammarus* (largely surface dwellers) and the genus *Niphargus* (largely subterranean), a frequent comparison in the older literature.

Poulson also made comparisons among stygobionts. Although the amount of eye and pigment degeneration was not the focus of his study, he used the relative degeneration of eyes and pigment cells in different species as an indicator of how long species had been isolated in caves (Poulson 1969). That is, he was using these characters as a morphological clock in the same way that differences in mtDNA sequences are currently often used as a molecular clock. He proposed that *T. subterraneus* had been isolated the shortest time, and *S. poulsoni* the longest. The implication is that, given enough time, all cave amblyopsids would evolve into something that looks similar to *S. poulsoni*. It is fair to say that nowadays a molecular clock should be used, as is now available (see Fig. 6.2)—such techniques were not available in the 1960s when Poulson was doing this work. There is no simple answer to the question of the age of isolation of the different populations.

At least some of the differences among stygobionts are probably the result of different subterranean habitats, rather than differences in age. Noltie and Wicks (2001) showed that the habitat of *A. rosae* is shallower than the typical habitat of *T. subterraneus*, and therefore there is likely to be more organic carbon available for *A. rosae*.

This kind of comparative approach and evolutionary theorizing was sharply criticized by two leading evolutionary theorists (Gould and Lewontin 1979). They criticized what they called the adaptationist programme because no matter what observations were made about the biology of organisms, it was always possible to create a scenario that the patterns observed were adaptive. In particular, what Poulson did not do, and could not have done at that time, was show that natural selection was actually occurring in cave fish populations. Two of his students (Culver and Kane), together with Fong, did embark on an extensive research programme to measure selection directly in subterranean populations (Culver *et al.* 1995).

6.4 Adaptation in the amphipod *Gammarus minus*

One of the problems with the adaptationist programme is that any theory that purports to generality contains the risk of becoming circular and not falsifiable as a scientific theory. To avoid this, Brandon (1990) suggested five requirements for a 'complete adaptation explanation'. Modified to fit a subterranean environment, they are as follows:

1. Evidence that selection has occurred, that is, that some morphological types, such as those with elongated appendages or reduced eyes, have higher reproductive rates in subterranean environments.

2. An ecological explanation of differential reproductive rates in terms of the selective environment in subterranean habitats.
3. Evidence that the traits in question, such as eye size and appendage length, are heritable, that is, they have a genetic component.
4. Information about gene flow and genetic distance among surface and subterranean populations.
5. Phylogenetic information concerning what has evolved from what, that is, which character states are ancestral (such as large eyes) and which are derived (such as vestigial eyes).

Culver *et al.* (1995) attempted a rigorous demonstration of natural selection of the kind proposed by Brandon, with the amphipod *Gammarus minus*.

G. minus is a widespread inhabitant of springs, occurring in a broad arc from Pennsylvania and Maryland to Arkansas and Missouri in the eastern and central United States. Spring populations are not troglomorphic and retain pigment and well-developed eyes (Holsinger and Culver 1970). It is only found in carbonate springs, or at least springs with pH greater than 6 and conductivity typically greater than 100 µS/cm (Glazier *et al.* 1992). It is a common inhabitant of caves throughout its range as well, but in most caves it is only slightly different morphologically from spring populations. In two areas in West Virginia (Greenbrier Valley) and Virginia (Ward's Cove), both with extensive cave development and caves of more than 20 km in length, morphologically distinct populations occur. Populations in these extensive cave systems were both large enough and isolated enough from surface populations that troglomorphy could evolve. Individuals in spring populations feed on detritus and associated biofilm, but individuals in cave populations are omnivores and include different sources of nitrogen than leaf litter in their diet (MacAvoy *et al.* 2016). One of the caves where it is found is Organ Cave, West Virginia (see Chapters 2 and 4). The selective environment is one of darkness, but without the extreme resource limitation characteristic of the amblyopsid fish populations. The streams have considerable organic input (Simon *et al.* 2007a) and the amphipods are either at the base of the food chain (Simon *et al.* 2003) or omnivores (MacAvoy et al. 2016). In both spring and caves there are often large populations, numbering more than 10^6 individuals.

There are many morphological differences between the two kinds of populations, and individuals in the cave populations are larger, pale but purplish in colour, have longer appendages, and have compound eyes reduced to blotches of pigment (Holsinger and Culver 1970). The area of the eye is much reduced, to differing degrees in different caves (Fig. 6.5). Neurological differences that parallel those in the amblyopsid fish (Fig. 6.4) are also present, namely an increase in the olfactory lobes and a decrease in the optic lobe of the brain (Fig. 6.6). There is no overlap at all in the size of the optic lobes of individuals from cave and spring populations. The morphologically

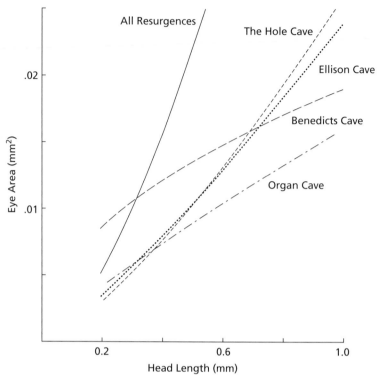

Fig. 6.5   Allometric curves ($y = ax^b$) of eye area and head length for four cave populations of *Gammarus minus* in four different karst basins in the Greenbrier Valley of West Virginia, USA. Allometric curves for the resurgences of the four karst basins are not significantly different from each other. Data from Culver (1987). Reprinted by permission of the publisher from *Adaptation and Natural Selection in Caves: the evolution of* Gammarus minus by David C. Culver, Thomas C. Kane, and Daniel W. Fong, p. 106. Cambridge, Mass.: Harvard University Press. Copyright © 1995 by the President and Fellows of Harvard College.

modified populations, with small eyes and large antennae, occurring in large caves in the Greenbrier Valley and Ward's Cove are easily distinguishable from spring populations, but they do not show extreme morphological change. That is, they still retain some components of the compound eye and some pigmentation. In fact, it is their variability that makes a detailed analysis possible.

Brandon's first requirement, differential fitness, requires some way to measure reproductive success. Ideally one could follow the fate of individual amphipods and know how long they survived and how many offspring they produced to measure lifetime fitness. This of course cannot be performed, but two components of fitness can be measured. One component is amplexus, the grasping and carrying of females by males before fertilization. Amplexus lasts for several weeks in *G. minus* and is an indicator of mating

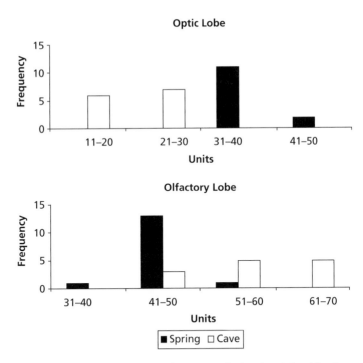

Fig. 6.6   Histograms of the size of the optic and olfactory ganglia (1 unit = 2.5 μm) for *Gammarus minus* from Organ Cave and the spring from which the water from Organ Cave resurges. Data from Culver *et al.* (1995).

success. The other component is the number of eggs carried by an ovigerous female in an external brood pouch. By comparing the morphology of individuals relative to their mating success and fertility, the intensity of selection on different characters can be estimated. There are numerous details about how this is done, the most important of which is that different characters are scaled to the same size so that selection gradients are standardized (Jones *et al.* 1992). Otherwise selection would seem to always be stronger on larger characters. They took a series of collections in several different caves and springs and estimated selection gradients for head length (a measure of overall body size), eye size, and antennal size. Overall, selection was occurring on all three character types (Table 6.3). The intensity and direction of selection did not differ among caves and did not differ among springs, but the direction of selection did differ between caves and springs—eyes were selected against in caves and selected for in springs.

Carlini and Fong (2017) provide evidence from genome-wide transcriptome analyses of widespread selection, and that upregulating selection was actually more prevalent in a cave population than a spring population. Of 104 630 transcripts identified, 1517 were significantly up-regulated in the cave population compared to the spring population, while only 551 were significantly

Table 6.3 Means and standard deviations (S.D.) for standardized selection gradients for *G. minus* from caves and springs.

Character	Habitat	N	Mean	S.D.
Head length	Cave	55	0.18[a]	0.19
Head length	Spring	36	0.19[a]	0.22
Antennae	Cave	130	0.06[b]	0.21
Antennae	Spring	90	0.06[b]	0.23
Eye	Cave	85	-0.08[a]	0.21
Eye	Spring	54	0.06[a]	0.22

Source: Data from Jones *et al.* (1992). Used with permission of Blackwell Publishing.
[a] $p < 0.01$
[b] $p < 0.05$
N = sample size

down-regulated in the cave population relative to the spring population (Fig. 6.7). Nucleotide diversity was reduced in the down-regulated transcripts in cave populations, and Carlini and Fong (2017) argue, on the basis of several measures of selection such as the ratio of synonymous to non-synonymous mutations, that this is due to natural selection. They also identify a small number of genes (five) in the cave population with premature termination codons, which they argue is evidence for relaxation of selection (neutral processes) since no functional transcript is produced.

Brandon's second criterion states that we should be able to make sense of the direction of selection observed, and that it fits with the morphological differences between populations. The direction of selection on different morphological features in cave populations makes sense in the general context of the subterranean environment—larger antennae are selected for to extend the zone of tactile perception in the darkness of the cave stream and smaller eyes are selected for as a result of energy economy or neurological efficiency. Larger animals are selected for, probably for a variety of reasons, including increased number of eggs produced by larger animals. The pattern in spring populations produced some surprises. Probably for the same reasons as in caves, larger body size was probably selected for but it was counteracted in some springs by differential predation and in some springs by the small fish *Cottus carolinensis* (Culver *et al.* 1995). Selection for larger eyes was expected, but selection for increased antennae size was not. Selection for larger antennae was an indication of lack of understanding of the selective environment of springs.

Brandon's third requirement is that the morphological differences that were important in selection were heritable, that is, that they had a genetic component. Fong (1989) estimated the percentage of morphological variance that was due to genetic variation by measuring what is called broad-sense heritability (Falconer and McKay 1996). Broad-sense heritability measures the

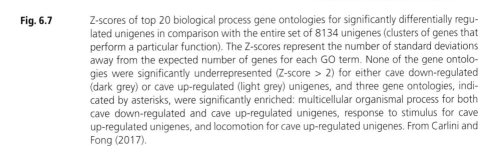

Fig. 6.7 Z-scores of top 20 biological process gene ontologies for significantly differentially regulated unigenes in comparison with the entire set of 8134 unigenes (clusters of genes that perform a particular function). The Z-scores represent the number of standard deviations away from the expected number of genes for each GO term. None of the gene ontologies were significantly underrepresented (Z-score > 2) for either cave down-regulated (dark grey) or cave up-regulated (light grey) unigenes, and three gene ontologies, indicated by asterisks, were significantly enriched: multicellular organismal process for both cave down-regulated and cave up-regulated unigenes, response to stimulus for cave up-regulated unigenes, and locomotion for cave up-regulated unigenes. From Carlini and Fong (2017).

fraction of the variance in a character due to additive genetic variance and maternal effects. Because of limitations in rearing *G. minus*, it was not possible to measure narrow-sense heritability, which includes only additive genetic variance. The broad-sense heritabilities (genetic plus maternal effects) were high with an overall average of 0.72 (Fig. 6.8). All but two of the 36 determinations of heritability were statistically significant and so the requirement that the traits have a genetic component was easily met.

Brandon's fourth requirement is that population structure is known from a genetic and selective point of view. The troglomorphic populations of *G. minus* occur in five underground drainage basins in the Greenbrier Valley of West Virginia and one such basin in Ward's Cove in Virginia. On the basis of an extensive survey of allozyme variation (Kane *et al.* 1992; Sârbu *et al.* 1993) and mitochondrial DNA gene sequencing (Carlini *et al.* 2009), cave populations within a basin were very similar to each other, but distinct from cave populations in other drainage basins. The differences are extensive, at the level of species differences (Culver *et al.* 1995). The morphologically modified cave populations in the different basins are the result of separate invasions upstream into subsurface basins, probably at different times, of spring populations. As the invasion of the subsurface is in the upstream direction, colonization was probably active, rather than the result of passive stranding (see Chapter 7). Active invasion may have been triggered by factors such as reduced predation pressure or reduced temperature fluctuations. Resurgence populations are quite similar, indicating gene flow among them. Genetic analysis thus indicates that *G. minus* is really a species complex with different cryptic species in different karst basins.

Brandon's final criterion is that the ancestral and derived state of morphological traits is known. This is generally an easy problem for subterranean populations, and one of the reasons why they are attractive models for the study of adaptation. Derived character states are usually the troglomorphic

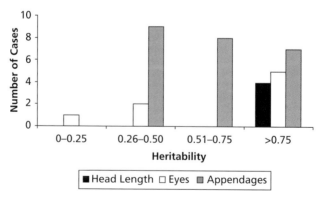

Fig. 6.8 Distribution of estimates of broad sense heritability (see text) for cave and spring populations of *Gammarus minus*. Data from Fong (1989).

states—reduced eyes and pigment and elongated appendages, although reversals are known, including cave scorpions (Prendini *et al.* 2010) and karst window populations of *Gammarus minus* (Culver *et al.* 1995).

The relationships between the physical environment (geographical distance, karst basin, and habitat) and genetic and morphological differences are summarized in Fig. 6.9. Genetic distance among cave populations is largely determined by which karst basins populations are in, rather than geographical distance or even habitat differences (caves vs springs). Morphological differences are largely determined by differences in selection in different habitats. Genetic and morphological differences are connected in part by the length of time populations have been isolated in caves.

All in all, the demonstration of selection and adaptation among cave populations of *G. minus* is both detailed and convincing. What is not so clear is what factors in addition to selection may be driving changes in cave populations, especially those 'regressive' features of eye and pigment loss. Culver *et al.* (1995) showed that selection was likely operating on eye size, but their study of adaptation did not consider in detail the possible role of neutral mutation. However, they did compare relative amounts of morphological change for eyes, appendages, and size in the population in Organ Cave to that of its resurgence (Fig. 6.10). To convert the differences to a rate of change, an estimate of time of divergence, based on genetic data, of 500 000 years was used. No matter how long the time of divergence, the differences between rate of change of antennae, body size, and eyes remain proportionally the same, and the change in eye size was at least one order of

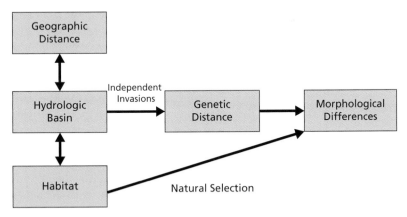

Fig. 6.9   Factors affecting morphological and genetic differentiation in *Gammarus minus*. Hydrological basin, habitat, and geographical distance are themselves all correlated and the remaining arrows indicate the major pathways. Modified and adapted from Culver *et al.* (1995). Reprinted by permission of the publisher from *Adaptation and Natural Selection in Caves: the evolution of* Gammarus minus by David C. Culver, Thomas C. Kane, and Daniel W. Fong, p. 155. Cambridge, Mass.: Harvard University Press. Copyright © 1995 by the President and Fellows of Harvard College.

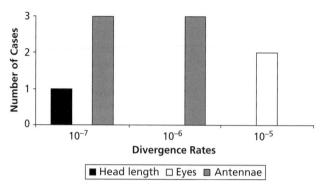

Fig. 6.10 Histograms of the rates of morphological change per year for standardized morphological variables (i.e., mean = 0, S.D. = 1) for the population of *Gammarus minus* in Organ Cave, West Virginia, USA. Rates, in standardized units per year, were calculated on the assumption that the cave population has been isolated from the spring population for 500 000 years. Data from Culver *et al.* (1995).

magnitude faster. The conclusion is that both neutral mutation and selection change eyes, while only selection changes antennae and body size.

6.5 Adaptation in the cave fish *Astyanax mexicanus*

Unlike the studies of amblyopsid cave fish and the amphipod *G. minus*, studies on the Mexican cave fish *A. mexicanus* have focused on the reduced features of its anatomy, especially the eye. While early work by Wilkens (1988) emphasized neutral mutation theory and other non-selectionist explanations, ideas about eye loss in *Astyanax* have come full circle, with Jeffery considering eye reduction an integral part of adaptation to the subterranean environment (Jeffery 2005a,b).

There have been many more papers devoted to the Mexican cave characin, *A. mexicanus*, than to any other subterranean species. Originally discovered in the 1930s (Hubbs and Innes 1936), it immediately attracted the interest of biologists because of the ease with which it could be transported and cultured in the laboratory. Approximately 30 cave populations of *A. mexicanus*, mostly from the Sierra de El Abra region of northeastern Mexico, are known (Mitchell *et al.* 1977). Unlike amblyopsid fish, by far the most obvious modifications of *A. mexicanus* for cave life are reduction of eyes and pigment (Fig. 6.11). Cave populations differed markedly in the degree of eye and pigment degeneration, and study of them was greatly facilitated by the fact that hybrids between cave populations and between cave and surface-dwelling populations could readily be produced (Tabin 2016). It is likely that they initially became isolated in caves when surface streams were captured by underground streams (Mitchell *et al.* 1977). They are found in residual

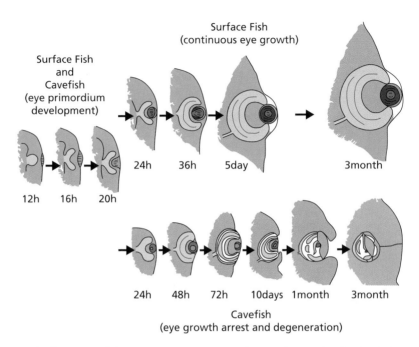

Fig. 6.11 Eye development and degeneration in *Astyanax mexicanus*. Diagram showing the timing of eye growth and development in surface fish (top) and eye degeneration in cavefish (bottom). Drawing by W. Jeffery, with permission. See Plate 15.

stream pools during the dry season and very little is known about their habitat and distribution during the wet season when caves are flooded. Organic carbon is probably not in short supply in these caves, and large amounts of organic matter enter the cave during floods. Rasquin (1947) noted that cave *Astyanax* carried large fat reserves, also indicating an ample food supply. As with *G. minus*, the main component of the selective environment is darkness.

Wilkens, a strong proponent of the neutral mutation theory of eye and pigment loss in subterranean organisms, embarked on an extensive research programme with breeding populations of *Astyanax* beginning in the late 1960s (see Wilkens 1971, 1988 and Wilkens and Strecker 2017). Several observations seemed to lend support to a neutral mutationist view. The high variability of eyes in many cave populations seemed to indicate that selection was relaxed; otherwise, there would not be so much variability (see also Hinaux *et al.* 2013). Wilkens' demonstration that different genes were involved in eye loss in different populations (Wilkens 1971) also supported the importance of mutations. He was able to demonstrate this because hybrids between different cave populations had larger eyes than either parental population. By examining the ratios of different kinds of offspring, he was able to show that the control of eyes in cave fish was multifactorial, and he thought around 10 genes were involved (the number seems to vary by population, see O'Quin and McGaugh [2016]). This research led Wilkens to propose a model of eye loss (elaborated more formally by Culver 1982) where a series of genes, acting independently, were needed to produce a fully developed eye. Mutations in any one of these genes would lead to at least eye reduction if not a loss of some eye components.

The availability of the genetic, cellular, and molecular tools necessary to study the actual genes involved and their role in development allowed Jeffery and colleagues to break new ground in the study of eye loss. Rather than hypothesize that a series of identically acting genes (what we would call quantitative trait loci, QTLs), they were able to identify the particular genes involved. The first breakthrough came with the identification of reduced levels of *pax6* expression and apoptosis (cell death) in the lens, which were considered as the key factors in eye degeneration (Jeffery and Martasian 1998). The developmental steps in eye regression were (1) *pax6* expression was reduced at the anterior midlines, (2) a smaller lens and optic vesicle probably the result of (1), (3) apoptosis rather than cell differentiation in the lens, (4) further eye structures failing to develop as a result of the absence of lens signalling, and (5) sinking of the eye into its orbit (Fig. 6.11). Evidence of the central role played by lens apoptosis comes from experiments where cave fish lenses were transplanted on to surface fish and vice versa. The lens vesicle of the eye of a cave fish with a surface fish lens underwent further differentiation and growth, while the lens vesicle in the eye of a surface fish with a cave fish lens died (Yamamoto and Jeffery 2000).

If this were the whole story of eye degeneration in *Astyanax*, it is consistent with a neutral mutation hypothesis—only losses and degeneration occur even at the cellular and genetic level. But it is not the whole story. Yamamoto *et al.* (2004) showed that another protein—sonic hedgehog (*shh*)—played a critical role, and unlike *pax6*, *shh* increased in expression in cave fish compared to surface fish. At the neural plate stage, its domain was 10 cells wide in cave fish and only 6 cells wide in surface fish (Yamamoto *et al.* 2004). One of its impacts was to induce apoptosis in the lens. They also mimicked the effect of increased *shh* activity by injecting *shh* messenger RNA in surface fish, resulting in reduced optic vesicles and cups. As they point out, this result cannot be explained by neutral mutations, which involve losses, because sonic hedgehog increases in expression. Even more telling was the discovery of an increase in function of the heat shock protein gene *HSP90* (Rohner *et al.* 2013). It masks standing eye variation in surface populations, but if it is suppressed in cave populations, more variation is exposed to selection. The authors suggest that reduced conductivity of the water in caves may trigger its inhibition, but whatever the mechanism, *HSP90* is likely to play an important role in the evolution of reduced eyes. This cannot be explained by neutral mutation theory. Jeffery (2005a) indicates that eye degeneration operates in a similar way in different *Astyanax* populations, indicating a non-random chain of convergent events (Table 6.4).

Table 6.4 Events and processes associated with eye degeneration in different cavefish populations.

Event or process	Cavefish populations				
	Pachón	Los Sabinos	Tinaja	Curva	Chica
Smaller eye primordium	+	+	+	+	+
Loss of lens and optic cup	+	+	+	+	+
Lens apoptosis	+	+	+	?	+
Eye restoration by lens transplantation	+	+	+	?	+
hsp90α activation	+	?	?	+	+
Pax6 down-regulation in optic vesicle fields	+	+	?	+	?
Hb expansion at embryonic midline	+	+	?	+	+

'+' indicates the event or process was detected and '?' indicates it has not yet been studied.
Source: Modified from Jeffery (2005a). Used with permission of Oxford University Press.

Jeffery (2005a,b) goes on to argue that energy conservation is not the likely selective agent. In lens apoptosis, cells first proliferate and then die, an energy inefficiency. Furthermore, populations under bat roosts, which are highly unlikely to be resource limited, also show reduced eyes. He suggests rather that eyes could be lost as a pleiotropic effect of selection for constructive traits, such as taste buds which may be positively regulated by *shh* signalling. Another potential case of pleiotropy is the retention of a light-sensitive rhodopsin-like protein in the pineal gland of cave fish, even though it is not present in the eye (Yoshizawa and Jeffery 2008).

The pigment system of *Astyanax* provides interesting similarities and contrasts with the cave fish eye. Most of the work has been on the production of melanin, a black pigment. Two other pigment cell types occur in cave fish but have been little studied (Jeffery 2006, Jeffery *et al*. 2016)—silver iridophores and orange xanthophores. In cave fish, melanoblasts are normally produced by the neural crest, but blocked in differentiation. Loss of pigment is caused by mutations in a single gene, *oca2* (formerly known as the *P* gene), and different mutations in different populations have resulted in albinism (Protas *et al*. 2006). In particular cave fish, melanoblasts are unable to convert L-tyrosine to L-DOPA (and melanin). Other genes are involved in pigment reduction (Protas *et al*. 2006, 2007) but *oca2* is always involved in albinism, although perhaps not universally so in other cave fish species (Trajano 2007). Bilandžija *et al*. (2013) suggest that knockdown of *oca2* expression allows for increases in the catecholamine system, which in turn controls feeding and sleep (Fig. 6.12).

In contrast to the pleiotropy of the albinism gene, *oca2*, Protas *et al*. (2007) suggest that selection may not be involved in the reduction of number of melanin-containing cells. They showed that the polarities of mutations for QTLs are different in eye and pigment systems. Polarity refers to whether the mutation increases or decreases gene expression. Assuming that mutations

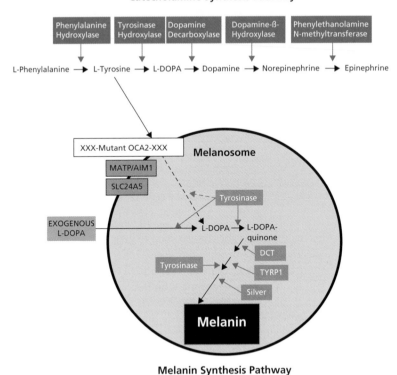

Fig. 6.12   The relationship between the catecholamine and melanin synthesis pathways in *Astyanax* cavefish. In albino cavefish, a mutated *oca2* gene (white box with XXX) affects the first step of the pathway prior to tyrosinase function and prevents melanin synthesis. The defect caused by *oca2* loss of function can be rescued by exogenous L-DOPA. Solid lines: steps that occur in surface fish and in cavefish after L-DOPA rescue of melanogenesis. Dashed lines: steps that are absent in cavefish. From Bilandžija *et al.* (2013). See Plate 16.

result in both increases and decreases, the presence of both positive and negative polarities for pigment QTLs is consistent with neutral mutation and genetic drift. In contrast, QTL polarity for eyes was negative, which they interpret as evidence for selection. However, whether there are many mutations that increase expression in an elaborate structure such as the eye is open to debate. For example, Wright (1964) argued that 'new alleles would on average tend to bring about reduction of the organ after its maintenance had ceased to be an object of natural selection'. Because pigments are less complex than eyes, this, rather than differences in the role of selection, may explain the different distribution of polarities.

While there is a convergence of pathways for albinism, this need not be the case when natural selection is involved, either directly or via pleiotropy.

For example, Kowalko *et al.* (2013) showed that the feeding angle in cave *Astyanax*, which is subject to strong selection, evolved via different pathways and different QTLs. In a transcriptomic analysis of two cave populations of *A. mexicanus*, Stahl and Gross (2017) show that a number of adaptive traits evolved through a combination of unique and shared gene expression patterns.

The emerging view of *A. mexicanus* is quite different from the view that it shows no troglomorphies except for losses of eyes and pigment (Romero 2001). There are both constructive changes and regressive changes (Table 3.4) and evidence that constructive and regressive changes could be linked by pleiotropy. The majority of workers in the field hold that selection has some role, even in eye and pigment regression. This is certainly the case for the contributors to the volume on *Biology and Evolution of the Mexican Cavefish* (Keene *et al.* 2016). This is not to say that members of the 'pleiotropy' school do not deny the importance of neutral processes and genetic drift (see Aspiras *et al.* 2015, Rétaux and Casane (2013), but they do find pleiotropy to be very widespread (Jeffery 2009). On the other hand, the minority who champions neutral mutation gives almost no role to pleiotropy in eye and pigment loss (Wilkens and Strecker 2017, Poulson 2017b). The differences between the neutralist and pleiotropy schools run deeper than this. In particular, Wilkens and Strecker (2017) dispute the pleiotropy school with respect to the number of times *Astyanax* has independently invaded caves, and dispute the basic taxonomy of the genus. Whether this dispute persists is open to question, but in section 6.8 we explore some possible new directions.

6.6 Experimental studies of adaptation in other subterranean organisms

Although the study of *Astyanax mexicanus* and to a lesser extent *Gammarus minus* and the Amblyopsidae have dominated studies of adaptation, other organisms have yielded useful insights, especially in the age of transcriptomics, the study of the expression of RNAs in a particular cell population or organism. This is a rapidly expanding field and we briefly mention several such studies.

Friedrich *et al.* (2011) looked at the transcripts for genes involved in vision and circadian rhythms in the beetle *Ptomaphagus hirtus* from Mammoth Cave, Kentucky. *P. hirtus* has only the vestige of an external eye; it is generally thought that such eyes have no function and occur in species that have not been isolated in subterranean habitats for a long enough time for the eye to entirely disappear. However, Friedrich and his colleagues demonstrated that *P. hirtus* was photophobic, and that the genes for phototransduction and circadian clock were present. They suggested a possible adaptive advantage to *P. hirtus* to be able to detect light. Mejía-Ortíz *et al.* (2018) suggest that troglomorphic species may be attracted to very weak light because there will be greater food resources in even weakly photic habitats. The whole field of the response of subterranean species to light is a complex one, but worthy of continuing study (see Fišer *et al.* 2016).

In common with *Astyanax mexicanus* and *Gammarus minus*, subterranean populations of *Asellus aquaticus* can interbreed with surface ancestors. In the Dinaric karst, several distinct subterranean subspecies occur, as well as the entirely subterranean *A. kosswigi* (Verovnik 2012). Stahl et al. (2015) produced a linkage map utilizing cave–surface hybrids and identified 742 single nucleotide polymorphisms (SNPs) and four new candidate genes to add to an existing linkage map. They also identified allele-specific expression differences in the hybrid individual, which should lead to a more detailed understanding of adaptive changes in the cave population. Given the complex history of multiple invasions of *A. aquaticus* into caves in the Dinaric karst, it holds considerable potential for further study (see Protas and Jeffery 2013).

The Chinese fish genus *Sinocyclocheilus* has more than 50 species, about equally divided between stygobionts and stygophiles (Ma and Zhao 2012, Zhao and Zhang 2009). Species in the genus have invaded caves multiple times (Chen *et al.* 2016), and based on a transcriptome analysis of eye genes, Meng *et al.* (2013) showed that eye reduction was lens-independent, in contrast with *Astyanax mexicanus*, and thus eye reduction can occur by very different pathways in different lineages. *Sinocyclocheilus* species show some unique modifications for subterranean life, especially dorsal horn-like structures that are apparently fat storage organs (Ma and Zhao 2012).

6.7 How long does adaptation to subterranean life take?

Time has entered into the consideration of adaptation and regressive evolution in indirect ways. For example, Poulson (1963) took a Darwinian gradualist view of adaptation, and held that the different degrees of troglomorphy of amblyopsid fish reflected different times of isolation (see also Eigenmann 1909). The working hypothesis at that time was that the isolation of cave fauna in North American caves happened sometime during the Pleistocene (e.g., Barr 1968). The implication was that adaptation to subterranean life takes several million years. Niemiller *et al.* (2012b) estimate the age of the subterranean invasion by Amblyopsidae is 10.3 Mya.

In contrast, Mitchell *et al.* (1977) suggested that *Astyanax* was isolated in caves about 10 000 years ago. This fits in with the view that *Astyanax* showed little in the way of adaptation to cave life. The major challenge was to understand how eye and pigment loss could occur within this time frame (Barr 1968, Culver 1982). In the case of *Astyanax*, good estimates of the age of different lineages are now available (Porter *et al.* 2007). They estimated time using two mitochondrial genes (*cytb* and *ND2*) as well as the fish fossil record. They looked at two lineages, one with surface populations still extant, and another without surface populations. They estimate that the time available for the evolution of troglomorphy in *A. mexicanus* is between 0.9 and 2.1 million years (Pleistocene), based on *ND2* and fossil calibrations, while it was between 1.5 and 5.2 million years (Pliocene) based on *cytb*. The actual time was probably on the lower end of the estimate, since *Astyanax* originated in South America and only passed northward into Mexico after the isthmus of Panama was formed in the Pliocene about 3 to 4 million years ago. Whatever the correct age of invasion (Ornelas-Garcià *et al.* 2008 provide similar estimates), it is two orders magnitude greater than Mitchell *et al.*'s estimate of 10 000 years. Using less extensive molecular data, Culver *et al.* (1995) estimated the time available for the evolution of troglomorphy in *G. minus* to be between 100 000 and 500 000 years, depending on the population. For other species, such as the salamander *P. anguinus*, the isopod *A. aquaticus* (see Chapter 7), and beetles in the genus *Leptodirus*, the time available for the evolution of troglomorphy can be much longer, in the range of 1–10 million years (Trontelj *et al.* 2007).

6.8 Revisiting the selection–neutrality controversy

Although most recent studies of eye and pigment loss, especially eye loss, ascribe some role to selection, usually in the form of pleiotropy (Keene *et al.* 2016), there remains vigorous opposition (Wilkens and Strecker 2017). There are several reasons for the continuing controversy, beyond attachment to one's own ideas. The first is that different kinds of evidence are typically used by the two schools of thought. Much of the support for pleiotropy comes from demonstration of the pleiotropic effects of a particular gene or pathway. A notable example of this is the bifurcation of catecholamine and L-DOPA pathways controlled in part by the *oca2* gene (Bilandžija *et al.* 2013). Much of the support for neutral mutation comes from observations of increased genetic variability during colonization of subterranean habitats, as reviewed by Wilkens (1988).

The second reason is that the different kinds of evidence have made it difficult to directly compare effects, even though many workers in the field acknowledge that both processes are happening (e.g., Rétaux and Casane 2013). Those studies that purport to be direct tests (especially Borowsky 2016) have not been widely accepted (Lande in Wilkens and Strecker 2017).

The third reason is that even both explanations together may be insufficient. Cartwright *et al.* (2017), based on population genetic models of gene frequency change in *A. mexicanus*, suggest that (1) neutral mutation cannot explain eye and pigment loss if there is continuing migration from surface populations, as there apparently is for *A. mexicanus* and (2) selection would have to be strong, perhaps unreasonably so, to explain the evolution and persistence of eyeless populations. Cartwright *et al.* (2017) suggest that differential migration from the cave, which is phenotype dependent, could offer an explanation. For example, if eyed individuals were photopositive and reduced eyed individuals photonegative, then this migration acts in the same way dynamically as selection. This harkens back to the ideas of Lankaster (1925) and Ludwig (1942). It will be interesting to see if these ideas gain traction.

6.9 Summary

The loss of characters, especially eyes and pigment, in subterranean animals has attracted the attention of biologists since their first discovery centuries ago. As the connection to adaptation was not immediately obvious, explanations often did not include natural selection, but rather Lamarckian, orthogenetic, and, more recently, neutralist explanations. Adaptationist ideas with regard to subterranean organisms were originally developed, not in connection with loss of eyes and pigment, but rather in connection with constructive changes such as appendage elongation and elaboration of extra-optic sensory structures. Three studies of adaptation epitomize adaptation as it applies to subterranean species. Poulson (1963) studied life history and metabolic and neurological changes in cave fish in response to darkness and low food availability. His basic approach was a comparative one, taking advantage of related surface-dwelling species. Culver *et al.* (1995) studied adaptation of populations of the amphipod *G. minus*, focusing on the demonstration of adaptation to the darkness of caves. Their basic approach relied on quantitative genetics. Jeffery (2005a) focused on the causes of eye and pigment degeneration in the Mexican cave fish *A. mexicanus*. Using an array of techniques from cell, molecular, and developmental biology, he demonstrated the critical role selection plays. Finally, the time available for adaptation, based on molecular clocks, is in the range of several million years.

7 Colonization and Speciation in Subterranean Environments

7.1 Introduction

Aside from taxonomic descriptions of subterranean species, more has been written about the biogeography of subterranean animals than any other topic in speleobiology. Of the approximately 20 000 described species of stygobionts and troglobionts, it is fair to say that the 'typical' species is known from only a handful of specimens from nearby localities (often only one), whose closest living relative is another stygobiotic or troglobiotic species. An example is shown in Fig. 7.1, where the ranges of four closely related troglobiotic beetles in the genus *Pseudanophthalmus* are shown. Each species is restricted to an isolated belt of cavernous limestone (Holsinger 2012). Given this limited information, the natural extension of the species description is to consider biogeographical questions rather than ecological or evolutionary ones for which there is little or no information. Two recurring questions have been: (1) why is the range so restricted?; and (2) is its closest relative a living subterranean species, an extinct surface species, or even a living surface species (in the example in Fig. 7.1, only the first two are possibilities)?

Looming over these two questions come other questions about relicts and relics (Humphreys 2000). Relic species, the last survivors of an ancient radiation, and relict species, species geographically separated from related species, figure prominently in subterranean biogeography. Perhaps our hypothetical 'typical' species is a relic and/or a relict (the actual frequency of relictualism is unknown). The coleopterist Jeannel (1943) entitled his book on cave biology, *Les fossils vivants des cavernes*—to emphasize what he thought was the relictual nature of the subterranean fauna. All these terms—living fossils, relics, and relicts—connote an evolutionary dead end, the absence of the potential for further adaptation, and the absence of dispersal. They fit in well with the non-adaptationist ideas of Banta, Vandel, and some other speleobiologists (see Chapter 6).

The Biology of Caves and Other Subterranean Habitats. Second Edition. David C. Culver and Tanja Pipan. Published 2019 by Oxford University Press. © David C. Culver and Tanja Pipan 2019. DOI: 10.1093/oso/9780198820765.001.0001

Fig. 7.1   Distribution of four troglobiotic beetle species (numbers 1–4) of the *hubbardi* group of *Pseudanophthalmus* in caves of the northern Shenandoah Valley of Virginia, USA. Cave localities are indicated by dots. From Holsinger (2012). Used with permission of Elsevier Ltd.

The broader discipline of biogeography has undergone changes in the past several decades (Lomolino and Heaney 2004) that have made for an explosion of information and ideas about subterranean biogeography. The development of cladistic and phylogenetic concepts in systematics has led to parallel developments in biogeography, and the most important concept to be developed was that of vicariance biogeography, based on the idea that splitting of a species' range is the result of the formation of biogeographical barriers due to historical events rather than dispersal. The historical events resulting from continental drift have provided hypotheses to explain distributions of subterranean species (see e.g., Fig. 3.6). An additional advance has been the availability of molecular techniques for the sequencing of DNA, both mitochondrial DNA (mtDNA) and nuclear DNA, coupled with computational methods for estimating times since divergence of populations (see Porter *et al.* 2007, Trontelj *et al.* 2007). Finally, there has been an explosion of data on species distributions, as a result of greater ease of travel, discovery and mapping of many more caves, and the availability of databases.

Conceptually, we divide the process of colonization and evolution in subterranean environments into four phases (see also Trontelj 2018). First, what causes animals to enter (colonize) subterranean environments? Second, what factors contribute to the success or failure of these colonizations? Third, what is the role of extinction of surface populations in isolation and adaptation of the subterranean populations (allopatric vs parapatric speciation)? Fourth, how much subsurface dispersal occurs? The question of age of invasions, already encountered in Chapter 6, will be reconsidered in a phylogeographical context. In the final sections, we review the evolutionary history of several lineages: (1) the isopod *Asellus aquaticus* in surface and subsurface habitats in Europe; (2) diving beetles in one calcrete aquifer (Sturt Meadows) in Western Australia; and (3) a trogloxenic rodent, *Leopoldamys neilli*, from Thailand. In this chapter, the focus is on historical biogeography. The other part of biogeography, ecological biogeography (and in particular island biogeography), will be considered in the context of subterranean habitats in Chapter 8.

7.2 Colonization of subterranean environments

In many subterranean habitats, there is a continuing flux of invaders and migrants. The fauna of surface streams gets swept into caves through swallets (sinking streams), and this is the likely path of the successful colonization of the Mexican cave fish *Astyanax mexicanus* (Mitchell *et al.* 1977). In epikarst, a shallow subterranean habitat, there is a constant rain of both aquatic and terrestrial species (Fig. 1.7), resulting in an average of one copepod/drip/day in Organ Cave, West Virginia (Pipan *et al.* 2006b). Epikarst itself receives a supply of invaders, especially copepods living in leaf litter above the epikarst (Reid 2001). Other shallow subterranean habitats, such as seeps and milieu souterrain superficiel (MSS), because of their proximity to surface habitats, are frequently entered by surface-dwelling species. Likewise, many stream invertebrates enter the hyporheic and groundwater habitats as a result of the vertical circulation of water in streams and rivers.

There is also frequently a steady supply of colonists of terrestrial cave habitats through entrances. Many species enter caves to avoid temperature extremes (both heat and cold) because cave temperatures are buffered (Fig. 1.2), and to avoid predators and competitors. For example, salamanders such as *Eurycea lucifuga* are more frequent in caves in the southeastern United States in summer (Camp and Jensen 2007), apparently to avoid high summer daytime temperatures. Balogová *et al.* (2017) document the overwintering of *Salamandra salamandra* in caves. These species are seeking refuge from environmental stress, and it is possible that the successful colonization of caves by species during the Pleistocene was the result of an initial colonization to avoid cold temperatures resulting from advancing ice sheets. Caves are thus refuges in this model. In North Temperate regions, glaciated areas have a depauperate cave fauna, and cave areas near glacial boundaries sometimes have an exceptionally diverse fauna. Barr (1960) provides an example of this with *Pseudanophthalmus* cave beetles in the

Mitchell Plain in Indiana, USA, a karst area on the edge of the Pleistocene ice sheet. This model of colonization is called the climatic relict hypothesis (CRH) (Peck and Finston 1993, Danielopol and Rouch 2012). Peck (1984) showed that the phylogeny of the *Ptomaphagus hirtus* group of leiodid beetles can be explained by assuming that colonization and isolation occurred during four succeeding interglacial periods. A similar explanation has been advanced for the aquatic cave fauna of Western Australia, but in this case it was aridity, not temperature, that was the environmental factor driving animals into caves (Leys *et al.* 2003, Cooper *et al.* 2007).

Subterranean species may also undergo what Howarth (1980, 1987) calls an adaptive shift. Adaptive shift is a phenomenon in which individuals from a population change to exploit a new habitat or food resource (Howarth and Hoch 2012). In his studies of the fauna of Hawaiian lava tubes, Howarth noted that food was in very short supply on the surface of lava flows, especially recent ones. In this environment, much of the organic carbon on the surface comes from wind-blown debris (Ashmole and Ashmole 2000) (see Chapter 2). In contrast, subterranean habitats in lava flows, including epikarst-like habitats which Howarth calls mesocaverns, as well as lava tubes enterable by humans, have many tree roots (Fig. 2.6) which are an abundant source of carbon. Furthermore, the lava tubes are a relatively benign environment without temperature extremes (Pipan *et al.* 2011). Howarth and Hoch (2012) suggest that adaptive shift can explain the presence of many troglobionts in tropical caves. We will consider this hypothesis in more detail when the question of the importance of isolation for speciation is considered later (in section 7.4).

One very interesting, and in many ways puzzling, source of stygobionts is the marine fauna. There are groups of subterranean organisms in predominately marine groups, and for many of these it is likely that they entered freshwater subterranean habitats via vertical migration from marine sediments into freshwater sediments at the sea margin. It is plausible that species colonized these freshwater sediments and were then stranded by the regression of the Adriatic Sea, especially during the Messinian Crisis approximately 6 million years ago, when the sea dried up (Gautier *et al.* 1994). More generally, such colonizations can occur whenever there is a lowering of sea level, and the end result, especially in the Mediterranean with a complex geological history, is a series of subterranean species, many living in interstitial sediments (Boutin and Coineau 2000), with different ages of isolation in subterranean habitats depending on their location (Fig. 7.2). The area cladogram shown in Fig. 7.2 represents not only the evolutionary history of the *Pseudoniphargus* amphipods but also their biogeographical history. The tips of the cladogram are the different strandings of *Pseudoniphargus* in the western Mediterranean Sea. The above scenario is just that—a plausible story, and it is extremely difficult to reconstruct the history of colonization of subsurface habitats in very old groups.

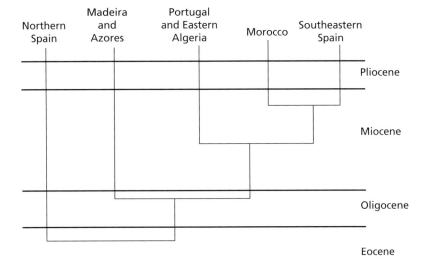

Fig. 7.2   Simplified area cladogram of the main groups of species of the genus *Pseudoniphargus* and dating of the colonization of continental groundwaters by the coastal ancestral populations of each lineage, from the Eocene to the Pliocene. Area cladograms are generated by substituting the area of occupancy for the species in a species cladogram (phylogeny). For example, there are separate species groups present in northern Spain, Madeira and the Azores, and so on, and their phylogenetic relationships are used to generate the area cladogram. Data from Boutin and Coineau (2000).

7.3 What determines success or failure of colonizations?

The fauna of any subterranean site is not a random sample of species occurring on the surface or even of the colonizing organisms. Any subterranean community is 'disharmonious' in this sense. Some of this is easy to understand. Aquatic insects, except for some aquatic beetles, are missing from the stygobiotic fauna, presumably because the winged adults would have difficulty mating. In general, strongly visually oriented organisms do not successfully colonize caves because they cannot initially overcome the penalty imposed by lack of visual orientation.

Speleobiologists have often used the word pre-adapted to describe successful colonists of subsurface water. Pre-adaptation is used in the sense of possession by an organism of the necessary properties to permit a shift into a new habitat. A structure is pre-adapted if it can assume a new function before it becomes modified itself. A similar concept and one without the connotation of destiny or neo-Lamarckism, is exaptation, an adaptation for one function serving for another function. An example of pre-adaptation (and exaptation) is that in temperate zone caves, the terrestrial fauna is largely, if not exclusively, derived from the forest litter fauna. Regions without forests, such as the Black Hills region in South Dakota, USA (Culver *et al.* 2003), have a depauperate fauna, at least in part because there are few surface species pre-adapted to caves. Species living in the dimly lit, humid environment of leaf litter are exapted for the aphotic, humid environment of caves even though there is little if any leaf litter present. Christiansen (1965) and Howarth and Hoch (2012) point out that behaviour as well as morphology may be important in determining colonization success, and behavioural changes may in fact precede morphological change.

In spite of the importance of exaptation and pre-adaptation, they remain rather elusive concepts. We can *a posteriori* explain why particular groups of animals are in caves. For example, omnivorous millipedes living in moist leaf litter would seem likely successful colonists, and they are. But we cannot always successfully make *a priori* predictions. For example, Symphyla, blind and eyeless inhabitants of leaf litter, would seem ideal candidates for subterranean life. Yet they are rarely found in caves (Juberthie-Jupeau 1994).

7.4 Allopatric and parapatric speciation

There are many successful colonists of subterranean habitats that have not evolved troglomorphic features and are often found in non-subterranean habitats. Careful examination of most lists of species found in subterranean habitats will include a number of such troglophilic and stygophilic species (recall the ecological classification of species discussed in Chapter 3). The next two stages in the evolutionary history of subterranean faunas determine how many stygobionts and troglobionts occur in a particular taxonomic group or in a particular region. For example, the two most common orders of insects[1] in subterranean habitats are Coleoptera and Diptera, but there are thousands of troglobiotic Coleoptera but only a handful of troglobiotic Diptera. The difference between Coleoptera and Diptera is not one of successful colonization (both have repeatedly been successful), but one of speciation.

The standard view of the evolution of troglobionts and stygobionts is that extinction of the surface-dwelling populations is required (Holsinger 2000, Sbordoni *et al.* 2000). Technically, this is allopatric speciation, occurring among populations that are geographically separated, in this case because the adjoining surface population is extinct. All things being equal, allopatric speciation is more likely to occur than when differentiating populations are

---

[1] Most modern taxonomic treatments do not put Collembola, which are wingless hexapods, in the Insecta but rather in the Entognatha. Collembola are one of the most common groups of troglobionts (Table 3.3).

adjoining (parapatric) or overlapping (sympatric) because there is no gene flow to retard the process of differentiation. Allopatric speciation fits nicely with the CRH, in which surface-dwelling populations and species go extinct because of climate changes, such as those that occurred during the Pleistocene. It also fits nicely with the observation that for many genera with many troglobiotic and stygobiotic species, there are few or no surface-dwelling species. A typical example is the beetle genus *Pseudanophthalmus* in North America, with more than 200 described species, all but one known only from caves, and the remaining species is only known from an MSS habitat (Barr 2004). In these cases, the most closely related species is another troglo-biont or stygobiont, typically in a nearby geographical region (Fig. 7.1).

The counterview has been put forward in connection with the adaptive shift hypothesis (ASH), in which active colonization of subterranean habitats occurs and speciation is parapatric, with differentiation occurring between contiguous, non-overlapping populations (Chapman 1982, Howarth 1987). Howarth and Hoch (2012) provide an impressive list of examples from Hawaiian lava tube fauna where speciation has apparently occurred parapa-trically, and the most closely related species of troglobionts in lava tubes are nearby surface-dwelling species (Table 7.1). Peck and Finston (1993) provide a similar list for the Galapagos Islands lava tube fauna. Riesch *et al.* (2011) suggest that sympatric speciation is not as unlikely as thought, because, at least in the case of the fish *Poecilia mexicana*, surface-dwelling females were unable to reproduce in darkness, effectively creating a reproductive barrier at the light/dark boundary.

Accumulating evidence strongly supports both hypotheses, although obvi-ously for different cases. We review first an example of the CRH from arid

Table 7.1 Parapatric cave and surface species pairs occurring on the island of Hawai'i.

Lava tube species	Surface-dwelling relative	Ancestral habitat
Isopoda: *Littorophiloscia* sp.	*Littorophiloscia hawaiiensis*	Marine littoral
Hemiptera: *Oliarus makaiki*	*Oliarus koanoa*	Mesic forest
Hemiptera: *Oliarus polyphemus*[1]	*Oliarus* sp.	Rain forest
Hemiptera: *Oliarus lorettae*	*Oliarus* sp.	Dry shrub land
Hemiptera: *Nesidiolestes ana*[1]	*Nesidiolestes selium*	Rain forest
Orthoptera: *Caconemobius varius*[1]	*Caconemobius fori*[1]	Barren lava flows
	Caconemobius sandwichensis	Marine littoral
Dermaptera: *Anisolabis howarthi*	*Anisolabis maritima*	Marine littoral
	Anisolabis hawaiiensis	Barren lava flows
Arnaneae: *Lycosa howarthi*	*Lycosa* sp.[1]	Barren lava flows

Source: From Howarth and Hoch (2012). Used with permission of Elsevier Ltd.

[1] Represented by several distinct populations or species. Polymorphic cave populations may represent separate invasions or may result from divergence and subterranean dispersal of a single lineage.

western Australia (Leys *et al.* 2003), and then examples of the ASH from lava tubes in Hawaii (Rivera *et al.* 2002) and the Canary Islands (Arnedo *et al.* 2007).

Leys *et al.* (2003) investigated a diverse assemblage of diving beetles in the family Dytisicidae found in calcrete aquifers in south Western Australia. Calcrete aquifers are a feature of arid landscapes in Australia (Fig. 7.3), and the details of their history provide a particularly useful site to contrast the

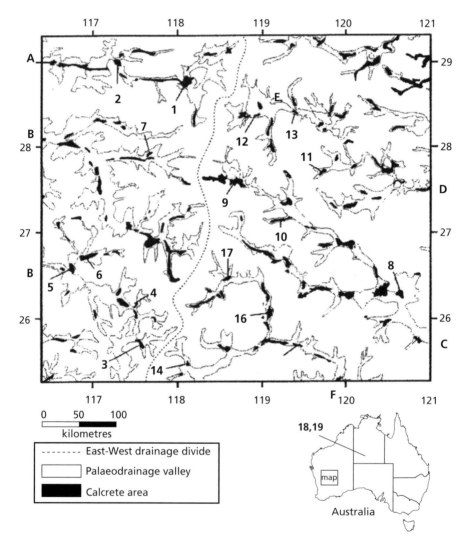

Fig. 7.3 A map of the sampled calcretes and their relative positions in paleodrainages in Western Australia. Letters and numbers refer to the paleodrainages (dotted lines) and calcretes (in black), respectively. From Leys *et al.* (2003). Used with permission of Blackwell Publishing.

CRH and ASH. They are formed by the precipitation of carbonates from shallow groundwater in climates with precipitation of less than 200 mm/ year (Mann and Horwitz 1979). They are upflow of salt lakes and are about 10 m thick, and thus are a superficial subterranean habitat (see Chapter 1), although they have less connection with the surface than other shallow subterranean habitats. The calcretes formed between 30 and 37 million years ago during a cool, dry period in the Eocene. From 30 million years ago until 10 million years ago, there was a warm temperate climate in this part of Australia. Beginning in the Miocene, there was a period of drying that began in the northwest and moved southeast over the next 5 million years. Leys *et al.* (2003) argue that if CRH is correct, species should become isolated in caves (as a result of the extinction of surface populations) only during this period of maximum aridification, and that if ASH is correct, species should become isolated in caves (as a result of parapatric speciation) throughout the time period since the formation of calcrete aquifers.

They analysed DNA sequences of *ND1* (820-bp fragment) and *COI* (822-bp fragment) mitochondrial genes for 60 species of aquatic dytiscid beetles in south Western Australia. To test the hypothesis, they need to calibrate the divergence rates. In the absence of a fossil record they used a 2.3 per cent pairwise divergence rate per million years. The resulting phylogeny (determined using Bayesian methods of tree building) together with a time scale is shown in Fig. 7.4. Leys *et al.* estimate that there have been at least 26 independent invasions of calcrete aquifers by dytiscid beetles. Except for eight pairs of species that are sympatric within a calcrete aquifer, there was no apparent geographical structure in the tree. For example, species from calcretes belonging to the same drainage system did not group together (see letters and numbers in Fig. 7.4). All of the times of isolation fall within the time of aridification from 5 to 10 million years ago, providing strong support for CRH. The sympatric species provide some of the most striking support for CRH. They are probably the result of a single invasion with subsequent speciation from subsurface dispersal within the aquifer (Cooper *et al.* 2002), and their divergence can be used to estimate the time of isolation in calcretes. These times range from 3.6 to 8.1 million years ago, and the differences in estimated time of divergence have a strong latitudinal component. Species from northwestern calcretes diverged earliest (Fig. 7.5) and it was in the northwest that aridification began. Overall, latitude accounted for 83 per cent of the variance in estimates of divergence times. Similar patterns have been shown for stygobiotic amphipods occurring in the same calcretes (Cooper *et al.* 2007).

Rivera *et al.* (2002) looked at a case where the ASH seems to hold. Two species in the terrestrial isopod genus *Littorophiloscia* occur in the island of Hawai'i—*Littorophiloscia hawaiiensis* occurring in the marine littoral and an undescribed troglobiotic species in lava tubes. They have a parapatric distribution and a phylogeny based on a 473-bp region of *COI* indicates that they

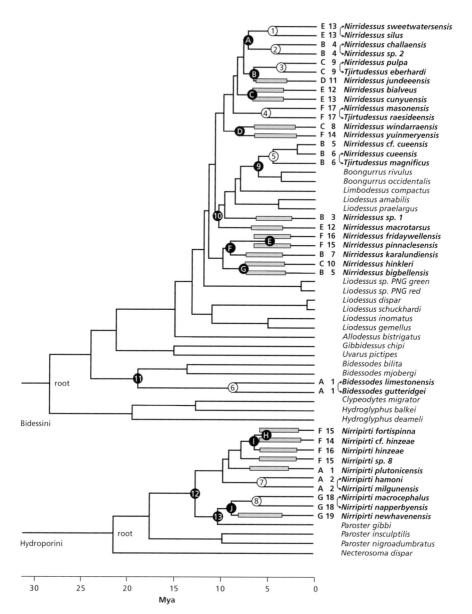

Fig. 7.4 Phylogenetic tree for dytiscid beetles in Western Australia calibrated with geological time. Stygobionts are in bold, and sympatric sister species are shown by double-headed arrows. Letters at the tips of the branches show drainage systems and numbers show calcrete aquifers. Aquifer numbers are shown in Fig. 7.3. Divergence time of the modes with black dots show maximum estimates of transition times to the subterranean environment and the open circles show minimum transition times, based on sympatric sister species. Bars on the branches represent 95 percent confidence intervals of predicted isolation times. From Leys et al. (2003). Used with permission of Blackwell Publishing.

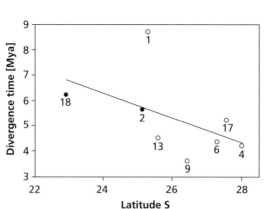

Fig. 7.5 Latitudinal variation in divergence times of eight sympatric sister pairs of stygobiotic dytiscid beetles in Western Australia. The open circles show species pairs belonging to the Bidessini; the black circles show species pairs belong to the Hydroporini. Numbers are shown in Fig. 7.4. From Leys *et al.* (2003). Used with permission of Blackwell Publishing.

are sister taxa, that is, that they are closely related. As they point out, ASH is most probable in places such as the Hawaiian Islands because climate has been relatively constant since the formation of the islands.

Arnedo *et al.* (2007) examined another case where the ASH seems to hold—spiders in the genus *Dysdera* on the Canary Islands. Eleven troglobiotic species and 34 surface-dwelling species are endemic to the Canaries. Using mtDNA sequences of *COI* and *rrnL*, they showed that the sister species of six of nine troglobionts was a surface-dwelling spider; one sister species was another lava tube species, and two had no close sister species (Table 7.2). This supports the ASH since surface–cave pairs are expected under this hypothesis. The other three pairs that do not fit ASH can be explained if there is subsequent extinction of the surface species unrelated to colonization or if subsequent speciation via subsurface dispersal took place.

A cautionary note is sounded by Villacorta *et al.* (2008) about assuming all tropical caves or even all island lava tubes have species that colonized as a result of adaptive shift. In a study of what appeared to be two species of terrestrial amphipods, one surface-dwelling and one cave-dwelling, in the genus *Palmorchestia* in the Canary Islands, actually consisted of a series of cryptic taxa. Caves seemed to serve as refugia for surface populations at various times in the evolutionary history of these populations.

There are some lineages where the ASH seems to hold although they have not been explicitly examined in this context. They include most of the examples of adaptation discussed in Chapter 6. Subterranean populations and subspecies of *Asellus aquaticus*, *Astyanax mexicanus*, and *Gammarus minus* all show considerable morphological differentiation even though there are parapatric surface populations. The same holds for the *Poecilia*

Table 7.2 Uncorrected genetic distances of *COI* sequences for sister pairs of troglobiotic lava tube *Dysdera* spiders on the Canary Islands.

Sister species	Number of pairs	Uncorrected genetic distance
Cave	1	0.110
Surface	6	0.113 (0.080 to 0.169)
Unknown	2	

Source: Adapted from Arnedo *et al.* (2007).
For surface sister species, median and range are given.

mexicana species complex of fish occupying both subterranean and sulfidic cave habitats (Tobler et al. 2018).

7.5 Vicariance and dispersal

Throughout the discussion of colonization and speciation in subterranean habitats in the previous section, the question of the relative role of vicariance[2] and dispersal has been present. Indeed, the central question for subterranean biogeography is the relative role of surface and subterranean dispersal (Holsinger 2012). The low subsurface dispersal hypothesis (Lefébure *et al.* 2006a) is that after a species disperses on the surface through a region and colonizes caves in the region, the surface population becomes extinct (the vicariant event). After species colonize caves and become isolated, relatively little dispersal occurs and that species occupies only a single cave or a few caves that are connected by impenetrable passages. The dispersal hypothesis is that after colonization and isolation, species not only occupied a single cave or a few connected caves, but also occasionally dispersed beyond this (perhaps through epikarst and MSS) and that these occasional migrants were themselves isolated and speciation occurred.

For a relatively long period of time, roughly until the middle of the twentieth century, the low subsurface dispersal hypothesis held sway. Caves, at least in most regions, were thought to be quite rare (the Dinaric karst being an exception) and so there was little opportunity for dispersal. The idea of most cave species being single cave endemics was quite common in the middle of the twentieth century, Valentine (1945) in North America being an example. The common practice among European taxonomists of naming subspecies, each from a single cave, is also a manifestation of this view (Culver *et al.* 2006b). Orthogeneticists such as Vandel (1964) allowed for little migration

---

[2] Vicariant species are ones that have evolved in place (no dispersal) with a common ancestor. Vicariance requires passive range fragmentation, which may result from climate change. In the example here, it is fragmentation resulting from the loss of the surface population. Another kind of vicariant event would be where the subterranean environment itself becomes fragmented.

since stygobionts and troglobionts were considered to be senescent lineages without adaptive features.

The antidispersalist view began to change following the discovery of non-cave subterranean habitats, including both interstitial and shallow subterranean habitats (see Chapter 1), encapsulated in the phrase 'global interstitial highway' (Ward and Palmer 1994). Americans were much slower to grasp the significance of non-cave subterranean habitats, but the continued discovery of caves (more than 44 000 in the United States by 1999; Culver *et al.* 1999) had the same effect of indicating increased possibilities for migration between caves, and it just seemed inconceivable that there could be thousands of species of troglobionts or stygobionts in a single genus. In a series of influential papers, Barr (1960, 1968, 1979) argued that subsurface migration was a significant factor and could lead to speciation.

The antidispersalist view gained new legitimacy beginning in the 1970s, due to both the development of vicariance biogeography (Nelson and Platnick 1981) and the use of molecular genetic techniques to determine population and species similarity. Vicariance biogeography is based on the hypothesis that distributions reflect the positions of continents before speciation, and thus offers explanations for 'strange' distributions such as ones reflecting the ancient southern supercontinent Gondwanaland. It stands in contrast to the older dispersalist biogeography where distributions reflect dispersal. The discovery of correlations of distributions of interstitial crustaceans with ancient shorelines (e.g., Fig. 7.2) supported a vicariant view with very little migration occurring. At a much smaller scale, genetic differentiation of cave populations only a few kilometres apart (Culver *et al.* 1995) suggested extremely low migration rates. Trontelj (2018) reviews the evidence for the vicariance model, and argues that it remains valid.

There are several ways to assess the relative importance of vicariance and dispersal. Early proponents of vicariance biogeography (e.g., Nelson and Platnick 1981) proposed that a vicariant framework be used to assess its goodness of fit to observed distributions, with dispersal accounting for anomalies. Alternatively, dispersal can be assessed using molecular markers, or indirectly through geological barriers or morphological features such as size (e.g., Culver *et al.* 2009).

In general, the discovery of cryptic species suggests that vicariance is important and that dispersal occurs over relatively short distances. There are few stygobionts or troglobionts anywhere with ranges of more than 200 km or so, and most have ranges much smaller.

The next issue is how to account for any dispersal that might occur. A model of speciation by surface vicariance events (multiple invasions of subterranean habitats and isolation) and by subsurface dispersal is shown in Fig. 7.6. With vicariance, different populations of a surface ancestor colonize separate

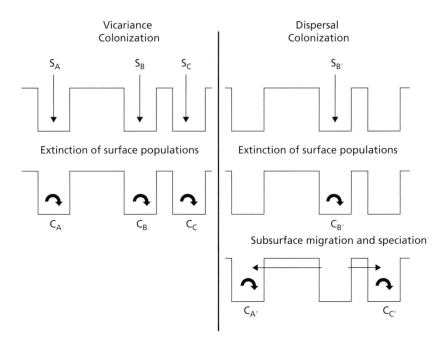

Fig. 7.6   Diagram of vicariance and dispersalist models of speciation in caves. In the vicariance model three surface-dwelling populations (S_A, S_B, and S_C) of the same surface-dwelling species enter caves. S_A is more geographically distant from the other two populations. Following the extinction of the surface populations, three cave populations speciate. In the dispersal model, one surface population (S_B) enters caves and the surface population goes extinct. Subsequently two other caves are colonized by subsurface dispersal and form separate species under the assumption that dispersal events are rare. From Culver *et al.* (2009). Used with permission of John Wiley and Sons Inc.

caves (or any subterranean habitat). Under the CRH and any other cases where the surface population goes extinct, a series of isolated populations are left (Fig. 7.1). Under the dispersal model, there is a single initial colonization of a subterranean habitat. After isolation of the subterranean population, it disperses to new sites where populations differentiate genetically. If migration is low enough, speciation occurs. The resulting phylogenies of these two processes—multiple isolation and subterranean dispersal—are shown in Fig. 7.7, and in fact the phylogenies are identical (Culver *et al.* 2009). Thus, an appeal to the power of cladistic analysis to sort out the hypotheses is insufficient. If surface populations persist, as happens in the ASH, the two scenarios can be distinguished (see also Desutter-Grandcolas and Grandcolas 1996), and if ages can be put on the nodes, then the two scenarios can be distinguished. This was the approach of Leys *et al.* (2003) in the context of separating CRH and ASH (see earlier in this section). Trontelj (2018) points out that teasing out the biogeographical scenario is complicated by the fact that dispersal can occur either at the time of colonization or after colonization, and it can occur either above ground or below ground. Non-cave shallow

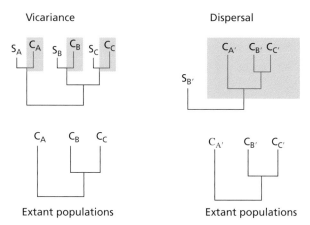

Fig. 7.7   Cladograms for the two models shown in Fig. 7.8. Populations are labelled as in Fig. 7.8. Shaded parts of the cladograms are those parts of the tree where evolution is occurring in caves. The resulting cladogram of extant populations is identical for both models. From Culver *et al.* (2009). Used with permission of John Wiley and Sons Inc.

subterranean habitats may serve as important dispersal corridors (Culver and Pipan 2014, Ortuño *et al.* 2013), making subsurface movement after colonization more feasible than previously thought.

Another approach to separating the two hypotheses is to take an ecological approach that relies on distribution patterns at different scales. One way to measure differences in dispersal among species is to measure how frequent they are in a small area. For example, if we assume that epikarst copepods can occur in the range of conditions found in drips in a single cave, then species found in many drips are likely better dispersers than ones found in few drips. The same argument can be used for the presence of macro-invertebrates in caves in a small area. If we use this localized situation as a measure of dispersal ability, then we can predict that good local dispersers will have larger ranges when larger areas are considered. If dispersal ability is unimportant, then there should be no connection between the patterns at small and large scales. In two tests of this idea, Culver *et al.* (2009) found that the number of drips occupied by copepods in Postojna–Planina Cave System (PPCS) was a good predictor of the number of caves occupied by these copepods in the epikarst of other Slovenian caves (Fig. 7.8), and that the number of caves in a 10-km² area in West Virginia in which different species of macro-Crustacea were found predicted the number of caves these macro-crustaceans occupied in a 400-km² region.

Christman *et al.* (2005) looked at the role of subsurface dispersal (or lack of it) in the evolution of endemism in the troglobiotic fauna in eastern North America. Endemism, the restriction of taxa to a particular geographical area, is very high in subterranean faunas, no matter what the size of the area used (Gibert and Deharveng 2002). Christman *et al.* (2005) chose the most

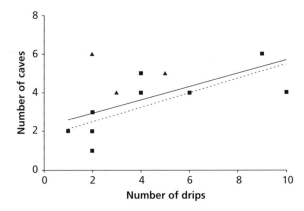

Fig. 7.8
Relationship between the number of drips in the Postojna–Planina Cave System and number of Slovenian caves occupied by epikarst copepods. Triangles depict stygophiles and squares stygobionts. The solid line is the regression line for all species (R^2=0.32, p<0.05) and the dashed line is the regression line for stygobionts (R^2=0.58, p<0.01). From Culver *et al.* (2009). Used with permission of John Wiley and Sons Inc.

restrictive definition of endemism possible—those known from a single cave. Even with this definition, endemism was remarkably high—211 of 467 troglobionts (45 per cent) were known from a single cave. They looked at the importance of vicariance and dispersal as causes of endemism. Their basic approach was to compare the pattern of species richness of endemics and non-endemics. They reasoned that if subterranean dispersal was highly restricted, then the patterns of the two should be similar. The pattern of endemism they tried to explain is shown in Fig. 7.9. There was a centre of endemism near the southern end of the distribution (northeast Alabama) of troglobionts and few endemics around the periphery of the distribution range of troglobionts. The number of non-endemics was a good predictor of the number of endemics, accounting for 69 per cent of the variance in number of endemics. This is the vicariance component because it requires no spatial context, i.e., the situation in nearby areas. However, the residual variance in number of endemic species, that remaining after the effect of non-endemics is accounted for, showed a pattern that indicated that dispersal was also important. The pattern of residuals had a strong geographical component, with excess endemism occurring in cave regions with high densities of caves and at the southern end of the distribution. Higher rates of secondary productivity may allow more successful colonizations and more successful survival of populations resulting from subsurface dispersal. Residual endemism is not higher in areas of more dissected limestone, which would reduce dispersal rates; if anything the reverse was true.

Ward and Palmer (1994) provide an especially interesting overview of dispersal in subterranean environments, from a wider perspective. They emphasize the meiofauna but their ideas may apply to the stygofauna in

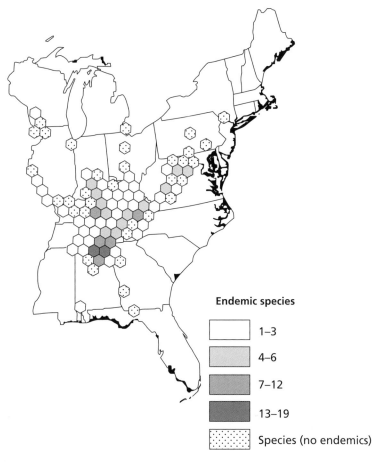

Fig. 7.9   Number of single cave endemic troglobionts in 5000 km² hexagons. Stippled hexagons are those with no single-cave troglobionts, but with non-endemic troglobionts. From Christman *et al.* (2005). Used with permission of John Wiley and Sons Inc.

general. They focus on the connections between different kinds of subterranean habitats in the broad sweep of evolutionary time. Alluvial aquifers, sand and gravel deposited by flowing waters, are the central core of what they call the 'global interstitial highway'. Alluvial aquifers constitute a more-or-less spatially continuous subterranean habitat, one that has had temporal continuity as well (Boutin and Coineau 2000). Eme *et al.* (2013) provide support for the interstitial highway in their analysis of *Proasellus* isopods. Rather than slow continuous dispersal, Eme *et al.* (2013) provide evidence for bursts of dispersal (Fig. 7.10). Ward and Palmer suggest that the stability and continuity of this habitat offer an explanation of distributions of groundwater animals that correspond to ancient geological events, such as the break-up of the supercontinent Pangaea in the early Mesozoic, nearly 200 million years ago (Schminke 1981). Other subterranean habitats are more

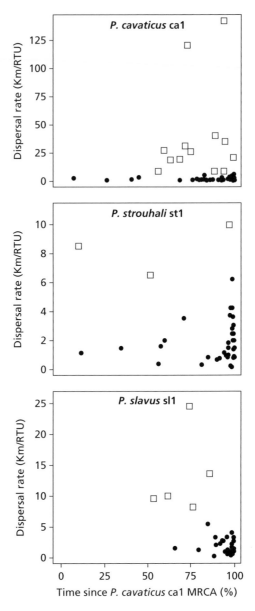

Fig. 7.10   Variation in dispersal rates over the course of a species' evolution for *Proasellus* isopods in central and northern Europe. For each branch the dispersal rate is plotted against the mean age of the branch. Relative time units (RTU) correspond to the time duration of a branch divided by the age of the *P. cavaticus* ca1 most recent common ancestor (MRCA). White squares are defined as outliers defined as branches with dispersal rates higher than 1.5 times the interquartile range (boxplot convention). From Eme *et al.* (2013). Used with permission of John Wiley and Sons Inc.

spatially and temporally discontinuous and are connected mainly through the alluvial aquifer system. They argue that these habitats, including caves, springs, and marine sands, are colonized from the alluvial aquifer system.

7.6 Cryptic speciation

To assess the potential for migration, it is important to know the actual range of species. While species definitions typically involve inferences about reproductive isolation (biological species concept) or monophyly, having arisen from one ancestral population (phylogenetic species concept), in practice subterranean species are defined on the basis of morphological differences, supplemented by genetic differences, that indicate monophyly.

Cryptic species are genetically distinct but morphologically indistinguishable species. Cryptic species can occur anywhere, but they appear to be especially common in subterranean habitats. Most studies of molecular variation in subterranean animals, from the early days when the only information available was different rates of movement of soluble proteins in an electric field (allozyme analysis) to the present, have indicated the presence of cryptic species (Sbordoni *et al.* 2000, Trontelj *et al.* 2009). With the exception of the cave shrimp *Troglocaris anophthalmus* (Zakšek *et al.* 2009), stygobiotic and troglobiotic species with large ranges (200 km in Trontelj *et al.*'s study) comprise several cryptic species. An example from the amphipod genus *Niphargus* is the study of Meleg *et al.* (2013). In 11 localities, they found 8 species, 4 of them cryptic (Table 7.3). Niemiller *et al.* (2012a) looked for cryptic species within the widespread cave fish *Typhlichthys subterraneus*, and concluded that there were between 4 and 21 cryptic species, the number being dependent on the number of individuals and the number of loci

Table 7.3 Cryptic molecular species in *Niphargus* from the Western Carpathians.

Nominal *Niphargus* species (morphological)	Putative cause of speciation	Cryptic species	Distribution
N. bihorensis	different catchments	N. bihorensis	single locality
		N. sp. 4	single locality
N. andropus	geological and ecological heterogeneity	N. andropus	single locality
		N. sp. 2	13 km (2 localities)
		N. sp. 3	single locality
N. laticaudatus	different catchments	N. laticaudatus	25 km
		N. sp. 1	20 km

Source: From Meleg *et al.* (2013)
Based on COI (mitochondrial) and 28S (nuclear)

sampled (Table 7.4). To understand why cryptic species are so common in subterranean habitats, the evolutionary processes, particularly the selective environment, that result in cryptic species need to be reviewed.

One kind of cryptic speciation occurs when an ancestral species splits into two descendent populations, for example, by extinction of the surface populations due to climate warming (a surface vicariant event) or a change in subsurface drainage (a subterranean vicariant event). The species differentiate genetically but not morphologically because the same selective environment acts on both species. The other kind of cryptic species result from the morphological resemblance of non-sister lineages as a result of convergent evolution. The presence of both kinds of cryptic species means that morphological similarity is not a good predictor of populations connected by dispersal.

Both kinds of cryptic speciation in subterranean environments result from the strongly convergent selective pressures found in subterranean environments, but the way they are detected is different. The first type can be detected from the amount of genetic differentiation among populations, but the difference must be rather large (Lefébure *et al.* 2006b) in order to be certain that the populations are reproductively isolated, and hence separate species. The second type can be detected from the topology of a phylogenetic tree, where the expected monophyly (having arisen from a single ancestral species) of the cryptic species is disrupted by other species.

A good example of cryptic speciation resulting from independent invasions is the familiar European salamander *Proteus anguinus*. On the basis of extensive sequencing of several regions of the mtDNA genome (Goricki 2006,

Table 7.4 Number of delimited species, number of best trees, and tree score for each *Typhlichthys* delimited species analysis using the non-parametric method of O'Meara (2010).

Loci	20 individuals			60 individuals			135 individuals		
	No. species	No. trees	Score	No. species	No. trees	Score	No. species	No. trees	Score
Three-gene	7 (7)	13	6.000	16 (16)	45	19.654	21 (19)	40	47.194
Six-gene	7 (6)	25	16.336	11 (11)	14	58.738			
Nine-gene	7 (6)	2	28.285						
Three-gene	4 (5, 0.53)			14 (2, 0.98)			15 (1, 1.00)		
Six-gene	4 (4, 0.46)			10 (2, 0.99)					
Nine-gene	6 (1, 1.00)								

Source: From Niemiller *et al.* (2012a)

The number of species used for subsequent analysis is indicated in parentheses after generating the 50% majority rule consensus tree of the best delimited species trees. In the bottom set, the number of species after Bayesian species delimitation is listed for each delimited species analysis. The number of different species delimitation models with posterior probabilities >0.01 and the posterior probability of the model with the highest posterior probability are listed in parentheses.

Gorički and Trontelj 2006), six groups of *P. anguinus* populations had greater genetic divergence levels between them than most species of salamanders within a genus (such as *Salamandra*) do (Trontelj *et al.* 2009). The greatest linear extent of any of these lineages was 200 km, compared to 500 km for the entire 'species' (Fig. 7.11). In this analysis, the black *Proteus anguinus*

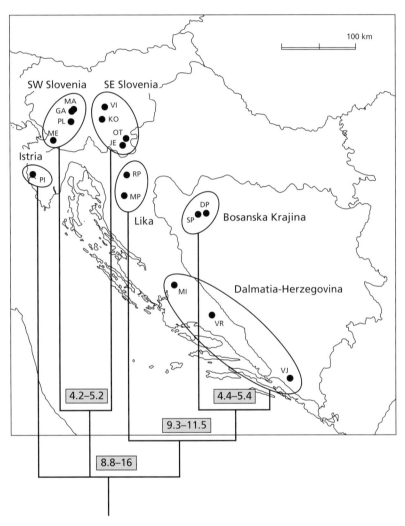

Fig. 7.11   Relationships among six monophyletic groups of *Proteus* inferred from mtDNA sequences (Gorički 2006, Gorički and Trontelj 2006) and their approximate divergence dates (in mY BP) estimated by applying the molecular clock for the 12S and 15S rRNA genes of sala-mandrid newts (Caccone et al. 1997). Figure by Š. Gorički, used with permission. Locality abbreviations are DP—Dabarska pećina, GA—Gašpinova jama, JE—Jelševnik, KO—Kompoljska jama; MA—Malo okence, ME—Mejama, MI—Miljacka pećina, MP—Markarova pećina, OT—Otovšski breg, PI—Pincinova jama, PL—Planinska jama, RP—Rupeciča, SP—Suvaja pećina, VI—Vir, VJ—Vjetrenica, and VR—Vrelo Stuba.

parkelji was found to be an intraclade variant, an example of divergence or perhaps plesiomorphy rather than convergence.

In general, the discovery of cryptic species suggests that vicariance is important and that dispersal occurs over relatively short distances. There are few stygobionts or troglobionts anywhere with ranges of more than 200 km or so, and most have ranges much smaller.

The power of convergent selective pressures in subterranean environments to confuse taxonomic relationships (the second kind of cryptic species) can be impressive. The subterranean stygobiotic salamanders of Texas, USA, are found in caves and deep phreatic wells that penetrate the Edwards Aquifer. Species are generally placed in two genera—*Eurycea* which is widespread in North America, and *Typhlomolge* has been used for the most troglomorphic species—*rathbuni* and *robusta* (Wiens *et al.* 2003). Wiens *et al.* (2003) created two phylogenetic trees, one based on 16 morphological characters

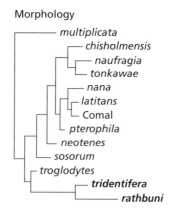

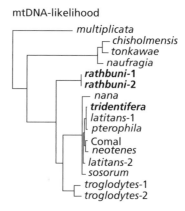

Fig. 7.12   Phylogenetic trees for central Texas (USA) salamanders in the genus *Eurycea* based on morphology (upper panel) and mtDNA (lower panel). The length of branches is proportional to the amount of estimated change for that lineage. See Wiens *et al.* (2003) for details. Used with permission of Oxford University Press.

which had been used in previous systematic studies, and another based on a 1141-bp sequence of the *cytb* mitochondrial gene. The two trees (Fig. 7.12) could hardly be more different. In the morphological phylogeny, *Eurycea tridentifera* and *Typhlomolge rathbuni* form a distinct clade with *Eurycea troglodytes* as a sister species. In the mtDNA phylogeny, the three species are now in separate clades. The morphological tree reflects selection and evolutionary relatedness, while the mitochondrial tree reflects only evolutionary relatedness.

7.7 Age of lineages

Throughout the history of speleobiology, there have been controversies concerning the age of the subterranean cave fauna. A neo-Lamarckian perspective is one where change happens very rapidly because of the efficiency of the process of evolution of acquired characters. Packard (1888) envisioned the process of isolation and evolution of troglomorphy as taking only a few dozen generations. On the other hand, neo-Darwinians and neutral mutationists both saw evolution as a slow process, because Darwinian evolution was gradual, and neutral mutation, as a process, was also slow. Neo-Darwinian proponents of the adaptive shift hypothesis (ASH) held that adaptation could be rather rapid, especially in lava tubes, which themselves were usually thousands or tens of thousands of years old.

Up until the last several decades, it was widely held that the events of the Pleistocene (beginning approximately 2 Mya) were the major force in determining the diversification and distribution of terrestrial cave species. The fauna of unglaciated areas is typically depauperate, especially with respect to the terrestrial fauna (e.g., Culver *et al.* 2003); but the glacial/interglacial cycles of the Pleistocene were widely held to result in diversity through the repeated process of expansion and contraction of ranges and the ability of species to survive on the surface (e.g., Guéorguiev 1977, Peck 1984). Aquatic species were thought to be older, but ages were uncertain. However, Mitchell *et al.* (1977) thought that *Astyanax mexicanus* became isolated in caves about 10 000 years ago.

The ability to estimate time since lineage splits by DNA sequence data has made it possible to at least obtain an independent estimate of the age of lineages. In general, lineages are much older than previously thought. For example, subterranean lineages of *Astyanax mexicanus* are estimated to be a million years old or more (Porter *et al.* 2007, Ornelas-Garcià *et al.* 2008). Espinasa and Espinasa (2016) suggest that *A. mexicanus* lineages are older than the caves they inhabit. For the much more troglomorphic cave fish in the Amblyopsidae, Niemiller *et al.* (2012b, 2013b) estimate the cave lineages are more than 10 million years old. Trontelj *et al.* (2007) suggest that species (including *Proteus anguinus*) became isolated in caves in the Dinaric karst between 2 and 5 million years ago. Some estimates of age of the terrestrial fauna fall within the Pleistocene, such as the hygropetric beetles in the genus *Hadesia* in Montenegro, which are about 4 million years old (Polak *et al.* 2016), but others, like the beetle genus *Geotrechus* from the eastern Pyrenees, had lineage splits up to 10 million years ago (Faille *et al.* 2015).

7.8 Phylogeography of three different subterranean lineages

In this final section, we consider the phylogenetic histories of three lineages that represent three very different circumstances. The first is a classic troglomorphic species from caves, *Asellus aquaticus*, one that presents a pattern of multiple invasions at different times in the same region. The second are the dytiscid diving beetles in the genus *Paroster* that occur in the Sturt Meadows calcrete aquifer in Western Australia. The third is the trogloxenic long-tailed giant rat *Leopoldamys neilli*, a species endemic to karst regions, but not to caves, in Thailand.

7.8.1 *Asellus aquaticus*

The evolutionary history of any subterranean lineage is likely to involve, at least at some scale, both dispersal and vicariance. The subterranean populations of the isopod *A. aquaticus* illustrate this well, but they also reveal a complexity due to colonizations at different times. *A. aquaticus* is an isopod common in lakes and rivers through all but westernmost Europe. Morphologically modified cave populations are known from much of its range (Fig. 7.13). On the basis of the sequence analysis of the mtDNA gene *COI* and the nuclear *28S rDNA* gene, Verovnik *et al.* (2005) suggest that *A. aquaticus* invaded Europe from Asia about 8–10 million years ago. It colonized most of Europe in several waves, including a pre-Pleistocene one. Populations in the Balkan Peninsula, in particular the Dinaric karst of Slovenia and north-east Italy, were much more differentiated and the only ones in Europe that showed variation in *28S rDNA*. In the heart of the Dinaric karst is PPCS, and troglomorphic populations have been known and studied from multiple streams in the system since at least the 1940s (Kosswig and Kosswig 1940). Populations in the cave stream differ among themselves in the degree of eye and pigment loss, and Kosswig and Kosswig (1940) attributed the high variability to relaxed selection or hybridization, a precursor to Wilkens' claim related to eye and pigment loss in *Astyanax mexicanus* and other species (Wilkens 1971, 1988).

Verovnik *et al.* (2003, 2004), using random amplified polymorphic DNA (RAPD), a technique that provides estimates of genetic differences without knowing sequences directly, and later supplemented by microsatellite analysis (Konec *et al.* 2015), found that populations of *A. aquaticus* in different

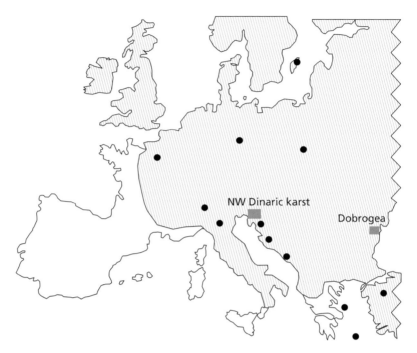

Fig. 7.13   Approximate distribution of *Asellus aquaticus* in Europe. Black dots denote troglomorphic populations; the squares, areas with specialized troglomorphic populations. From
Verovnik (2012). Used with permission of Elsevier Ltd.

streams in PPCS were genetically distinct (Fig. 7.14). The Planina polje
population is the most distinct and in fact is in a separate subspecies,
A. aquaticus carniolicus (Verovnik 2012). The morphologically variable
population in Pivka Cave is in fact genetically homogeneous. Previous
workers, such as Kosswig and Kosswig (1940) attributed this to hybridization. The Pivka Cave population likely represents a later colonization than
that found in the Pivka channel and the Rak channel.

The PPCS populations were different one from the other, different from
surface populations in the same area, and very different from populations
in Abisso di Trebiciano in the closely related species *Asellus kosswigi*, in a
separate karst drainage basin about 30 km away in Italy. The genetic dis-
tinctiveness of the PPCS populations indicated that they were separate
colonizations of PPCS, probably three in all. These same populations, as
well as populations from Abisso di Trebiciano, were also morphologically
distinct for a wide variety of characters (Prevorčnik *et al.* 2004, Konec
et al. 2015). They found morphological characters subject to parallel and
convergent selection, especially in size and shape of appendages, but with
differences because of the differences in age of colonization, but there were
even more characters that were subject to divergent selection, once again

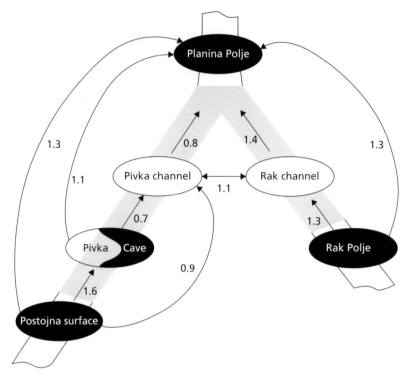

Fig. 7.14   Analysis of gene flow between *Asellus aquaticus* populations along two confluent sinking rivers in the Postojna–Planina Cave System. The values on the arrows are estimated
numbers of effective migrants per generation. Black symbols indicate populations with
pigmented individuals belonging to *Asellus aquaticus aquaticus* (Planina surface) and
Asellus aquaticus carniolicus (Planina and Rak polje). White symbols indicate troglomorphic populations belonging to *A. aquaticus cavernicolous*. The black-and-white symbol
stands for the morphologically mixed population. The darker area represents the subterranean river sections. From Verovnik (2012). Used with permission of Elsevier Ltd.

emphasizing the role of interspecific competition in moulding morphology
(see Chapter 5).

The reconstructed evolutionary history of *Asellus aquaticus* and *A. kosswigi*
in the northwestern Dinaric karst is shown in Fig. 7.15. The descendent
subterranean populations have remained genetically distinct. Subsurface
dispersal along 30 km of underground river took place after the invasions,
while longer distance dispersal probably occurred during the surface-dwelling
phase. *A. aquaticus* indicates the unexpected complexity of the biogeographical history of a subterranean animal, especially with regard to multiple
invasions through time rather than a single invasion or multiple invasions at
the same time. This history of *A. aquaticus* suggests that the colonization of
subterranean habitats may be more common and frequent than previously
thought.

Plate 1   Photo of the karst landscape of Halong Bay, Vietnam. Karst landscapes take many different shapes and forms in different regions. Among the most spectacular are the towers
and pinnacles of Halong Bay, a UNESCO World Heritage site. The remaining limestone is
slowly being dissolved away. See page 5 in text.

Plate 2   Photo of Pivka River sinking at the entrance to Postojnska jama, Slovenia. Photo by
M. Petrič, used with permission. See page 11 in text.

Plate 3   Photo of Unica Spring, the resurgence of the Postojna–Planina Cave System, Slovenia.
Photo by M. Blatnik, used with permission. See page 12 in text.

Plate 4   Photograph of the authors at a hypotelminorheic site at Scotts Run Park, near Washington, DC, USA. Photo by W.K. Jones, with permission. See page 18 in text.

Plate 5   Main trunk passage in Lower Kane Cave, Wyoming, USA. White, filamentous microbial mats dominated by sulfur-oxidizing bacteria are present in shallow sulfidic water, beginning at the lower right corner (water flows from the lower right to upper left). The microbial mat extends for approximately 20 m with an average thickness of 5 cm. From Engel (2012). Photo by A.S. Engel, with permission. See page 26 in text.

Plate 6   Photograph of drip water in Organ Cave, West Virginia, USA which percolates into the cave from the epikarst. Photo by H.H. Hobbs III, with permission. See page 31 in text.

Plate 7   Dead raccoon at the base (10 m depth) of Sunnyday Pit, West Virginia, USA. Photo by H.H. Hobbs, with permission. See page 37 in text.

Plate 8   Roots of *Metrosideros polymorpha* coming through the ceiling of Lanikai Cave, Hawai'i. Photo by H. Hoch, with permission. See page 40 in text.

Plate 9   Photo of the remipede *Lasionectes entrichoma* showing male and female reproductive systems, ventral view. Photo by D. Williams and J. Yager, with permission. See page 57 in text.

Plate 10   Photograph of *Proteus anguinus*. Photo by G. Aljančič, with permission. See page 71 in text.

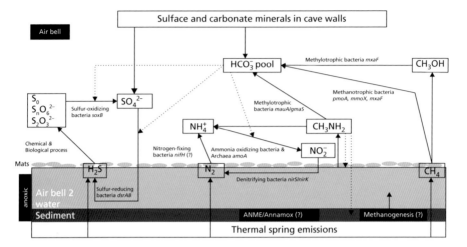

Plate 11   Schematic representation of microbial sulfur, carbon, and nitrogen cycling in Peștera Movile, Romania. Evidence for metabolic pathways comes from functional gene analyses. From Kumaresan *et al.* (2014). Used with permission of Walter de Gruyter GmBH. See page 93 in text.

Plate 12   Concentration of cave-crickets, *Ceuthophilus stygius*, on the ceiling of Dogwood Cave, Hart Co., Kentucky, USA. Photo by H. H. Hobbs, with permission. See page 101 in text.

Plate 13   Typical (A) *Proteus anguinus anguinus* and the pigmented, eyed subspecies (B) *Proteus anguinus parkelj*. Photographs by G. Aljančič, with permission. See page 121 in text.

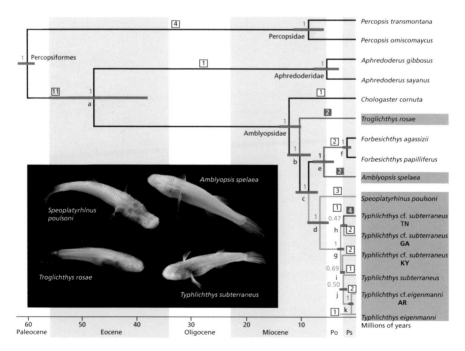

Plate 14   Molecular phylogeny of Amblyopsidae, based on Niemiller *et al.* (2012b). Used with permission of John Wiley and Sons Inc. See page 125 in text.

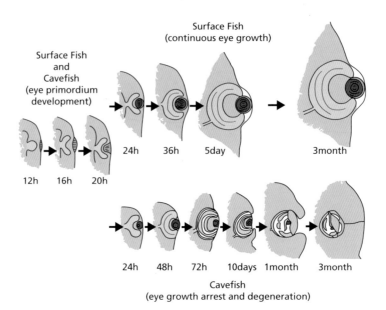

Plate 15   Eye development and degeneration in *Astyanax mexicanus*. Diagram showing the timing of eye growth and development in surface fish (top) and eye degeneration in cavefish (bottom). Drawing by W. Jeffery, with permission. See page 139 in text.

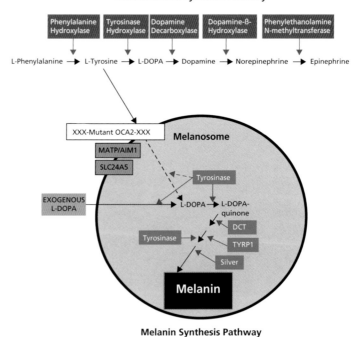

Plate 16   The relationship between the catecholamine and melanin synthesis pathways in *Astyanax* cavefish. In albino cavefish, a mutated *oca2* gene (white box with XXX) affects the first step of the pathway prior to tyrosinase function and prevents melanin synthesis. The defect caused by *oca2* loss of function can be rescued by exogenous L-DOPA. Solid lines: steps that occur in surface fish and in cavefish after L-DOPA rescue of melanogenesis. Dashed lines: steps that are absent in cavefish. From Bilandžija *et al.* (2013). See page 142 in text.

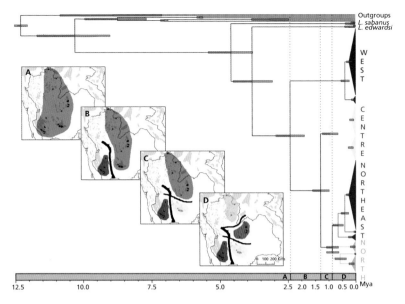

Plate 17 Dating of the most recent common ancestors with 95 percent HPD (highest posterior density), and graphical representation of the biogeographical scenario of *Leopoldamys neilli* according to four time periods A, B, C, and D. The four maps depict the hypothetical of *L. neilli* ancestral population (mid grey), western (mauve), central (yellow), northern (light blue), and northeastern (red) groups and the locations of barriers (black lines) leading to three vicariant events. From Latinne *et al.* (2012). See page 177 in text.

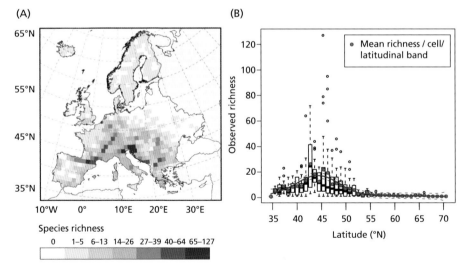

Plate 18 Map of species richness patterns of European stygobionts. A. Species richness of 10 000 km² cells. B. Relationship between the cell average of species richness per 0.09′ latitudinal band and latitude. Black horizontal bars and boxes show the median and interquartile range, respectively, for latitudinal bands. The maximum length of each whisker is 1.5 times the interquartile range and open circles represent outliers. The thick red line is the fit of generalized additive model to the averages of latitudinal bands. From Zagmajster *et al.* (2014). See page 200 in text.

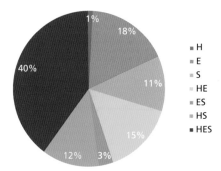

Plate 19   Pie chart of relative contributions of different combinations of drivers of species richness to the explained variance. H is historical climate stability, E is productive energy, and S is spatial heterogeneity. Data from Eme *et al.* (2014). See page 200 in text.

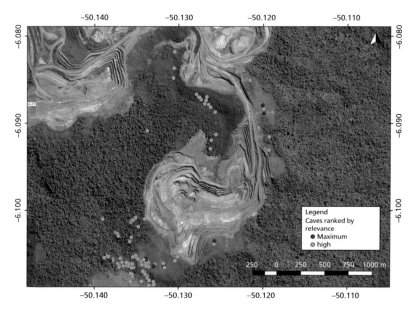

Plate 20   Iron-ore mine (N5, Serra Norte, Carajás, Brazil) showing the location of caves coloured by their classification. Caves with maximum relevance have at least one rare troglobiont; caves with high relevance have at least one troglobiont. From Jaffé *et al.* (2018). See page 215 in text.

Plate 21   Gate at the entrance to Fisher Cave, Missouri, USA, designed to allow unimpeded access for bats. Photo by H. Hobbs III, with permission. See page 234 in text.

Plate 22   Channelized Trebišnija watercourse in Popovo polje, Bosnia & Herzegovina in 2005. Photograph by M. Zagmajster, with permission. See page 237 in text.

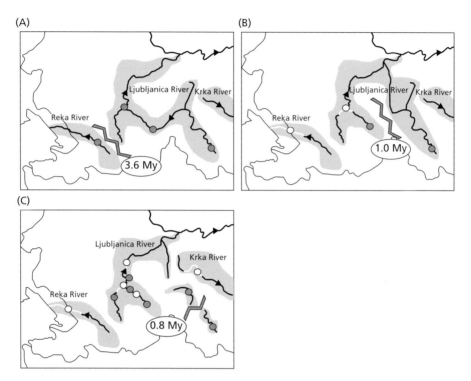

Fig. 7.15 Hypothesized chain of events that best explains the current distribution of haplotypes of *Asellus aquaticus* and *Asellus kosswigi* in the northwest Dinaric karst. Zig-zag line indicates fragmentation events that separate the drainages. The pale line denotes subterranean river passages; white circles subterranean populations; and grey circles isolated karst polje surface populations. The maximum age of the events was estimated by applying the COI clock rate of stenasellids. From Verovnik (2012). Used with permission of Elsevier Ltd.

7.8.2 *Paroster* in calcrete aquifers

The stygobiotic diving beetles isolated in calcrete aquifers in Western Australia are perhaps the best example of the climate relict hypothesis (CRH); in this case the driving force was continental drying, rather than Pleistocene glaciation. In 47 calcrete aquifers, at least one dytiscid diving beetle has become isolated (Fig. 7.3) sometime between 3 and 10 Mya (Fig. 7.4). Each calcrete aquifer is 'an island under the desert' (Cooper *et al.* 2002) with unique species present. In 29 of these aquifers, at least two species are present, and in 13 of these the species present are sister species (Guzik *et al.* 2009). The question that Guzik *et al.* asked was what the likely nature was of speciation within a calcrete aquifer. Speciation could be sympatric, which is in accord with the adaptive shift hypothesis (ASH), micro-allopatric, in accord with CRH, or the result of colonization and isolation by the same ancestral species at different times. This last possibility is similar to what Verovnik (2012) demonstrated for *Asellus aquaticus*.

In order to examine the question of the nature of speciation, Guzik *et al.* (2009) looked at the microgeographic pattern of three sister species (*Paroster macrosturtensis, P. mesosturtensis*, and *P. microsturtensis*) in a 3.5 km² grid in the Sturt Meadows calcrete aquifer, itself with an area of 43 km². Utilizing a grid of boreholes between 100 to 200 m apart, they sequenced the mitochondrial gene CO1 (*cox1*), and grouped adjoining boreholes together based on genetic similarity. Analysis of Molecular Variance (AMOVA) indicated that *P. macrosturtensis* and *P. mesosturtensis* had significant spatial structuring, and isolation by distance, even within their relatively small study area (Table 7.5). On the other hand, *P. microsturtensis* showed no structuring. They attribute this difference to the small size of *P. microsturtensis* (1.8 mm), which presumably enhances their dispersal ability. Bayesian models based on a coalescent approach yielded times of the most recent common ancestor of between 0.66 and 2.17 Mya, depending on the species and the model used (Guzik *et al.* 2009). As they point out, their results are consistent with either micro-allopatric speciation or isolation at different times by a common surface ancestor.

Leijs *et al.* (2012) looked at the likely nature of speciation in *Paroster* at a broader scale. In the Yilgarn, they found 11 aquifers with three sister species of dytiscids, 16 with two sister species, and 18 with one. They point out that this sympatry of sister species could the result of speciation within the aquifer (sympatric, parapatric, or micro-allopatric) or the result of multiple colonizations from the same surface ancestral species. Because there were multiple cases of sister species sympatry at the calcrete aquifer scale, Leijs *et al.* (2012) were able to devise a statistical test to separate the two hypotheses. The multiple invasion model fails to account for the observed fractions

Table 7.5 Analysis of Molecular Variance (AMOVA) using SAMOVA groups in two stygobiotic *Paroster* species and for individual bores with n>7 in *P. microsturtensis*.

Species	Source of variation	Sum of squares	Variance components	% of variation
P. mesosturtensis	Among SAMOVA bore groups	228.0	2.8 Va	40.1
	Among bores within SAMOVA bore groups	190.1	0.03 Vb	0.5
	Within individual bores	814.9	4.2 Vc	59.4
	Total	1233.1	7.0	100
P. macrosturtensis	Among SAMOVA bore groups	9.9	3.1 Va	44.8
	Among bores within SAMOVA bore groups	169.3	0.7 Vb	10.9
	Within individual bores	233.3	3.0 Vc	44.4
	Total	412.5	6.8	100
P. microsturtensis	Among bores	11.4	0.02 Va	0.7
	Within bores	149.2	2.6 Vb	99.3

Source: From Guzik *et al.* (2009)

of aquifers with pairs or triplets, except when the number of species in the ancestral species pool was very low. On the other hand, a within-aquifer speciation model could account for the observed fractions for a wide range of niche colonization probabilities, the main parameter of the within-aquifer model, based on the pattern of limiting similarities of dytiscids (Vergnon *et al.* 2013). This system provided a unique opportunity for a statistical test of modes of speciation, but as Leijs *et al.* (2012) pointed out, speciation within the aquifer could be sympatric, parapatric, or micro-allopatric.

7.8.3 *Leopoldamys neilli* in karst in Thailand

Neill's long-tailed giant rat (*L. neilli)* is limited to karst areas in Thailand (Fig. 7.16). The reasons for its limitation to karst are not entirely clear but it uses caves as nesting and breeding sites (Latinne *et al.* 2012). The species is a trogloxene, but Latinne *et al.* use the context of isolation in caves to explain the isolation and phylogeography of *L. neilli*. They used a wide array of genetic information, including two mitochondrial genes, two nuclear gene fragments, and 12 microsatellite loci. Their comparison of vicariance versus dispersal indicated that a vicariance model fits the data much better, and that the species was isolated from its sister species, *L. edwardsi*, about 3.8 Mya. During the geological history of the Central Plain of Thailand, there were periods of burying of karst which were vicariant events, leading to the isolation of four main biogeographical regions of distribution of *L. neilli* (Fig. 7.17; Latinne *et al.* 2012). Whether the split with *L. edwardsi* was due to adaptive shift or climate refuge is unclear. At present the ranges of the two species overlap. The use of caves for breeding could have been an adaptive shift, or the overlap between ranges could have been the result of secondary overlap. This study is remarkable is several aspects. The maps of distribution and explanations are nearly identical in kind to the maps and explanations for the evolution of troglobionts and stygobionts. On the one hand, this suggests that ASH and CRH have an applicability beyond the isolation of species in caves. But on the other hand, it suggests that colonization and isolation is not different in kind from colonization of different surface habitats and that genetic modifications need not be that profound. This echoes some of Romero's (2004, 2009) views on the evolution of troglomorphy, which for him has a large non-genetic phenotypic component.

7.9 Summary

Colonization and speciation in subterranean environments can be conveniently divided into four stages. The first step is colonization of subsurface environments. There is a constant flux of colonists into most subterranean habitats, and both changes in environmental conditions (such as warming

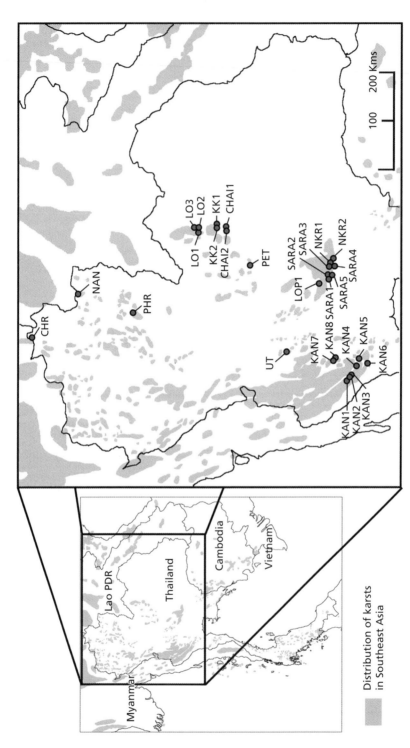

Fig. 7.16 Distribution of karst and of sampling localities for *Leopoldamys neilli*, Neill's long-tailed giant rat, in Thailand. From Latinne et al. (2012).

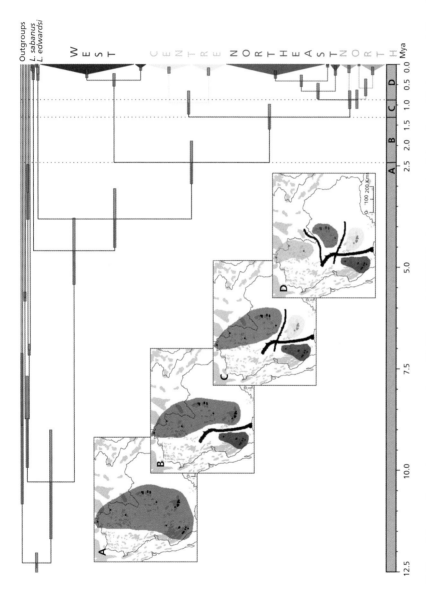

Fig. 7.17 Dating of the most recent common ancestors with 95 percent HPD (highest posterior density), and graphical representation of the biogeographical scenario of *Leopoldamys neilli* according to four time periods A, B, C, and D. The four maps depict the hypothetical of *L. neilli* ancestral population (mid grey), western (darkest grey/mauve), central (palest grey/yellow), northern (pale grey/light blue), and northeastern (dark grey/red) groups and the locations of barriers (black lines) leading to three vicariant events. From Latinne *et al.* (2012). See Plate 17.

or aridification) and active colonization by individuals searching for new resources can increase colonization rates. The second step is the success (or failure) of these colonizations. Success varies from group to group, and in general depends on exaptation (pre-adaptation) to subterranean conditions. The third step is speciation. Under the CRH surface populations go extinct but under the ASH they do not necessarily do so, and speciation can be parapatric between adjoining populations. There is strong evidence for the CRH among the temperate zone subterranean fauna, and growing evidence for the ASH in tropical caves, especially lava tubes. The final step is possible further speciation as a result of subsurface dispersal. In the absence of surviving surface populations, it is difficult to distinguish between a group of species with a single subterranean origin and a group of species all of which colonized subterranean habitats separately. The presence of cryptic species reduces the potential amount of dispersal because ranges of cryptic species are much smaller. On the other hand, there is ecological evidence supporting dispersal, such as the correspondence between range size and percentage occupancy of habitats such as epikarst drips. In the broad sweep of evolutionary time, alluvial aquifers may serve as dispersal routes (the interstitial highway). Detailed analysis of the evolutionary history of the isopod *A. aquaticus* in the Dinaric karst, the diving beetles *Paroster* in a calcrete aquifer in Western Australia, and the trogloxenic *Leopoldamys neilli* in Thailand reveal some of the complexities of species' phylogeography.

8 Geography of Subterranean Biodiversity

8.1 Introduction

In the previous chapter, the emphasis was on the evolutionary history and geographical distribution of individual lineages. This is the focus of historical biogeography. However, there is another way to look at biogeographical patterns, and this is to consider the sum of all of the geographical patterns and evolutionary history of subterranean lineages. From this point of view, it is the number of species (species richness) that is the object of analysis. The advantage of this method is that it is a summary approach. The patterns that result are the consequence of independent events of colonization, isolation, and speciation. This makes it easier to make and test hypotheses about the causes of these patterns. The disadvantage of this approach is that individual evolutionary histories and corresponding details are lost. A lowly beetle nearly indistinguishable from many other beetles in the genus, such as one of the more than 200 species of *Pseudanophthalmus* in North America is as important in this kind of analysis as a species that is the only member of its genus, such as the beetle *Neaphaenops tellkampfi*, also found in some of the same North American caves.

In spite of the considerable interest in caves as ecological, evolutionary, and microbiological laboratories, many examples of which we have discussed in previous chapters, the heart of speleobiology remains the description and explanation of species diversity. While the study of biodiversity is important in any habitat, it seems especially so for subterranean habitats. There are undoubtedly several reasons for this, but certainly the bizarre morphology of subterranean animals combined with the seemingly inexhaustible supply of undescribed species is a major component.

This chapter has three main parts. The first is a consideration of the problems associated with sampling completeness and adequacy, the struggle to measure subterranean biodiversity. Owing to the high levels of endemism

The Biology of Caves and Other Subterranean Habitats. Second Edition. David C. Culver and Tanja Pipan. Published 2019 by Oxford University Press. © David C. Culver and Tanja Pipan 2019. DOI: 10.1093/oso/9780198820765.001.0001

of the stygobiotic and troglobiotic fauna (Christman *et al.* 2005) and the large number of cryptic species (Trontelj *et al.* 2009), a complete enumeration of species seems far from complete. We consider ways to compensate for this incompleteness. The second part is an examination of the analogy between islands and subterranean habitats, particularly caves. The theory of island biogeography developed by MacArthur and Wilson (1967) provides an explanatory framework to explain differences in species richness on islands and island-like habitats such as caves. We pay special attention to the scale at which the island analogy holds. In the third part, we look at the emerging global patterns of subterranean biogeography, including regional and continental patterns.

The concept of species diversity, itself part of the more general concept of biodiversity, includes not only the number of species but also their relative abundance. Some information on relative abundance is available for subterranean communities but only in those relatively rare cases where sampling was quantitative, when Bou–Rouch pumps, epikarst filters, or other quantitative samplers were used (see Chapter 3). Nearly all of the ensuing discussion will be about species richness rather than relative abundance.

8.2 The struggle to measure subterranean biodiversity

There is probably not even a single relatively species-rich cave, let alone a region for which we can be confident that all cave-limited species (aquatic stygobionts and terrestrial troglobionts) have been discovered and described. In the two of the six most species-rich caves known (see Table 8.3) and among the best studied—Vjetrenica in Bosnia and Herzegovina and Mammoth Cave in Kentucky—there is no indication that all species have been discovered, based on the accumulation of described species up to 2000 (Fig. 8.1). The apparent asymptote for Vjetrenica results from the Bosnian War, and new species are still being discovered (Lučić and Sket 2003, Ozimec and Lučić 2009). The same holds for the stygobiotic fauna of France, probably the best studied country (Ferreira *et al.* 2007). Even in the world's most diverse cave—the Postojna–Planina Cave System—new species are being discovered.

While there is a tendency on the part of some speleobiologists to conclude that no generalizations are possible because sampling is incomplete, this approach ignores the techniques available to estimate total species richness from incomplete data. We all need to have the best information available, but given that new caves and other subterranean habitats are continually being discovered and sampled, it will never be possible to generalize if one waits for total sampling completeness. Except for some karst regions in the British Isles (Proudlove 2001), there are no reports of sampling of more than 25 per cent of known caves. Sampling completeness is even more problematic

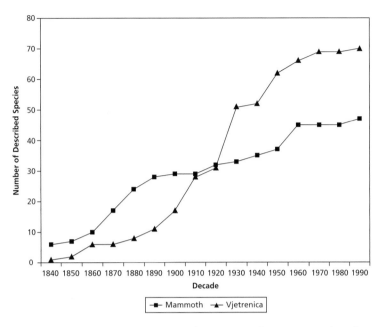

Fig. 8.1   Accumulation of described species known from Mammoth Cave, Kentucky, USA and Vjetrenica, Bosnia & Herzegovina. Data for Vjetrenica are from Lučić and Sket (2003) and data for Mammoth Cave are from Culver and Hobbs (2017).

for interstitial habitats. Species composition varies over small geographical scales (Rouch 1991; Ward and Palmer 1994), but relatively few sites have been sampled along any river, and the coverage of different rivers is very incomplete. More relevant than 'site sampling completeness' is 'species sampling completeness', whether all species have been found.

The standard tool to answer this question is the species accumulation curve, which is created by randomizing caves (or quadrats) and selecting 1, 2, 3, . . . sites at random and resampling 100 or more times. This is not the same as the actual accumulation of described species (see Fig. 8.1), but is a procedure that allows for statistical assessment of the accumulation curve. For the most part, such accumulation curves for subterranean fauna have not reached an asymptote (see Schneider and Culver 2004 for one of the first such studies in caves and Rouch and Danielopol 1997 for an example from a hyporheic habitat). Exceptions to the lack of an asymptote have been either very intensive sampling at small scales such as epikarst drips within a cave (Pipan and Culver 2007a; Pipan et al. 2018) or intensively sampled areas aggregated into equal-sized sampling areas (Culver et al. 2006b). The epikarst drip study is especially instructive because Pipan and colleagues looked at completeness at several nested scales—different samples of the same epikarst drip, different drips in the same cave, and different passages in a cave system (Fig. 8.2). For all of these scales, Pipan and Culver (2007a) found that

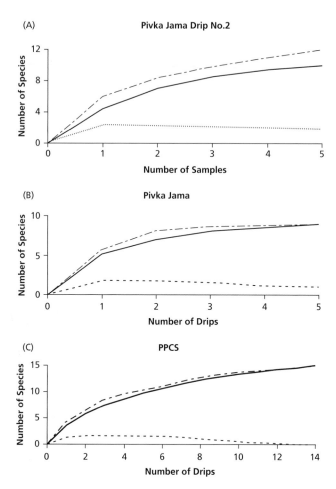

(A) Pivka Jama Drip No.2

(B) Pivka Jama

(C) PPCS

Fig. 8.2   Species accumulation curves based on 100 randomizations of 1, 2, 3, … samples for different scales in Postojna–Planina Cave System: (A) monthly samples of a drip in Pivka jama; (B) different drips in Pivka jama, approximately 100 m apart; and (C) different drips in several stream passages in Postojna–Planina Cave System (PPCS), up to 1 km apart. Modified from Pipan and Culver (2007a). Used with permission of Blackwell Publishing.

their sampling was sufficient for the accumulation curves to reach an asymptote. In their analysis of caves in the Dinaric karst, Pipan and Culver (2007a) found that it took approximately four samples of one drip, three to four samples of different drips in a stream passage, and ten drips in different stream passages to accumulate 90 per cent of the total species found. Pipan *et al.* (2018) did find some caves where the equivalent sampling was incomplete. At the next larger scale, that of a region (in Pipan *et al.*'s study, the Dinaric karst and the Alpine karst), the accumulation curve did not reach an asymptote (Fig. 8.3).

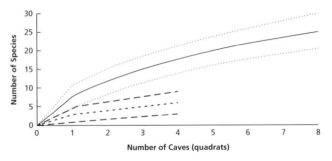

Fig. 8.3   Accumulation curves for epikarst copepod species in the Alpine (dotted line) and Dinaric (solid line) karst. From Pipan *et al.* (2018).

In those cases where accumulation curves did not reach an asymptote, there are group of procedures, the best known of which are the Chao1 and Chao2 estimates (Chao 1984; Colwell *et al.* 2004; Colwell *et al.* 2012; Colwell 2016), which allow an estimate of the 'missing' species richness. For example, Chao2 estimates are based on the ratio of the number of individuals known from a single site compared to the number of species known from two sites. As sampling becomes more complete, the number of species known from a single site declines, and correction due to Chao2 estimates becomes smaller. In a study of overall diversity of stygobionts from both karst and interstitial habitats in six countries in Europe, Deharveng *et al.* (2009) found that the percentage of expected additional species ranged from 30 per cent for Belgium to 83 per cent for Portugal (Table 8.1). This percentage was based on an estimator similar to Chao2—Jackknife1 that uses the number of species only found in a single site (Deharveng *et al.* 2009). Differences in estimated number of species were also related to differences in shapes of the accumulation curves—curves close to an asymptote had lower expected numbers of new species.

What remains to be determined is which of these estimators, for example, Chao2, Jackknife1, and so on, are most appropriate for cave data. What makes cave data unique is that, unlike most samples in surface environments, the number of species known from a single subterranean site often does not decline to zero as is typical of most samples of surface-dwelling fauna analysed at the same spatial scale, but rather reaches a non-zero asymptote.

The importance of thorough sampling was investigated by Dole-Olivier *et al.* (2009a). They investigated the asymptotes of sampling curves for six regions of about 400 km^2 in size that were the study sites for the most ambitious and intensive sampling of subterranean fauna to date—*P*rotocols for the *A*ssessment and *C*onservation of *A*quatic *L*ife *I*n the *S*ubsurface (PASCALIS). In each region, several drainage basins, both karstic and interstitial aquifers, and different habitats were sampled (Gibert 2005). Dole-Olivier *et al.* (2009a)

Table 8.1 Groundwater biodiversity measures in six European countries.

Country	Number of Sampled Cells	Number of Sampled Sites	Number of Species	Additional Species Predicted by Jackknife 1	Proportion (%)
Belgium	17	155	33	10	30
France	566	1712	320	114	36
Italy	337	1580	288	106	37
Portugal	24	34	48	40	83
Slovenia	54	491	183	63	34
Spain	241	737	216	92	43
Total	1228	4709	930	361	39

The number and proportion of expected additional species refers to the estimates of species discovered if further samples are taken. Jackknife1 is a procedure for estimating missing species richness, analogous to Chao2. Cell size is 2 km × 2 km.
Source: Deharveng et al. (2009).

produced accumulation curves for each of the six regions. For each of these regions, between 187 and 206 samples were taken, a very intensive sampling effort (Fig. 8.4). They showed that if only 10 samples were taken, conclusions about which regions had the highest species richness would be wrong. This is because the accumulation curves cross each other. With 100 samples, the rank order of the sites in terms of species richness is nearly the same as that for 180 samples [only the order of Jura (France) and Cantabrica (Spain) is reversed]. This clearly shows that intensive sampling is necessary. They also investigated whether adding new drainage basins, aquifer types, or habitats increased the efficiency of sampling.

Generalizations were difficult because unfortunately, the results depend on the region. For example, in some regions, the inclusion of both interstitial and karst aquifer samples significantly increased the number of species (and hence the efficiency of sampling) and in others it made no difference. In a parallel study, Stoch *et al.* (2009) investigated whether a subset of the stygo-fauna could be used to predict the species richness of the rest of the fauna. They found that it was also dependent on region; no single indicator or set of taxonomic indicators could predict overall stygobiotic richness.

A final example of the importance of thorough sampling is the study of Eberhard *et al.* (2009) on sampling of wells in the Pilbara region of Western Australia. A series of wells are the only sampling access to a rich groundwater fauna of over 300 species, including both stygobionts and non-stygobionts. A single sample contained on average less than half of the species known from a particular well, and so multiple samples of the same well were essential. They also point out that inadequate sampling can lead to conclusions about high levels of endemism that are artefacts of incomplete sampling.

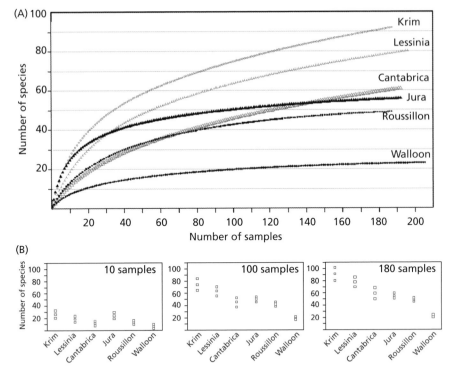

Fig. 8.4   Species accumulation curves for stygobionts in the six European PASCALIS regions (A) and 95% confidence for the observed number of species (Mao-Tau estimate of Colwell *et al.* 2004) for three different sampling efforts: 10, 100, and 180 samples (B). Sampling regions are Krim Mountain (Slovenia), Lessinian Mountains (Italy), Cantabrica (Spain), Jura Mountains (France), Roussillon (France), and Walloon (Belgium). From Dole-Olivier *et al.* (2009a). Used with permission of Blackwell Publishing.

Culver *et al.* (2004b) took a very different approach to assessing sampling adequacy. Using the extensive database on Slovenian caves maintained by the Karst Research Institute ZRC SAZU and the Speleological Association of Slovenia, they looked at three time 'snapshots'—1940, 1970, and 2000. In 1940, about half of the known stygobionts and troglobionts had been described and about a third of the caves sampled by 2000 had been sampled. By 1970, these numbers had reached about 80 per cent of what was known in 2000. They showed that many spatial relationships between species richness and cave density and geographical location remained constant from 1940 to 2000. For example, both hotspots of stygobiotic species richness were known by 1940 and two of four hotspots of troglobiotic species richness were known by 1940. All were known by 1970. Their study provides a strongly optimistic note in the rather depressing consideration of the need for sampling thoroughness.

8.3 Caves as islands

At a very simple level, there is an analogy between real oceanic islands and 'virtual islands' of caves. Similar to islands, caves are isolated habitats, but the 'ocean' for caves is the surface, with the incumbent dangers of predators, sunlight, and environmental variation. For cave-adapted organisms, these barriers may be as formidable as the open ocean is to terrestrial organisms on islands (Culver and Pipan 2008b). The reality is more complicated because of other connected subterranean habitats such as fissures, MSS, and epikarst, but the reason that it is useful to explore the island analogy is that, if it holds, then the theory of island biogeography developed by MacArthur and Wilson (1967) can be applied to subterranean habitats. The crux of island biogeography theory is that the number of species on an island is in dynamic equilibrium between the addition of new species through immigration and the loss of species through extinction. MacArthur and Wilson's book has been cited by thousands of papers and remains both a powerful theory and a controversial one (Losos and Ricklefs 2010).

Perhaps the most widely used prediction from island biogeography theory is that z in the relationship between species number and island area is clustered around 0.25 (e.g., Sugihara 1981):

$$S = CA^z$$

where S is the number of species, A is the area, and C and z are fitted constants.

The physical analogy between caves and islands is actually rather weak because of the complexities of the karst subsurface (see Fig. 1.6), and the analogy at both larger and smaller scales seems more apt. In almost all cases, the rock in which caves occur has fractures and small solution tubes that allow subsurface connections between caves. In particular, epikarst and the associated vertical percolation of water are more or less continuous in karst areas at the landscape scale (Williams 2008). In addition, caves often occur in relatively dense clusters. In some places in Slovenia and Tennessee (USA), cave passages reach a density of between 150 and 250 m/km². Lava tube passages may reach densities ten times that (Culver and Pipan 2014). At densities such as these, the analogy with islands weakens. It is not surprising that the first two attempts to find an area effect among the invertebrate fauna in caves have failed to do so (Culver 1970, Vuilleumier 1973).

However, the analogy of caves and islands does seem to hold in the case of bats' use of caves. In this case, the small interconnections between caves are irrelevant and the landscape, from the point of view of bats, consists of a series of roosting chambers (caves) separated from each other. Brunet and Medellín (2001) found that the number of bat species roosting in caves in central Mexico was strongly affected by ceiling area (Fig. 8.5), with a z-value of 0.31, well within the range of values found for fauna on oceanic islands. They suggest that area is also a surrogate for habitat diversity, especially solution pockets in the ceiling that act to retain the body heat of roosting bats. The maximum distance between any two caves was 13 km, well within the flight range of bats, so isolation was not a factor.

In addition, Souza-Silva *et al.* (2011) showed that the total number of invertebrates, including all that were found regardless of their ecological status, has a strong linear dependence on cave length. Further analysis of their data, using only troglomorphic species (S_{TM}) in iron-ore caves, also demonstrated a strong linear dependence on length (L) ($S_{TM} = 1.34 + 0.10$ L, $R^2 = 0.48$, Fig. 8.6). Their results thus directly contradict the earlier findings of Culver (1970) and Vuilleumier (1973). More studies of this type would be informative.

Perhaps the best analogy between caves and islands in ecological time is that of cave drip pools fed by percolating water, a habitat that bears a superficial

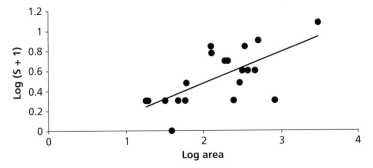

Fig. 8.5   Relationship between ceiling area in a cave and number of bat species roosting in central Mexican caves. Data from Brunet and Medellín (2001).

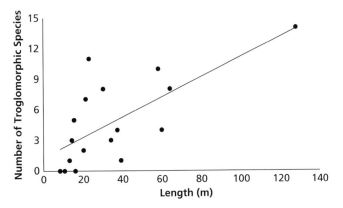

Fig. 8.6   Relationship between number of troglomorphic species and cave length of iron-ore caves in the Atlantic Coastal Rain Forest of Brazil. Data from Souza-Silva *et al.* (2011).

similarity to water trapped in bromeliads, also an island-like habitat. These drip pools have a diverse fauna that comes from water dripping from epikarst. The analogy with islands is that epikarst (the source of dripping water) is the mainland, and pools are islands (Fig. 8.7). In a study of this system, Pipan *et al.* (2010) found between five and seven stygobiotic copepods (mean = 6) in drip pools in six Slovenian caves and between four and twelve (mean = 8) in drips that fed these pools. The number of species immigrating over the course of a year was the same order of magnitude as number of the resident species. Drips, however, were not the only source of migrants to pools because drip pools also had between one and seven non-stygobiotic species (mean = 4.2) while drips had between zero and three non-styobiotic species (mean = 0.8).

This little-studied system has advantages as a model system of virtual islands. They are highly replicated; it is possible to completely census and manipulate

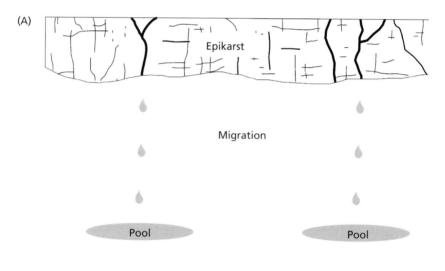

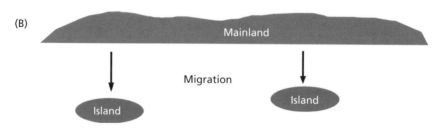

Fig. 8.7   Sketch of the relationship between epikarst, ceiling drips, and pools (A) and the analogy with continents and islands (B).

immigration over long periods of time; and, given the high migration rates, it is likely to be in equilibrium between migration and extinction rates.

Other superficial subterranean habitats have an apparent island-like structure and may be analogous to islands in ecological time. Seeps are an obvious example. In the lower Potomac drainage near Washington, DC, USA, seeps can occur as close as 10 m apart and yet have different species composition (Culver *et al.* 2012). Dispersal between seeps can occur when heavy rains result in sheet flow of water, potentially moving animals between seeps.

For subsurface habitats outside of cave regions, such as the hyporheic along streams and rivers, the analogy to islands completely breaks down because the habitat is branch-like and linear.

There may also be an analogy between subterranean habitats and real islands in evolutionary time, a timescale at which the equilibrium processes of immigration and extinction are dominated by isolation and speciation. Different karst drainage basins (see Chapter 4) may be analogous to islands because movement between them is highly restricted. This was the situation with the amphipod *Gammarus minus* in West Virginia where different basins had genetically distinct populations (Figs 6.4 and 6.7). The correlation of ranges of the *Monolistra* isopods with paleodrainages also points to a physical analogy of karst basins and islands (Sket 2002). The present-day drainage patterns are different and the distribution of *Monolistra* corresponds to the paleodrainages (islands), not present-day drainages (islands).

Stygobionts and troglobionts also share characteristics of isolated oceanic archipelagos. Dispersal abilities are reduced; endemism at the scale of single islands or single caves is common; and morphologies are often obviously adaptive. What is different is that isolated oceanic archipelagos often have groups that have undergone adaptive radiation within the archipelago, the classic example being the finches of the Galapagos Islands (Grant 1986). It is likely that there have been adaptive radiations in caves, including the species-rich trechine beetle fauna of the Pyrenees (Faille *et al.* 2010) and the species-rich *Niphargus* amphipod fauna of the Dinaric karst (Trontelj *et al.* 2012). More generally, it may be that dispersal rates are too low or opportunities for niche differentiation are too restricted to allow for frequent co-occurrence of similar species.

At yet a larger scale, that of contiguous karst regions, there is also a physical and biological analogy with islands. The major karst areas in the United States are shown in Fig. 8.8. The number of stygobionts ranges from zero in the Black Hills to 80 in the Appalachians, and the number of troglobionts ranges from zero in the Florida Lime Sinks to 257 in the Interior Low Plateau. Species overlap was very low and even at the generic level there were differ- ences. For example, the contiguous karst areas of the Interior Low Plateau and the Appalachians shared only half of the stygobiotic and troglobiotic

Fig. 8.8   Map of the karst regions in the United States for which biological data are available. The Interior Low Plateau is shown in stippling to differentiate it from the Appalachians and the Ozarks. From Culver *et al.* (2003). Used with permission of Springer Publishing.

genera. Culver *et al.* (2003) looked at seven possible variables that would predict the number of stygobionts and troglobionts. Four of these were measures of the physical environment: size of karst area, number of known caves, number of caves greater than 1.6 km, and number of caves deeper than 120 m. These variables captured various aspects of the different karst areas. The other three variables measured opportunities or stresses leading to cave colonization: distance from the Pleistocene ice sheet, distance from the embayment of the late Cretaceous Seas, and vegetation type. Vegetation type was an indirect measure of both surface productivity (see Chapter 2) and availability of exapted ancestors in the forest litter. Of all these variables, only the number of caves was a significant predictor of species number, both for stygobionts and troglobionts. There was no area effect because of the differences in amount of cave development in different regions. Because of the size of the areas, the analogy with islands is strained and this example leads us to a consideration of more global aspects of diversity, as does the likelihood that factors controlling species richness vary both geographically and according to the scale of analysis (Eme *et al.* 2015).

8.4 Global species richness

Beginning in the 1960s there was growing interest among ecologists generally about the patterns of diversity of different taxonomic groups, especially with regard to tropic–temperate gradients. Subterranean biologists were no exception and there were some early discussions about why subterranean species diversity was lower in the tropics (Mitchell 1969). However, there was no quantitative data to inform such a discussion, and discoveries of a rich troglobiotic fauna in the Hawaiian Islands (Howarth 1972) raised doubts about whether tropical caves actually had fewer stygobionts and troglobionts. Data on stygobionts and troglobionts is widely scattered in the taxonomic literature, and species lists for most countries were not attempted until the 1990s, largely spurred on by the impetus of Juberthie and Decu's *Encyclopaedia Biospeologica*. Such lists are particularly difficult in tropical countries where the frequency of troglomorphism among stygobionts and troglobionts is lower than in temperate areas (Deharveng and Bedos 2012) and even most surface-dwelling species are undescribed (Trajano 2001; Gallaõ and Bichuette 2018). There is still no compilation of troglobiotic diversity on a worldwide basis but an estimate of stygobiotic diversity on a continental basis is shown in Table 8.2. Such data must be treated with caution, not only because they are incomplete and there remain many undiscovered and undescribed species (see earlier in this chapter), but also because areas differ. Nevertheless, a couple of points do emerge. There are stygobionts (and troglobionts) on all continents, and species richness is much higher for Europe than for any other continent. Also of interest is the relatively low stygobiotic richness in North America, even compared to the less well-studied continent of Asia. Many of the numbers in Table 8.2 are outdated because additional compilations have been done since 2005, e.g. Kayo *et al.* (2012) for Africa, but using partially updated lists introduces bias towards the most recently updated list (Culver *et al.* 2012).

Realizing that regional lists of stygobionts and troglobionts were a long way from completion, Culver and Sket (2000) decided to approach the problem of subterranean diversity patterns from a different direction. In many cave regions, at least a few caves are relatively well studied biologically, and these caves tend to be the most interesting ones with a rich fauna. This is especially true in tropical areas, where only a few caves have been studied, but the ones studied have often been visited repeatedly (Deharveng and Bedos 2012). Culver and Sket (2000) compiled a list of 20 'hotspot' caves and wells throughout the world that had at least 20 stygobionts and troglobionts. Culver and Sket's analysis was limited to caves and wells in karst. No equivalent study of interstitial sites has been carried out, but it would be interesting to do so. Relatively few interstitial sites have been studied, so an analysis of α-diversity would be an appropriate starting place.

Since their publication, a number of sites have been reported to have at least 20 stygobionts and troglobionts, both as a result of new research (e.g., Deharveng and Bedos 2012; Niemiller and Zigler 2013; Souza-Silva and Ferreira 2016; Por 2007; Pipan 2005) and as the result of being overlooked in Culver and Sket's original study (especially the Canary Islands, see Culver

Table 8.2 The number of stygobionts known from six continents and selected countries.

Continents	Countries	Number of Stygobionts	
Europe		**2000**	
	Dinaric karst (parts of Italy, Slovenia, Croatia, Serbia, Bosnia & Herzegovina, Montenegro)		396
	France		380
	Italy		265
	Romania		193
Asia		**561**	
	Japan		210
	Indian, Indonesia, Cambodia, Laos, Thailand, Vietnam		380
	Middle East		265
	Turkey		193
North and Central America		**500**	
	United States		300
	Central America		200
Africa			
	Northern Africa (Morocco, Algeria, Tunisia, Libya, Egypt)	**500**	
	Southern Africa (South Africa, Namibia, Botswana, Zambia, Zimbabwe)		300
	Madagascar		200
South America		**200**	
Oceania		**226**	
	Australasia (Australia, Tasmania, New Zealand)		170
	Melanesia, Micronesia, Polynesia		56
Antarctica		**0**	

Source: Modified from Ferreira (2005).

and Pipan [2013]). While Culver and Sket's original publication generated a great deal of research and interest in patterns of subterranean species richness, it is clear that it had several limitations. First, it did not distinguish between aquatic and terrestrial species, and they often have different patterns of distribution (Culver *et al.* 2000). Second, because of the growing number of cases (more than 50 are known to us), its utility in identifying global hotspot areas is limited.

Therefore, Culver and Pipan (2013) examined the much smaller number of sites with either 25 stygobionts or 25 troglobionts (Fig. 8.9). Aquatic and

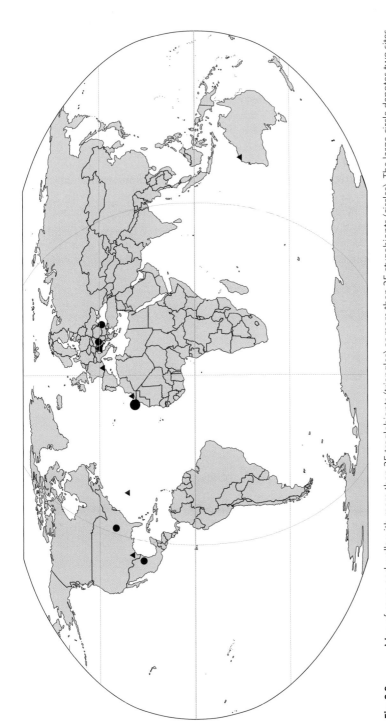

Fig. 8.9 Map of caves and wells with more than 25 troglobionts (triangles) or more than 25 stygobionts (circles). The large circle denotes two sites in the Canary Islands, and the large triangle denotes four sites in Slovenia. From data in Culver and Pipan (2013). Map by Magda Aljančič, used with permission.

terrestrial sites present very different patterns, but share several features (Table 8.3). First, there are no known hotspots in Africa, South America, or Asia. While it may be that there are undiscovered hotspots in these continents, all three have been studied and investigated in some detail. Second, the concentration of sites is in southern Europe, especially the Dinaric karst.

With one exception, terrestrial hotspots are either in the Canary Islands or southeast Europe, especially the Dinaric karst. A number of explanations have been put forward to explain the high species richness of the Dinaric karst, including high productivity of the surface landscape (Culver *et al.* 2006b; Eme *et al.* 2015). Peştera Movile (Romania) has high subterranean productivity since it is chemoautotrophic (Kumaresan *et al.* 2014). The Canary Islands are an interesting second hotspot. Howarth (1972, 1987) championed the view that terrestrial faunas in tropical lava tubes, especially in Hawaii, were sites of high species richness. It turned out that the Canary Islands were much richer in species than the Hawaiian Islands (Oromí 2004; Culver and Pipan 2014). The reasons for the difference between the two

Table 8.3 Caves with more than 25 stygobionts (A) or 25 troglobionts (B).

Site name	Country	Number of Species	Region/Ecology
A. Stygobionts			
Postojna–Planina Cave System	Slovenia	48	Dinarides
Vjetrenica	Bosnia & Herzegovina	40	Dinarides
Walsingham Cave	Bermuda	37	Anchialine/ chemoautotrophic
Triadou Aquifer	France	34	Phreatic
Robe River	Australia	32	Phreatic
Jameos del Aqua	Lanzarote, Canary Islands, Spain	32	Anchialine/ chemoautotrophic
Križna jama	Slovenia	29	Dinarides
Logarček	Slovenia	28	Dinarides
Šica–Krka System	Slovenia	27	Dinarides
Edwards Aquifer	Texas, USA	27	Phreatic/ chemoautotrophic
B. Troglobionts			
Postojna–Planina Cave System	Slovenia	36	Dinarides
Cueva de Felipe Reventón	Tenerife, Canary Islands, Spain	36	Lava tube
Vjetrenica	Bosnia & Herzegovina	30	Dinarides
Peştera Movile	Romania	29	Chemoautotrophic
Cueva del Viento	Canary Islands, Spain	28	Lava tube
Sistema Purificación	Mexico	28	Tropics
Mammoth Cave	Kentucky, USA	26	Longest cave

Source: modified from Culver and Pipan (2013) and Deharveng and Bedos (2019).

island systems include the older age of the Canaries, and the fact that Canarian lava tubes are shallower. Both have root systems penetrating the ceiling that are important food sources (Stone *et al.* 2012). Mammoth Cave is the remaining site, and it is distinguished by its great length (>500 km) and diversity of habitats, a real outlier (Toomey *et al.* 2017).

The pattern of stygobiotic hotspots is more diffuse. As was the case for troglobionts, the Dinaric karst is a hotspot, with five of the ten sites being in the Dinaric (Table 8.3). While productivity may be important, the long geological history of the Dinaric karst and its proximity to the Adriatic Sea, a source of colonists during the Messinian Salinity Crisis, may also be important (Sket 1999). The remaining sites, scattered across the continents, are usually phreatic or chemoautotrophic, or both. Chemoautotrophic sites are high productivity sites, and it may turn out that most phreatic sites are also chemoautotrophic.

The only tropical or subtropical hotspots are the lava tubes of the Canary Islands and a single deep well in the Robe River of Australia, and only the Robe River is a mainland site, one that is in highly arid western Australia. There may be additional Australian hotspots in the calcrete aquifers of the arid west (Guzik *et al.* 2011) but species lists for individual sites have not been published.

A further analysis of tropical cave patterns by Deharveng and Bedos (2012) showed that subterranean species richness is higher in the Oriental and Australasian regions than in the Neotropics or Africa (Fig. 8.10). The reasons for this are unclear, but the Oriental and Australasian regions are more fragmented, perhaps promoting speciation. Deharveng and Bedos exclude chemoautotrophic and anchialine caves from their analysis and point out that they likely have a different pattern.

Aside from their conservation importance (see Chapter 10), hotspots may serve as a surrogate for overall subterranean species richness. Species diversity can be divided into three components—a local, point diversity (α-diversity), a between-site diversity (β-diversity), and an overall regional diversity (γ-diversity). Hotspot caves are a measure of the maximum α-diversity, and if the components of diversity are correlated, then hotspot diversity should reflect overall diversity. If this were always true, then species accumulation curves would not intersect. At least in some cases they do intersect (Fig. 8.4), but the frequency of this occurrence is unknown.

8.5 Regional species richness

Considerably more information about troglobiotic species richness in caves is available for many areas of Europe and North America. Since α-diversity

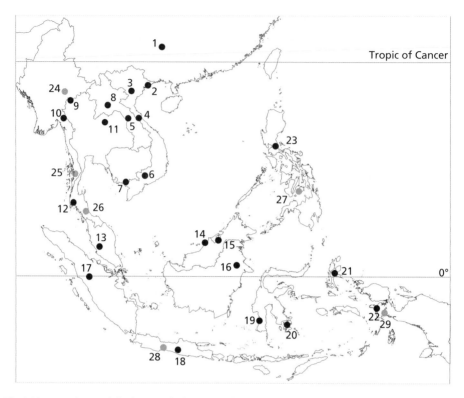

Fig. 8.10   Some of the best studied caves or karsts with more than four obligate cave species in Southeast Asia. 1: Ma San Dong cave (Guangxi); 2: Halong Bay karst (northern Vietnam); 3: Pu Luong karst (northern Vietnam); 4: Ke Bang karst (central Vietnam); 5: Tham Thon cave (Khammouane); 6: Tan Phu lava tubes (southern Vietnam); 7: Hon Chong karst (southern Vietnam); 8: Vang Vieng karst (central Laos); 9: Tham Chiang Dao cave (northern Thailand); 10: Farm caves (southern Burma); 11: Tham Phulu cave (eastern Thailand); 12: Phangnga caves (southern Thailand); 13: Batu Caves (Malaya); 14: Niah Cave (Sarawak); 15: Air Jernih system (Mulu karst in Sarawak); 16: Baai system (Sangkulirang); 17: Sangki system (Gunung Seribu); 18: Gunung Sewu karst (Java); 19: Salukkan Kallang–Tanette system (Maros karst); 20: Muna karst (Sulawesi); 21: Batu Lubang cave (Halmahera); 22: Fakfak karst (Papua); 23: Montalban cave (Luzon); 24: Shan State karst (eastern Burma); 25: Tanintharyi caves (southern Burma); 26: Songkhla karst (southern Thailand); 27: Bohol and Panglao karst (central Philippines); 28: Nusakambangan karst (Java); 29: Lengguru karst (Papua). Many caves and karsts recently sampled in southern China are not given on the map. Map courtesy of L. Deharveng and A. Bedos.

may not be a good indicator of β- and γ-diversity, Culver *et al.* (2006b) analysed seven karst areas, three in Europe and four in North America, ranging in size from 2000 to 6300 km², each with more than 120 sampled caves and more than 350 records of troglobionts. To reduce variability of the data and minimize the impact of incomplete sampling of individual caves, caves were associated with 100 km² hexagons. These hexagons were then used for species accumulation curves. As was the case for the analysis of stygobiotic diversity in the PASCALIS study (Fig. 8.4; Dole-Olivier *et al.* 2009a), some but not all of the accumulation curves reached an asymptote. All seven sites could be compared at a level of 20 hexagons (Fig. 8.11), the minimum number of hexagons in each karst area. The seven sites fell into two groups: three sites—the Ariége region of France, the Dinaric karst of Slovenia, and northeast Alabama (USA)—had an estimated total species richness (Chao2 estimate) of about 120 and an observed species richness of between 50 and 80. The other four sites had lower values, with estimates of total species richness

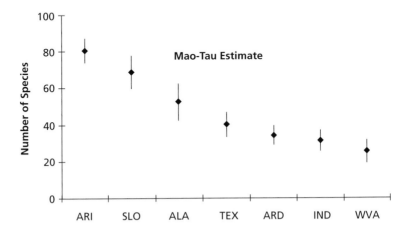

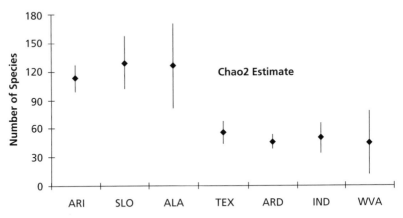

Fig. 8.11 Species richness estimates with standard errors for observed number of species, using the Mao-Tau procedure (Colwell *et al.* 2004) and for expected total number of species, using the Chao2 procedure (Colwell 2016) for 20 randomized 100 km² hexagons. Abbreviations for sites are as follows. ARI: Ariège, France; SLO: Dinaric karst of Slovenia; ALA: northeast Alabama, USA; TEX: Balcones Escarpment of Texas, USA; ARD: Ardèche, France; IND: Mitchell Plain of Indiana, USA; and WVA: Greenbrier Valley of West Virginia, USA. From Culver *et al.* (2006b). Used with permission of Blackwell Publishing.

of about 50. They used these results to calibrate information to categorize species richness in six other regions in North America and 10 other regions in Europe. When these 23 regions were plotted on a map, a striking pattern emerged, one with a ridge of species richness at about 42°N to 46°N in Europe (Fig. 8.12) and 34°N in North America. In North America, the 'ridge' consists of a single region in NE Alabama and adjacent areas of Tennessee (see Niemiller and Zigler 2013) bounded on the north and west by karst areas with lower diversity. To the south is mostly non-carbonate rock but the karst area in Florida, an area submerged during the Pleistocene, has no troglobionts (Fig. 8.8; Culver *et al.* 2003). The Alabama/Tennessee hotspot is the only area in the U.S. that rivals the species richness of the European ridge.

Regions within the ridge of high troglobiotic diversity share two features. One is that cave density is higher in these regions than in non-ridge regions. This suggests that habitat availability is important. The other is that they are also on a ridge of long-term high surface productivity, as measured by high temperature and high rainfall. Regions to the north and the south are not as

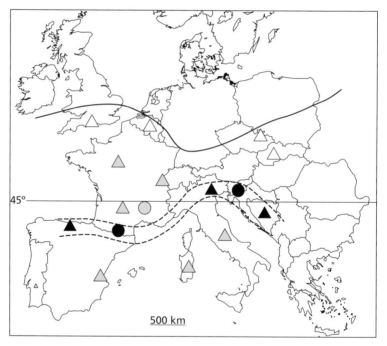

Fig. 8.12   Map of species richness patterns of European troglobionts. The open triangles are areas with few if any troglobionts, the grey triangles are areas with <50 species, and the grey circle is Ardèche with <50 species in 5000 km² of area. The black circles are the diversity hotspots in Slovenia and Ariège. Black triangles are other possible diversity hotspots. The boundary of the Pleistocene ice sheet is shown as a solid line. A pair of dashed lines indicates the hypothesized position of the high-diversity ridge. From Culver *et al.* (2006b). Used with permission of Blackwell Publishing.

productive. Thus productivity may be a major determinant of species richness (see Chapter 2). The ridge may continue around the globe, and Culver *et al.* (2006b) suggest that cave regions in the western Caucasus of Georgia and Shikoku Island in Japan are also along this hypothetical ridge.

All other quantitative studies of subterranean biodiversity to date have considered a single continent, or region of a continent. Zagmajster *et al.* (2014) and Eme *et al.* (2015) analysed a massive dataset on European stygobionts (21 700 records of 1570 species) for both species richness patterns and explanations of the observed patterns. They also found a ridge of stygobiotic species richness that was coincident with the one observed by Culver *et al.* (2006b) for troglobionts (Fig. 8.13). Eme *et al.* (2015) did a much more extensive analysis of possible causes of the pattern, simultaneously analysing the impact of historical climate stability, spatial heterogeneity, and productive energy. They mapped the independent and overlapping effects of mechanisms using partial geographically weighted regressions. When analysed separately, the three mechanisms explained the same amount of variation in species richness, but in the joint analysis, the influence of historical climate stability became hidden in the variation shared with the other mechanisms (Fig. 8.14). Spatial non-stationarity in the independent and overlapping effects of the three mechanisms was the most plausible explanation for the hump-shaped latitudinal pattern of crustacean species richness (Fig. 8.13). Productive energy and spatial heterogeneity were important predictors at mid and southern latitudes, whereas historical climate stability overlapped with the two other mechanisms in northern Europe and productive energy in southern Europe. Spatial heterogeneity is important in determining species richness and there is also a strong dependence on food supply from the surface.

Malard *et al.* (2009) analysed the crustacean stygofauna in eight regions in Europe to determine the relationship of diversity in interstitial habitats and karst habitats, and the contribution to species richness of different geographical scales. Individual samples (α_1) from karst and interstitial sites were associated with a catchment (α_2) and with a region (α_3). At each scale the total diversity S_T equals $\alpha + \beta$ (Lande 1996) where β (the between-sample component) is the difference between local diversity (α) and total diversity. Overall the total diversity, γ, is:

$$\beta_{1\ \text{aquifers}} + \beta_{1\ \text{aquifers}} + \beta_{2\ \text{catchments}} + \beta_{3\ \text{regions}}$$

Of these components, $\beta_{3\ \text{regions}}$ is by far the largest, accounting for 71.8 per cent of the variance, followed by $\beta_{2\ \text{catchments}}$ with 21.1 per cent of the variance, $\beta_{1\ \text{aquifers}}$ with 11.8 per cent of the variance, and $\alpha_{1\ \text{aquifers}}$ accounting for 9.4 per cent of the variance (Malard *et al.* 2009). All of the components were significantly different from zero, except for $\beta_{1\ \text{aquifers}}$. In the four regions where this complete analysis was possible, there was little difference in the components among regions (Fig. 8.15). These results provide striking confirmation

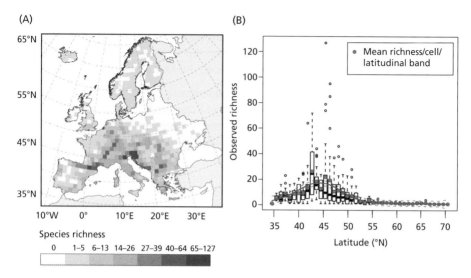

Fig. 8.13   Map of species richness patterns of European stygobionts. A. Species richness of 10 000 km² cells. B. Relationship between the cell average of species richness per 0.09´ latitudinal band and latitude. Black horizontal bars and boxes show the median and interquartile range, respectively, for latitudinal bands. The maximum length of each whisker is 1.5 times the interquartile range and open circles represent outliers. The thick red line is the fit of generalized additive model to the averages of latitudinal bands. From Zagmajster *et al.* (2014). See Plate 18.

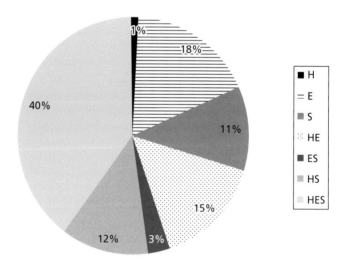

Fig. 8.14   Pie chart of relative contributions of different combinations of drivers of species richness to the explained variance. H is historical climate stability, E is productive energy, and S is spatial heterogeneity. Data from Eme *et al.* (2014). See Plate 19.

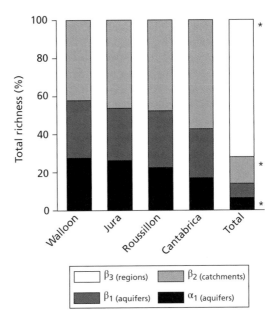

that local species richness is a relatively small component of overall species richness. Somewhat surprising was the lack of importance of the contribution of different habitat types (karstic vs porous aquifers), $\beta_{1\,\text{aquifers}}$. Only 29 per cent of the stygobionts studied that were found in more than a single site were limited to one habitat type. In fact, species richness in one aquifer type was a good predictor of species richness in the other aquifer type (Fig. 8.16), with a Spearman correlation coefficient of 0.94. Malard *et al.* also reported that the local species richness in karst aquifers, as shown in Fig. 8.16, is linearly correlated with overall regional species richness, giving some justification to using single cave hotspots as a surrogate for overall species richness.

On a smaller geographical scale, Zagmajster *et al.* (2008) and Bregović and Zagmajster (2016) analysed species richness patterns of the troglobiotic beetle fauna of the Dinaric karst. In a study area of 56 000 km^2 a total of 276 troglobiotic species, and an estimated 400 total species, based on Chao2 estimates, from 1709 localities were investigated. An idea of how species-rich this region is can be gained by comparing it to the Interior Low Plateau karst area of the United States. With an area of 60 000 km^2, and including well-studied caves such as Mammoth Cave, only 103 species of beetles are known from the Interior Low Plateau. Zagmajster *et al.* (2008) looked at

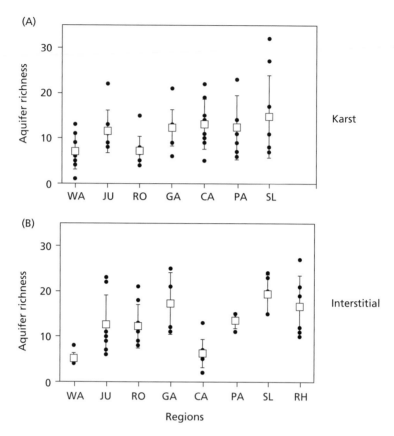

(A)

Karst

(B)

Interstitial

Regions

Fig. 8.16   Differences in average species richness (± standard deviation) of karst (A) and porous (B) aquifers between regions: WA (Walloon, Belgium); JU (Jura, France); RO (Roussillon, France); GA (Garonne, France); CA (Cantabrica, Spain); PA (Padano-alpine, Italy); SL (Slovenia); and RH (Rhône River corridor, France). From Malard *et al.* (2009). Used with permission of Blackwell Publishing.

diversity at several scales and ultimately settled on a grid size of 20 km by 20 km because this grid size had the fewest gaps in coverage and the highest spatial autocorrelation of all investigated grid sizes. After considering the effect of differing sampling intensities (Zagmajster *et al.* 2010), they showed that there were actually two hotspots of beetle diversity—one in the northwest in Slovenia (largely Carabidae) and one in the southeast in Montenegro and Bosnia & Herzegovina (largely Cholevidae; Fig. 8.17). Sket *et al.* (2004) had previously hypothesized a centre of beetle biodiversity in the southeast Balkans but the presence of two hotspots rather than one was unexpected. Their results show that although the relative importance of species richness drivers differed between the two families, in most cases habitat heterogeneity had a bigger influence than historical climate stability and productive energy.

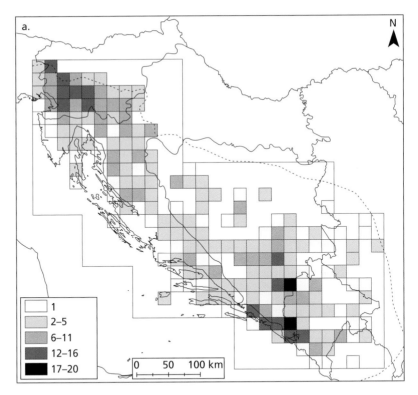

Fig. 8.17   Number of troglobiotic trechine and leptodirine beetles in 20 × 20 km grid cells in the Dinaric karst. Dotted line delimits the extent of the Dinarides. From Zagmajster *et al.* (2008). Used with permission of Blackwell Publishing.

The final regional study we consider is that of Christman on the stygofauna of the eastern United States (Christman and Culver 2001; Christman and Zagmajster 2012). Habitat availability, measured by the number of caves, is a candidate to be a good predictor of the number of stygobionts (and troglobionts for that matter) in an area. For 1622 counties in the study area, the number of caves and the number of stygobionts was analysed. There were a total of 263 stygobionts and 27 280 caves. The obvious way to proceed was to consider whether the number of caves in a county could predict the number of stygobionts in a county. Christman and Culver (2001) showed that the two were connected by the equation:

$$\ln(\text{Number of stygobionts}) = -0.023 + 0.259 \ \ln(\text{Number of caves}),$$

which accounted for 35 per cent of the variance in the log of species number.[1] Christman and Zagmajster (2012) went on to analyse regional variation that

---

[1] Logs were taken to make the data fit a normal distribution. They also found that the quadratic equation did not result in an improved fit.

cannot be explained by the overall relationship between number of caves and number of stygobionts. For example, a county with a large number of caves surrounded by counties also with a large number of caves may have a larger number of stygobionts than a county also with a large number of caves but surrounded by counties with few caves. If such regional contexts are important, then subterranean dispersal must also be important. Christman decomposed the number of stygobionts in a county into three components: the overall linear relationship between the logs of species numbers and number of caves; the spatial correlation component due to regional effects; and an unexplained component (Fig. 8.18). Her analysis makes it clear that habitat availability is an important determinant of species number and that there is both a local (number of caves in a county) and a regional (number of caves in adjoining counties) effect of habitat availability.

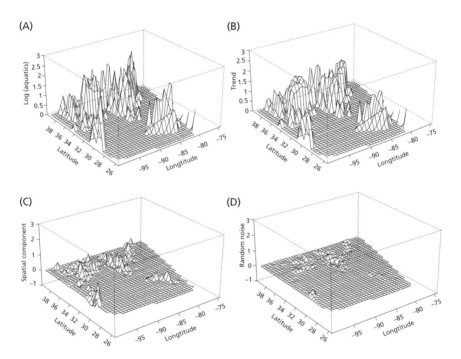

Fig. 8.18   Perspective plots showing (A) geographical distribution of the observed counts of stygobionts (log-transformed) in counties in the southeastern part of the United States; (B) the fitted regression plane for the local component of model in which the log-transformed number of stygobionts is related to the number of caves (log-transformed); (C) the spatial autocorrelation regional component in which the logged number of stygobionts is predicted conditionally using the logged counts in surrounding counties; and (D) the residuals or unaccounted for variation in these data. From Christman and Zagmajster (2012). Used with permission of Elsevier Ltd.

8.6 Summary

The focus of this chapter was on the patterns and explanations for the patterns of species richness of stygobionts and troglobionts over broad geographical scales. Because of the large number of subterranean sites, the frequent numerical rarity of species, and the high levels of endemism, species lists are always incomplete. Techniques, especially randomized species accumulation curves and estimators, such as Chao2, of total richness, based on the number of single site species, provide powerful tools for the analysis of patterns.

The analogy of caves as islands has relevance in both ecological and evolutionary time. The analogy with islands in ecological time is most appropriate at scales smaller than caves, such as seeps or epikarst drips, and the analogy with caves in evolutionary time is more appropriate at larger scales, such as karst basins or contiguous karst areas. Interstitial habitats are generally not island-like but rather linear. There are some similarities with isolated oceanic islands but there are no known examples of adaptive radiation in subterranean habitats.

The only estimates of global patterns available are estimates of the number of stygobionts in each continent and the distribution of high diversity individual caves. For troglobionts, southern Europe, especially the Dinaric karst, and the Canary Islands are regions of high richness. For stygobionts, southern Europe, especially the Dinaric karst, is a hotspot. Other sites are typically chemoautotrophic and/or phreatic. In the tropics, subterranean species richness is highest in the Oriental and Australasian regions. Other patterns were revealed in studies with a more limited geographical scope. In Europe and North America, there appears to be a ridge of high troglobiotic and stygobiotic diversity in southern Europe and the southeast United States that corresponds to an area of long-term high surface productivity. Several patterns emerged from the multinational European project, PASCALIS. In particular, local diversity is a small component of regional stygobiotic diversity and the importance of spatial heterogeneity, historical climate stability, and productivity are both scale and spatially dependent. The geological attributes of the habitat are an important determinant of species numbers. Habitat availability seemed especially important at smaller scales, both for the beetle fauna of the Dinaric karst and the stygofauna of the southeastern United States.

9 Some Representative Subterranean Communities

9.1 Introduction

In previous chapters we introduced the subterranean environment and its inputs (Chapters 1 and 2), its inhabitants (Chapter 3), the processes that mould the species and communities (Chapters 4–7), and the resulting geographical pattern (Chapters 7 and 8). In this chapter, we put the component parts and processes together from a naturalist's point of view, and describe some representative subterranean communities and habitats. The range of habitats is large (see Chapter 1) and the inhabitants are diverse, even bizarre (see Chapter 3). Besides the absence of light and resulting absence of photoautotrophy, subterranean habitats share another feature— they are difficult to sample. Some must be sampled indirectly, such as epikarst; some must be nearly destroyed to sample, such as the hypotelminorheic; some require specialized sampling devices, such as interstitial aquifers; and some seem to have no life at all. With rare exceptions (such as chemoautotrophic caves), caves seem, at first glance, nearly devoid of life. Unencumbered by vegetation, the geologist's task of deciphering the story in the rocks is easier, and may explain the relatively higher popularity of caves among geologists compared with biologists. However, with a little bit of time and effort, it is possible to find the biota of these apparently lifeless places.

In this chapter, we provide an overview of representative communities both of shallow subterranean habitats (SSHs) and of caves and other deeper subterranean habitats. SSHs are habitats less than 10 m below the surface (Culver and Pipan 2014). We have relied heavily on our own experiences of studying these different communities, and for those we have not directly studied, we have chosen well-studied examples from the literature.

The Biology of Caves and Other Subterranean Habitats. Second Edition. David C. Culver and Tanja Pipan. Published 2019 by Oxford University Press. © David C. Culver and Tanja Pipan 2019. DOI: 10.1093/oso/9780198820765.001.0001

9.2 Shallow subterranean habitats

The most superficial of the shallow habitats is the *hypotelminorheic*, small seeps of water (Culver and Pipan 2014). Characterized as a persistent wet spot, typically with blackened but not skeletonized leaves, and underlain by an impervious layer, usually clay (see Fig. 1.15), they have been most thoroughly studied in the lower Potomac River basin near Washington, DC (USA). They can be difficult to distinguish from small vernal pools and other temporary small bodies of water. Conductivity is higher (400–500 μS/cm) than for other waters and temperature in hypotelminorheic habitats is generally close to the mean annual temperature but with less fluctuation. Perhaps the best indicator of the hypotelminorheic is the presence of stygobiotic species, with the proviso of course that not all hypotelminorheic habitats have stygobionts at all times.

A total of eight stygobionts (amphipods, isopods, and snails) have been found in lower Potomac River basin seeps, including five that are apparently hypotelminorheic specialists (Table 9.1). Five of the stygobiotic amphipods (three of them found only in seeps) are members of a single genus, *Stygobromus*. Outside of the Edwards Aquifer in Texas, a deep extensive aquifer, seeps along the Potomac have the richest *Stygobromus* assemblage known. Two of the other seep specialists, the snail *Fontigens bottimeri* and the isopod *Caecidotea kenki*, retain some pigment and a small eye, but are limited to these sites and habitat. The final stygobiont, *Caecidotea jeffersoni*, is the only strongly troglomorphic isopod found in these seeps. There are an additional five species of amphipods and one species of isopod that are found in seeps, but are also found in surface habitats, especially streams, and only some of them have been found to reproduce in seeps (stygophiles in Table 9.1). The amphipods, isopods, and snails can be found by turning over leaves in the seep and looking at the undersides of the leaves. An individual seep may have two, or rarely three, species of amphipods, as well as isopods and snails. For the most part, it is possible to easily find animals from late winter to late spring, before the tree canopy leafs out. An especially good time to find hypotelminorheic species is during snow melt—areas around the seeps melt first and so they are quite visible against a white background. Some of these species have ranges of only a few kilometres (especially *Stygobromus hayi* and *Stygobromus kenki*) while others (especially *Stygobromus pizzinii* and *Stygobromus tenuis potomacus*) are known from several states. All together it is a diverse assemblage, especially of amphipods. Hypotelminorheic sites in Croatia and Slovenia have a similar mixture of specialized and non-specialized amphipods, although with fewer numbers of species (Culver *et al.* 2006a).

Epikarst communities have been most extensively studied in the Postojna–Planina Cave System (PPCS) in Slovenia (Pipan 2005; Pipan *et al.* 2006a; Pipan and Culver 2007a, b). The most efficient way of sampling epikarst

Table 9.1 Species of amphipods, isopods, and snails found in seeps in the lower Potomac River drainage in the environs of Washington, DC.

	Species	Stygobiont, stygophile, or accidental	Hypotelminorheic specialist	Troglomorphic
Amphipoda:	Stygobromus tenuis potomacus	stygobiont	no	yes
	Stygobromus pizzinii	stygobiont	no	yes
	Stygobromus hayi	stygobiont	yes	yes
	Stygobromus kenki	stygobiont	yes	yes
	Stygobromus sextarius	stygobiont	yes	yes
	Crangonyx floridanus	stygophile	no	no
	Crangonyx palustris	accidental	no	no
	Crangonyx serratus	accidental	no	no
	Crangonyx shoemakeri	stygophile	no	no
	Crangonyx stagnicolous	accidental	no	no
	Gammarus fasciatus	accidental	no	no
	Gammarus minus	stygophile	no	no
Isopoda:	Caecidotea kenki	stygobiont	yes	weakly
	Caecidotea jeffersoni	stygobiont	yes	yes
	Caecidotea nodulus	stygophile	no	no
Gastropoda:	Fontigens bottimeri	stygobiont	yes	weakly

Source: Data from Culver *et al.* (2012).

communities, and it is still indirect, is to filter the water from ceiling drips in a cave (see Chapter 3). Directly underneath epikarst is a zone with small cracks and fissures through which the water vertically percolates. In essence, the collections are of animals that have fallen out of their habitat. There are drips that are only active seasonally and then at very slow rates and there are ones that are steady streams of water throughout the year. Drips with intermediate drip rates from thin ceilings were often the most productive in terms of the number of animals collected. The total input of organisms from 20 drips over a 12-month period is summarized in Table 9.2. A wide variety of animals was found, and on average 3.3 animals were collected in a drip/day. Most of them were copepods (2.5/drip/day). The copepod fauna was diverse, with 12 stygobiotic species and 4 other species. The number of stygobiotic copepod species found in epikarst in PPCS is greater than the number of stygobionts from all habitats in most caves and other subterranean habitats. The copepod fauna itself added another component to spatial heterogeneity—the maximum linear extent of copepod populations was about 100 m so that species composition changed over short distances (Pipan and Culver 2007b). This likely reflects the semi-isolated nature of the small cavities that comprise the epikarst (Figs 1.7 and 1.8).

Table 9.2 List of fauna collected from 20 drips in PPCS over a 12-month period from March 2000 to March 2001.

	Group	Abundance
Aquatic:	Turbellaria	5
	Nematoda	192
	Oligochaeta	30
	Ostracoda	5
	Bathynellacea	1
	Isopoda	1
	Amphipoda	2
	Copepoda—adults	534
	Copepoda—nauplii	368
	Diptera—larvae	15
Terrestrial:	Araneae	2
	Acarina	20
	Diplopoda	4
	Gastropoda	1
	Collembola	11
	Coleoptera	3

Source: Data from Pipan (2005)

The terrestrial epikarst fauna was a small but significant part of the animals collected in drips, as it was in other studies (Gibert 1986; Pipan and Culver 2005; Culver and Pipan 2014). Collembola and mites predominate, and other studies (especially Gibert 1986) have demonstrated that some are troglobionts. It is exceedingly difficult to study the terrestrial epikarst community until some more direct way of accessing the habitat is found.

The *milieu souterrain superficiel* (MSS) is commonly found in scree slopes where soil has not filled in the spaces. It bears some resemblance to dry epikarst, although connectivity is likely higher. Sampling is usually done with specialized baited pitfall traps lowered into holes dug in the MSS (López and Oromí 2010), or from Berlese extractions of litter and debris taken from the MSS (Juberthie *et al.* 1980). The geographical extent of the MSS is probably more restricted than that of seeps and epikarst. It occurs only in areas of moderate to high slope with some sort of rock fracturing or rubble. Three MSS sites in the central Pyrenees (Tour Laffont, Bellissens, and Col des Marrous) are the most intensively studied (Crouau-Roy *et al.* 1992; Gers 1992). They found that the environmental variation of the habitat was intermediate between that of the surface and nearby caves—temperatures in the MSS fluctuated about 10°C annually compared to about 20°C on the surface. Troglobiotic beetles in the subfamily Bathysciinae (especially *Speonomus hydrophilus*) predominated, accounting for nearly 77 per cent of the 8141 animals collected in the 12-month study. Troglophilic Diptera in the family Phoridae accounted for most of the remaining animals—22 per cent. Several other troglobionts were present in low numbers, including millipedes, carabid beetles, Diplura, and Collembola. In many cases, there were no adult beetles or flies in winter samples. The sites were at an elevation of between 960 and 1350 m and Crouau-Roy *et al.* (1992) argued that a combination of low temperature and humidity made the immediate habitat unsuitable, and that presumably at least some adults survived deeper in the MSS.

Formed by a meander arm, the Lobau wetlands, an alluvial aquifer, are part of the floodplain of the Danube River near Vienna, Austria, and comprise the Danube Flood Plain National Park. This UNESCO Biosphere Reserve, with an area of 0.8 km², has been extensively sampled for decades (Pospisil 1994; Danielopol *et al.* 2000, 2001; Danielopol and Pospisil 2001). Bou–Rouch pumps (Chapter 3) and minivideo cameras in shallow wells revealed a complex habitat with areas of differing porosity and permeability, variable oxygen levels, and a rich fauna. A small 900-m² component of this flood plain, called 'Lobau C' (Pospisil 1994), was monitored and sampled intensively. It is a recent terrace of the backwater system 'Eberschüttwasser–Mittelwasser' and is a self-contained ecosystem with clear inputs and outputs because of its position between two channels and a dam (Danielopol *et al.* 2001). Loosely packed gravel, alternating with a thin layer of finer sediments, extends from 4 to 8 m beneath a thin soil cover. Both oxygen and dissolved organic matter concentrations are very heterogeneous, even at scales of 1 m or less. Animals were found throughout the depth of gravel, but were most common 0.5 m beneath the surface, and rare below 2 m (Pospisil 1994). More than 100 species, 35 of them stygobionts, have been recorded from the National Park (Danielopol and Pospisil 2001), and in tiny Lobau C at least 27 species were found, 11 of them stygobionts (Table 9.3).

Calcrete aquifers, often less than 10 m in vertical extent, are formed by evaporation of paleodrainages, and can only be sampled by boreholes. They only occur in areas of low rainfall and high evaporation, and are best known from western Australia. What makes them remarkable is the high levels of both species richness and endemism (Humphreys 2001; Guzik *et al.* 2011). It is an ancient landscape with a fauna that dates back to the Tertiary and before. Examples of the species richness of calcrete aquifers are shown in Table 9.4.

Even though soil is typically excluded from a discussion of subterranean habitats (see Chapter 1), it shouldn't be. The habitat, especially the deeper

Table 9.3 Number of species and stygobionts from the 'Lobau C' area of Danube Flood Plain National Park, Austria.

Group	Number of species	Number of stygobionts
Rotatoria	1+	1
Mollusca	2+	2
Copepoda: Cyclopoida	14	3
Copepoda: Harpacticoida	7	2
Amphipoda	1	1
Isopoda	2	2

Source: Data from Pospisil (1994), Danielopol *et al.* (2001) and Danielopol and Pospisil (2001).

Table 9.4 Stygobiotic species in the three calcrete aquifers associated with Lake Way, Western Australia.

Order	Family	Species	Hinkler Well	Lake Violet	Uramurdah Lake
Bathynellacea	Bathynellidae	Gen. nov. sp. 1	●		●
	Parabathynellidae	Gen. nov. sp. 1	●	●	●
		Gen. nov. sp. 1		●	
Coleoptera	Dytiscidae	Limbodessus macrohinkleri	●		●
		L. hinkleri	●		●
		L. raeae	●	●	
		L. wilunaensis			
		L. hahni	●		
		L. morgani	●		
Cyclopoidea	Cyclopidae	Fierscyclopes fiersi	●	●	●
		Mesocyclops brooksi			
		Metacyclops laurentiisae	●	●	●
		Halicyclops ambiguus		●	●
		H. kieferi		●	
Harpacticoida	Ameiridae	Haifameira pori	●		
		Nitocrella trajani		●	●
		Parapseudoleptomesochra karamani			●
		P. rouchi			●
	Diosaccidae	Schizopera austindownsi			●
		S. uramurdahi			●
Oniescidea	Philosciidae	Andricophiloscia pedisetosa			●
	Scyphacidae	Haloniscus longiantennatus			●
		H. stilifer			●
		Haloniscus sp. nov.			●
Podocopida	Candonidae	Candonosis dani		●	
		Gomphodella sp.			●
TOTALS			10	9	17

Source: Data from Humphreys et al. (2009). Used with permission of Springer Publishing.

layers, is aphotic, and many soil specialists are without eyes and pigment, albeit with a more compact morphology than the subterranean species occupying larger spaces (Culver and Pipan 2014). Gers (1992, 1998) has emphasized the close connections between MSS habitats and the soil, and a significant fraction of the species found in MSS are, on the basis of their morphology, soil specialists (e.g., Pipan *et al.* 2011). One of the most extensive surveys of taxonomic diversity of the soil fauna is that of Coiffait (1958), who worked in the same region of the Pyrenees as did Gers and Crouau-Roy in their pioneering study of the MSS. He found that of 88 beetle species collected in the soil, 26 were what he called edaphobionts, species limited to the soil. Most of these showed signs of eye reduction and appendage shortening. The overall diversity of the Pyrenean soil fauna is shown in Fig. 9.1.

Lava tubes, typically formed very close to the surface (see Chapter 1) can harbour a very diverse troglofauna, and in the case of the Canary Islands they are among the most diverse subterranean sites in the world (Table 8.3). On islands such as Hawai'i and Tenerife, densities of lava tube passages exceed cave passage densities of the most developed karst areas, such as the Dinaric karst of Slovenia (Culver and Pipan 2014). They can be quite long (Kazumura Cave in Hawai'i is more than 65 km long), and typically have a uniform cross-sectional passage dimension. Lava tube troglobionts are often from different taxonomic groups than troglobionts of karst caves. In Hawaii, planthoppers in the genus *Oliarus*, which feed on plant roots, form the base of the food web (Stone *et al.* 2012). The taxonomic composition of the troglobionts of Cueva de Felipe Reventón is shown in Fig. 9.2.

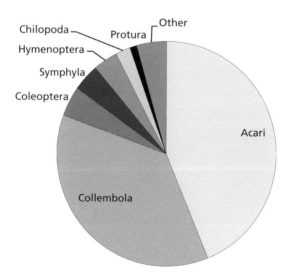

Fig. 9.1   Pie diagram of species composition of soil samples from 100 sites in the Pyrenees Mountains. Compiled from Coiffait (1958).

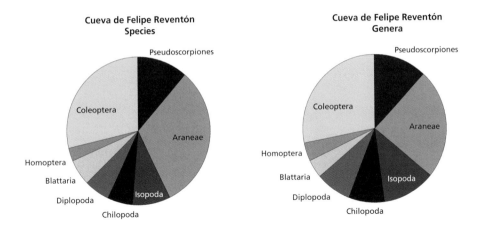

Fig. 9.2   Pie diagrams of troglobiotic species, troglobiotic genera, and troglophilic/trogloxenic species from Cueva de Felipe Reventón in Tenerife, Canary Islands. Data from Arechavaleta *et al.* (1999) and unpublished updates by Oromí.

Lavoie *et al.* (2017) studied microbial diversity in lava tubes in Lava Tube National Monument, California (USA). Both because most microbes cannot be cultured and because of the large numbers of species, microbial diversity is typically analysed using DNA sequencing. In this study, Lavoie *et al.* (2017) used sequences of the small subunit of ribosomal RNA to compare microbes found in microbial mats in lava tubes with those of the surrounding soil (Fig. 9.3). Overall, there was only an 11.2 per cent overlap between soil and lava tube samples, indicating that very different communities exist in the two adjoining habitats. Lava tubes were relatively enriched in Actinobacteria and Nitrospirae, and many sequences were impossible to assign to genus.

SURFACE

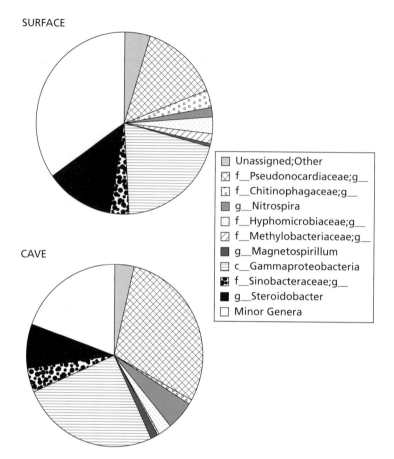

CAVE

Unassigned;Other
f__Pseudonocardiaceae;g__
f__Chitinophagaceae;g__
g__Nitrospira
f__Hyphomicrobiaceae;g__
f__Methylobacteriaceae;g__
g__Magnetospirillum
c__Gammaproteobacteria
f__Sinobacteraceae;g__
g__Steroidobacter
Minor Genera

Fig. 9.3   Percent Operational Taxonomic Units (OTUs) by lowest level of classification (family or genus). (A) mean surface soils and (B) mean cave microbial mats in Lava Beds National Monument, USA. From Lavoie et al. (2017).

Iron-ore caves, formed by bioreduction of Fe^{3+}, are very common in iron-ore regions (Fig. 9.4) of Brazil, although most of the caves are very short, often less than 50 m in length. The protection of the fauna of these iron-ore caves poses formidable challenges, given the proximity of iron-ore mining and its economic importance to the Brazilian economy (Jaffé et al. 2016). Jaffé et al. (2018) list the 30 most common troglobionts in Brazilian iron-ore caves from Eastern Amazonia (Table 9.5). Using multiple logistic regression, they show that altitude, presence of guano, and cave slope are the most frequent predictors of occurrence of the 14 common troglobionts. Altitude is an indicator of geological position, and the other two reflect availability of food. Souza-Silva et al. (2011) found that cave length was also an important determinant of species richness in iron-ore caves.

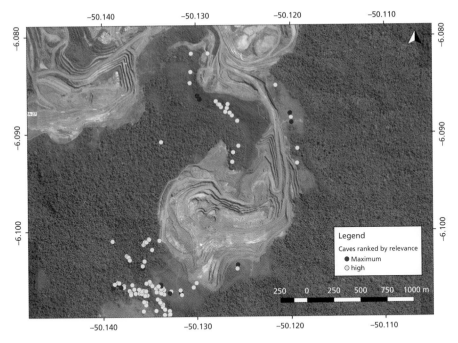

Fig. 9.4   Iron-ore mine (N5, Serra Norte, Carajás, Brazil) showing the location of caves coloured by their classification. Caves with maximum relevance have at least one rare troglobiont; caves with high relevance have at least one troglobiont. From Jaffé *et al.* (2018). See Plate 20.

Table 9.5 List of troglobiotic species occurring in at least 30 iron-ore caves in Eastern Amazonia in Brazil.

Species/OTU	Order/Class
Xyccarph	Araneae
Carajas	Araneae
Charinus	Amblypygi
Circoniscus	Isopoda
Cyphoderidae (2 species)	Collembola
Entomobryidae	Collembola
Entomobryomorpha	Collembola
Isotomidae	Collembola
Matta	Araneae
Paranellidae	Collembola
Pyrgodesmidae	Polydesmidae
Systrophiidae	Gastropoda

Source: Data from Jaffé *et al.* (2018).

9.3 Deep non-cave habitats

Wells often provide the only access to deeper subterranean aquatic habitats. The habitat of the community found in a 13 m deep well (with a water level of 6 m) in the alluvial plain of the Robe River in the Pilbara region of Western Australia is actually a mixture of a shallow subterranean habitat—calcretes— and an interstitial aquifer of the coarse alluvial gravels (Eberhard *et al.* 2009). The primary means of sampling is a phreatobiological net (Cvetkov 1968; Malard 2003), the lower end of which consists of a container closed with a valve that prevents the animals from escaping once they are caught. The net is moved up and down to capture animals in the well. Environmental conditions reflect the arid landscape above—31°C, salinity of 480 mg/L, and dissolved oxygen of 4.1 mg/L. Total stygobiotic richness was 32 stygobionts, 29 of which were Crustacea (Amphipoda, Copepoda, Isopoda, and Ostracoda). The high diversity results from the combination of habitats (interstitial and ones formed by dissolution of limestone), long-term stability of the habitat, a connection with the sea which provides an increased number of potential colonists, and aridification during the Mesozoic that forced species into subterranean environments (Leys *et al.* 2003).

The Edwards Aquifer, a *phreatic aquifer* that occupies an area of 10000 km², is located in central Texas, USA, along the Balcones Escarpment (Schindel *et al.* 2004). It is a complex aquifer, with four surface water divides, and artesian springs and wells (Fig. 9.5). To the south and east is the 'bad water' zone, where total dissolved solids exceed 1000 mg/L, largely the result of the presence of saltwater brines. In some areas the freshwater zone reaches a depth of 1000 m and wells reach depths of 600 m. Overall, the aquifer has 45 species of stygobionts (Table 9.6), including 13 species of snails (Hershler and Longley 1986) and 12 species of amphipods (Holsinger and Longley 1980). Some of the stygobionts were first known from caves that intersected the groundwater, such as the salamander *Eurycea rathbuni* in Ezell's Cave, Texas. It may be the most diverse aquifer in the world (Longley 1981, 2004). The subterranean amphipod fauna in this system is the most taxonomically diverse of its kind in the world and includes eight genera in four families (Holsinger and Longley 1980). It is likely that the resource base for this ecosystem is organic matter in the form of petroleum and peat. In the 'bad water' zone there is evidence of chemoautotrophy (Hutchins *et al.* 2013, 2016), and the unique sucker-like mouth of the stygobiotic catfish, *Trogloglanis pattersoni*, may be an adaptation for feeding on sulfur-oxidizing bacteria (Langecker and Longley 1993; Engel 2012b). The macrofauna of the 'bad water' zone itself has been little studied. The aquifer is threatened by excessive draw-down, both by agricultural interests and by the city of San Antonio, Texas. Lowering of the water table of the aquifer also threatens species limited to the springs of the Edwards Aquifer, including two fish (*Etheostoma fonticola* and *Gambusia georgei*) and two aquatic beetles

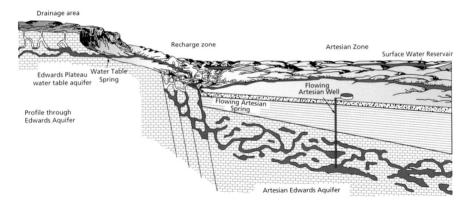

Fig. 9.5   Cross section of the Edwards Aquifer near San Antonio, Texas, USA. From Schindel *et al.* (2004).

Table 9.6   The number of stygobionts known from the Edwards Aquifer of Texas, from a variety of wells.

Group	Number of stygobionts
Nematoda[1]	1
Platyhelminthes: Turbellaria	1
Annelida: Hirudinea[1]	1
Mollusca: Gastropoda	13
Crustacea: Ostracoda	1
Crustacea: Isopoda	5
Crustacea: Amphipoda	12
Crustacea: Thermosbaenacea	1
Crustacea: Decapoda	2
Insecta: Coleoptera	3
Pisces	2
Amphibia	4[2]
TOTAL	45

[1]Parasites on *Eurycea rathbuni*
[2]Longley indicates 13 species but Wiens *et al.* (2003) indicate only 4 *Eurycea* are stygobionts.
Source: Data from Longley (2004).

(*Heterelmis comalensis* and *Stygoparnus comalensis*) on the US Endangered Species List.

In contrast to the Edwards Aquifer are deep phreatic aquifers in the Columbia River Basalt in Washington state, USA (Stevens and McKinley 1995; Krumholz 2000). Formed during the Miocene between 6 and 17 million years ago, there is no obvious external source of organic carbon, and the water in confined aquifers between basaltic flows can be more than 35 000 years old (Stevens and McKinley 1995). Somewhat surprisingly, 2–5 mg/L of dissolved organic carbon was found, as well as sometimes a dense anaerobic bacterial population with up to 104 cells/mL (Krumholz 2000). A variety of anaerobic bacteria were present, including iron reducers, sulfate reducers, methanogenic bacteria, and acetogenic bacteria. The relatively high levels of molecular hydrogen (20–100 µM), combined with other evidence, suggests that the primary producers are hydrogenotrophic bacteria, using hydrogen as an electron donor. All in all, it is very odd ecosystem even from the point of view of the generally 'odd' subterranean ecosystems.

9.4 Cave habitats

The tropical terrestrial cave community in Gua Salukkan Kallang in Sulawesi, Indonesia, provides insights into the organization and structure of tropical cave communities (Deharveng and Bedos 2000, 2004, 2012). The cave has 24 km of passage, including 10 km of an underground river course. The two major sources of nutrients are flood debris and droppings from bats, swiftlets, and crickets. Unlike many tropical caves Gua Salukkan Kallang does not have large bat populations, but guano still forms the basis for the giant arthropod community (Chapter 2). Deharveng and Bedos (2000, 2004) report a total of 93 species from the cave, including an especially rich arachnid (17 species) and collembolan (24 species) fauna. Of the 93 species, 21 are troglobionts and another 17 are found only in guano in caves. By convention (Chapter 3), guano specialists are not included as troglobionts because of their specialized habitat. Deharveng and Bedos (2000) point out that it is not necessarily the case that all species originated by adaptive shift and that some species may be relicts. There are also some species that are highly troglomorphic, even though the general trend is for tropical species to be less troglomorphic. An example of a highly troglomorphic species (see also Fig. 3.1) is the palpigrade *Eukoenenia maros*, a genus common in caves in the Mediterranean region. One oddity in the cave is the pterostichid beetle *Mateullus troglobioticus*, the only subterranean member of its tribe.

Mammoth Cave in Kentucky is, with the exception of PPCS, probably the best-studied cave system in the world. Over 1.7 million people visit the cave annually, using 15 km of trails open to the public. Protected by an impermeable sandstone caprock, erosion of upper levels of the cave has been slowed, so that five or six levels of the cave have been preserved. It was the uppermost levels of Mammoth Cave where the beetle–cricket interaction discussed in Chapter 5 occurred. Because of its immense size, there are habitats in the cave that are rare elsewhere. These include vertical shafts on the edge of the sandstone caprock, the drains of the shafts that integrate the system, and pond-like areas in slow-moving streams. The first biologist to visit the cave was Rafinesque in 1882, and by 1888 the fauna was relatively well known (Packard 1888). There is an enormous literature on the biology of the cave, summarized by Barr (1967), Poulson (1992, 2017a) and Helf and Olson (2017).

The terrestrial cave communities of Mammoth Cave are especially diverse, with 26 troglobionts (Culver and Hobbs 2017; Chapter 8). Part of this diversity is probably the result of the immense size and habitat variability of Mammoth Cave, but it is also positioned at the boundary of several physiographic provinces that may allow for a greater species pool to disperse into the cave (Barr 1967). The terrestrial communities can be readily subdivided into three types. The first is the beetle–cricket interaction (Chapter 5) that occurs in upper-level sandy areas where crickets lay their eggs in the substrate. The second is that associated with the organic matter carried in by water entering sinkholes and vertical shafts. Among the species that are part of this riparian community are the carabid beetles, *Pseudanophthalmus striatus* and *Pseudanophthalmus menetriesii*, which feed on oligochaetes and other invertebrates in the mud banks of the stream (Poulson 2017a). However, this community does not occur throughout the cave since only the lowest levels of passages have streams or flood regularly. More widespread are communities with a resource base of guano deposited by animals moving in and out of the cave and guano of animals feeding on those animals. Poulson (1992, 2012) divided guano types into five categories—bats, woodrats, raccoons, crickets, and beetles (Table 2.5)—and found that the majority of species were common on only one type of guano. For example, woodrat (*Neotoma magister*) guano is dominated by *Psychoda* flies, *Bradysia* fungus gnats, leptodirid beetles (*Ptomaphagus hirtus*), and predaceous rove beetles (*Quedius*). Bat and woodrat populations in the cave have declined in historic times (Barr 1967; Olson 2004) and so the species dependent on their guano have also likely declined in numbers.

Sket (2004a) identified cave hygropetric habitats in caves in the Dinaric karst. Thin sheets of permanently flowing water over vertical rock make up the hygropetric habitat. The sheets are no more than 2 mm thick and flow is laminar rather than turbulent. It has been most extensively studied in the cave Vjetrenica in Bosnia and Herzegovina, a moderately sized (8 km) cave on the edge of Popovo Polje about 20 km from the Adriatic Sea. Two species of Coleoptera in the subfamily Leptodirinae of the Cholevidae (*Hadesia vasiceki vasiceki* and *Nauticiella stygivaga*) are specialists in the hygropetric in Vejtrenica. Four other species are also hygropetric specialists in other caves in the Dinaric karst. The species, from different leptodirine lineages, have a convergent morphology with elongated, narrow prothorax, inflated abdomen, thick femurs, large claws, and mouthparts with many setae used for filter feeding. Other species are sometimes found in the cave hygropetric. The amphipod *Typhlogammarus mrazeki* may use the hygropetric for dispersal (Sket 2004a). The cave hygropetric has not been investigated outside of the Dinaric karst, and habitats such as the walls of vertical shafts in Mammoth Cave are likely candidates for the site of hygropetric communities.

The first known and best-studied chemoautotrophic community is from Peștera Movile, Romania (see Chapters 2 and 5). Originally discovered in 1986 by Lascu as a result of the drilling of test shafts for a new power station,

Peștera Movile provides the most convincing case of chemoautotrophy in caves and is one of the most diverse caves in the world, with 48 stygobionts and troglobionts (Culver and Pipan 2013; Lascu 2004). The cave would appear to be an unlikely hotspot of biodiversity. The cave is only 240 m long with no natural entrances; the smell of sulfur dioxide and elevated CO_2 levels make it a most unpleasant environment for humans. On the other hand, the abundance of life in the cave makes it obvious that this is a very special place. An 18 m shaft gives access to the cave. Most of the cave is a dry upper-level passage with a small number of troglobionts, especially pseudoscorpions. At the lower level, one encounters water with a partial microbial mat covering the water surface. Water-filled passages provide access to three additional air-filled bells about 2 m in height and 1 m wide. The air is enriched in CO_2 (up to 3.5 per cent) and depleted in O_2 (as low as 7.0 per cent). On the surface of the water in the air bells is a microbial mat with a diverse bacterial population, including sulfur oxidizers (Vlăsceanu *et al.* 1997; Kumaresan *et al.* 2014; Chapter 2). The cave is special in that there is a terrestrial as well as aquatic community dependent on chemoautotrophic production (Table 9.7). The vast majority of the 46 species are either endemic to the cave or to the aquifer to which Peștera Movile provides access. Among

Table 9.7 Styogobiotic and troglobiotic species from Peștera Movile in Romania.

Taxonomic group	Stygobionts	Troglobionts
Nematoda	3	
Rotatoria	2	
Platyhelminthes: Turbellaria	1	
Annelida: Hirudinea	1	
Annelida: Aphanoneura	2	
Annelida: Oligochaeta		1
Mollusca: Gastropoda	1	
Crustacea: Ostracoda	1	
Crustacea: Copepoda	3	
Crustacea: Amphipoda	2	
Crustacea: Isopoda	1	4
Arachnida: Pseudoscorpionida		3
Arachnida: Araneae		5
Arachnida: Acarina		1
Chilopoda		4
Diplopoda		1
Hexapoda: Collembola		3
Hexapoda: Diplura		2
Insecta: Coleoptera		4
Insecta: Hemiptera	1	

Source: Data from Sârbu (2000).

the more unusual species is the water scorpion *Nepa anophthalma*, the only stygobiotic water scorpion in the world. The fauna is estimated to be several million years old (Sârbu 2000). Peştera Movile provides access to a sulfidic aquifer that extends tens of kilometres north and south. The Black Sea area in general is understudied with respect to these habitats and there are likely more caves such as Peştera Movile waiting to be found.

Anchialine communities are found in caves with a connection to the sea and with the presence of saltwater. Caves near the sea and connecting with the sea have been known for a long time, but it was not until the work of Sket (1986) that its ecological and biological interest became evident. The first and one of the best-studied anchialine habitats is in the cave Šipun, at the extreme southern end of Croatia just outside the village of Cavtat (Sket 2004b, 2012). A pool about 100 m from the entrance has a highly stratified water column with differing salinities (Fig. 9.6). Stratification by salinity is characteristic of anchialine caves—different densities prevent mixing. The pool, about 10 m in length, has nine species of crustaceans, including a thermosbaenacean, *Monodella halophila*, and six species from other groups. The fauna shows a remarkable vertical stratification, with most species found in the upper, well-oxygenated freshwater layer (Fig. 9.6). Besides Thermosbaenacea, several other

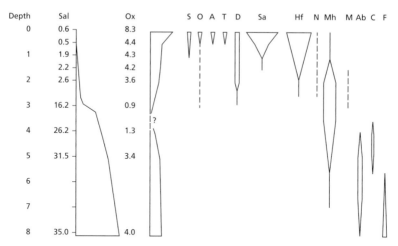

Fig. 9.6   Ecological stratification in the anchialine lake of the cave Šipun near Cavtat, Croatia in September, 1975. Depth in metres, temperature (temp) in °C, salinity (sal) in ppt, oxygen content (ox) in mg/L. Species abbreviations are S—*Saxurinator sketi* (Gastropoda), O—Oligochaeta; A—*Acanthocyclops venustus* (Copepoda); T—*Thermocyclops dybowskii* (Copepoda); D—*Diacyclops antrincola* (Copepoda); Sa—*Salentinella angelieri* (Amphipoda); Hf—*Hadzia fragilis* (Amphipoda); N—*Niphargus salonitanus* (Amphipoda); Mh—*Monodella halophila* (Thermosbaenacea); M—*Metacyclops trisetosus* (Copepoda); Ab—*Ammonia beccarii* (Foraminifera); C—*Caecum glabrum* (Gastropoda); and F—*Filogranula annulata* (Gastropoda). The population density is measured in estimated relative values. From Sket (1986, 2004c).

relatively obscure crustacean groups are primarily anchialine, including the order Mictacea and the class Remipedia. Sket (1996) estimated that, worldwide, about 400 stygobionts are specialized for anchialine habitats. These include several fish, including what is probably the world's longest stygobiont, *Ophisternon candidum* from Tantabiddi Well, an anchialine well in the Cape Range of Western Australia, reaching a length of 375 mm (Humphreys and Freinberg 1995). The anchialine pool in Šipun is not unusual among anchialine habitats in having a high species diversity. One anchialine cave—Walsingham Cave in Bermuda—has more than 20 stygobionts (see Chapter 8). Anchialine habitats tend to have high productivity, including chemoautotrophic production at the freshwater–saltwater boundary (Engel 2012) (Table 2.2).

The most frequently encountered aquatic cave communities are those in streams. Organ Cave in West Virginia has over 60 km of passage, most of it with active streams in a small (8.1 km^2) drainage basin. It is unusual in that a dendritic pattern of streams is accessible (Culver *et al.* 1994). The stream fauna is moderately diverse (Table 9.8), with a total of 15 species, six of them stygobionts. The amphipod and isopod communities are especially interesting, with a total of five species. Densities of the amphipods and isopods ranged from 20 to 120 per m^2. However, all amphipod and isopod species are not uniformly or even generally distributed throughout the cave. The streams can be classified by stream order—first-order streams are unbranched, second-order streams result from the joining of two first-order streams, and so on. First-order streams in Organ Cave can originate either from sinking streams or from percolating water (see Chapter 1). Although one might expect the highest diversity in the largest streams (third-order streams in Organ Cave), in fact there is no overall difference in median number of amphipods and isopods in streams of different orders (Table 9.9)—the median number is two for all stream orders. However, most low-diversity and most high-diversity streams are first-order streams. The low-diversity first-order streams originate from sinking streams and have mostly surface-dwelling species, and the high-diversity second-order streams originate from percolating water and have more stygobionts (Fong and Culver 1994). The only amphipod species that is more common in higher-order streams is *Gammarus minus*, a consequence of its origin from spring populations. The other amphipods are more common in epikarst-associated habitats. Only the isopod *Caecidotea holsingeri* is ubiquitous, although it is not present in epikarst itself. The other two stygobionts—the snail *Fontigens tartarea* and the flatworm *Macrocotyla hoffmasteri*—are less well known. *M. hoffmasteri* is found in stream pools through the cave and *F. tartarea* is locally very abundant (more than 100 per m^2) on smooth-sided rocks, often with thin black (manganese) coating. In many caves such as Organ Cave, it is the small, insignificant streams (trickles really) that harbour more stygobionts than the larger, more obvious streams.

Cave streams in glaciated areas provide interesting similarities and differences to cave streams, like the ones in Organ Cave, that are in unglaciated areas. Knight (2011) studied the aquatic fauna of Swildon's Hole, in the

Table 9.8 List of species found in streams in Organ Cave, West Virginia.

Group (phylum, class, order)	Species
Platyhelminthes: Turbellaria: Tricladida	*Macrocotyla hoffmasteri* *
Mollusca: Gastropoda: Mesogastropoda	*Fontigens tartarea* *
	Physa
Arthropoda: Maxillopoda: Harpacticoida	*Bryocamptus*
	Bryocamptus nivalis
Arthropoda: Malacostraca: Amphipoda	*Crangonyx gracilis*
	Gammarus minus var. *tenuipes* *
	Stygobromus emarginatus *
	Stygobromus spinatus *
Arthropoda: Malacostraca: Isopoda	*Caecidotea holsingeri* *
Arthropoda: Malacostraca: Decapoda	*Cambarus bartonii*
Arthropoda: Insecta: Plecoptera	*Taeniopteryx*
Chordata: Amphibia: Urodela	*Eurycea bislineata*
	Eurycea lucifuga
	Gyrinophilus porphyriticus

Stygobionts are indicated by an asterisk.
Source: From Culver et al. (1994). Used with permission of Elsevier Ltd.

Table 9.9 The number of stygobiotic amphipods and isopods in streams of different order in Organ Cave.

	Number of species			
	1	2	3	4
First-order stream	4	2	4	1
Second-order stream		8	1	
Third-order stream		3		

Source: From Fong and Culver (1994).

Mendip Hills of UK. As is typical of glaciated areas, few stygobionts occur. In the main stream, he found a total of 38 taxa, mostly Trichoptera (8 taxa) and Diptera (9 taxa). Only two amphipod and no isopod species were present. Of these 38 species, Knight (2011) lists only three potential stygophiles—the two amphipods plus the flatworm *Polycelis felina*. Tributary streams presented a different picture, more reminiscent of the case in Organ Cave. He lists 10 taxa, including two amphipods and two isopods, including three stygobionts. He suggests that predation by *Gammarus pulex* in the main stream prevents the occurrence of stygobiotic species, but it may also be the case that the stygobiotic species' primary habitat is epikarst.

The birthplace of the concept of the karst basin as ecosystem was a small drainage basin in the Pyrenees located 10 km southwest of Saint-Girons,

Table 9.10 Stygobiotic crustaceans collected in a continuous filtering net at the resurgence (La Hountas) and by hand collecting in the cave (Grotte de Sainte Catherine) in the Baget basin, France.

Group (phylum, class, order)	Species	Las Hountas	Grotte de Sainte Catherine
Arthropoda: Maxillopoda: Harpacticoida	*Nitocrella subterranea*	X	X
	N. gracilis	X	X
	N. delayi	X	
	Moraria catalana	X	X
	Ceuthonectes gallicus	X	X
	Elaphoidella coiffaiti	X	X
	Antrocamptus catherinae	X	X
Arthropoda: Malacostraca: Syncarida	*Bathynella* sp.	X	
Arthropoda: Malacostraca: Isopoda	*Stenasellus virei hussoni*	X	X
	Microcharon rouchi	X	
Arthropoda: Malacostraca: Amphipoda	*Niphargus kochianus*	X	
	Salentinella petiti	X	
	Parasalentinella rouchi	X	
	Ingolfiella thibaudi	X	

Source: Data from Rouch (1970).

France (Rouch 1977). A basin roughly the size (11.4 km²) of the Organ Cave basin, the main exit point of the subsurface drainage is the spring 'La Hountas' (Rouch 1970). Unlike the Organ Cave system, there are no large caves in the basin—only a relatively small cave, Grotte de Sainte Catherine, that does not have a permanent stream. Having difficulty in finding many harpacticoid copepods to study, Rouch utilized the technique of continuously filtering water emerging from the spring. Continuous filtering of springs was first performed near Cluj, Romania, by Chappuis (1925), but it was Rouch who utilized the technique for ecological study. What he found was astonishing (Table 9.10). A total of seven stygobionts (six of which were harpacticoid copepods) were known from Grotte de Sainte Catherine. After 19 months of continuous filtering with a 110-μm mesh net, he found a total of 13 stygobionts, 7 of which were harpacticoids. An additional 13 species of surface-dwelling harpacticoids were collected, having washed through the system. Out of total of 18 342 individuals collected in a 19-month period, 99 per cent were harpacticoids, and 73 per cent of these were stygobionts. He called the stream of crustaceans coming out of the spring the 'haemorrhaging' of the system. The name is very appropriate because the habitat for the 27 species collected was rarely the spring itself, just as the habitat of copepods collected in dripping water is obviously not the dripping water itself. The spring was a much better collecting site than the cave, a site where fauna from different components of the ecosystem could be collected, and a site where the different components of the karst ecosystem could be collected.

Rouch went on to publish more than 20 papers on the functioning of the Baget ecosystem.

9.5 Summary

In parallel with Chapter 3, which focused on individual species, this chapter has focused on communities. Among shallow subterranean habitats, representative communities of hypotelminorheic (Lower Potomac seeps, Washington, DC), epikarst (PPCS, Slovenia), MSS (central Pyrenees, France), soil (central Pyrenees, France), calcrete aquifers (Pilbara, Western Australia), lava tubes (Tenerife and Lava Beds National Monument, California), fluvial aquifers (Lobau wetlands, Austria) and iron-ore caves (Brazil) were described. Among non-cave deeper habitats, communities of phreatic aquifers (Edward Aquifer, Texas), and deep phreatic aquifers (basalt aquifers, Washington) were described. Among cave habitats, representative tropical terrestrial (Gua Salukkan Kallang, Sulawesi, Indonesia), temperate terrestrial (Mammoth Cave, Kentucky), chemoautotrophic (Peştera Movile, Romanian), hygropetric (Vjetrenica, Bosnia & Herzegovina), anchialine (Šipun, Croatia), cave streams (West Virginia and U.K.) and spring (Las Hountas, Baget basin, France) communities were discussed.

10 Conservation and Protection of Subterranean Habitats

10.1 Introduction

Subterranean habitats and species, especially those involving cave habitats, are attracting increasing interest and concern among conservationists, cavers, and speleobiologists, and for good reason. Most stygobionts and troglobionts are highly restricted geographically and often are numerically rare, making them vulnerable to even relatively minor disturbances. An examination of the concept of rarity (Rabinowitz *et al.* 1986), how it applies to the subterranean fauna, and why being rare increases vulnerability are the subjects of the first part of the chapter. There are other biological factors that may put the obligate subterranean fauna and many bats that utilize caves at increased risk of extinction, including low reproductive rates, high susceptibility to environmental change, and inability to withstand disturbance. Taken together, these factors reduce the ability of subterranean species to respond to environmental stress. In light of these biological attributes, including rarity, we consider the threats to subterranean communities. These include universal threats such as global warming and groundwater pollution, which should be recognized as a universal threat too. Other important threats are more local or regional. For example, mining and quarrying are big threats to caves and aquifers in Brazil and Western Australia (Auler and Piló 2015; Hamilton-Smith and Eberhard 2000) but little threat at present to caves and aquifers in central Texas, where development poses a much larger threat (Hutchins 2017). Water diversion and control projects are major problems in Bosnia & Herzegovina but not at present in Slovenia [although they were a threat in the past (Sket 1979)]. In caves, a special threat exists—human visitors to caves, and we take up the question of the impacts of human visitation. In the fourth section, we consider what should be protected, where it should be protected, and how this might be accomplished. With the availability of increasingly accurate maps of the geography of subterranean species richness

The Biology of Caves and Other Subterranean Habitats. Second Edition. David C. Culver and Tanja Pipan. Published 2019 by Oxford University Press. © David C. Culver and Tanja Pipan 2019. DOI: 10.1093/oso/9780198820765.001.0001

(Chapter 8), it is possible to decide where preserves should be. Site selection and protection cannot be performed in a vacuum, and in the fifth section we review the international, national, and local agencies, both governmental and non-governmental, that have been part of protection efforts for subterranean habitats and species. In the final section, what in many ways is the most difficult problem of all is considered—how a local site is protected and managed. The time is past when it is appropriate to protect a cave or well by only protecting a few metres around the site. The vexing and controversial strategy of gating of caves as a conservation tool will be reviewed. There can be no doubt that in some cases, appropriately designed cave gates have protected populations and probably species such as *Myotis grisescens* and *Myotis sodalis* from extinction, but there can also be no doubt that inappropriately designed gates have caused the disappearance of some bat populations (Elliott 2012).

For many biologists and others interested in subterranean habitats and communities, subterranean species are worth protecting in their own right—a biocentric view of species protection. But subterranean species also provide ecosystem services—services to human populations provided by ecosystems that would otherwise have to be accomplished in some other way (Daily 1997). The ability of groundwater microorganisms to decompose organic matter provides a number of significant ecosystem services, including rendering harmless human pathogens, breaking down organic wastes, and the net result—purifying groundwater (Herman *et al.* 2001; Griebler and Avramov 2015). Groundwater microorganisms often play a significant role in the clean-up of groundwater contaminants, sometimes called natural attenuation. Groundwater clean-up is a multibillion dollar industry. Additionally, groundwater provides mitigation of floods and droughts and nutrient cycling (Griebler and Avramov 2015). Bats also provide important ecosystem services, including insect control, seed dispersal, and pollination (Kunz *et al.* 2011). Bat guano is still an important source of fertilizer in some countries. Fenolio (2016) makes the more general case that subterranean species can provide new drugs, a general and powerful argument for the protection of biodiversity. Indeed, there have been several studies of groundwater and cave microbiota in the context of new drugs (Bhullar *et al.* 2012).

10.2   Rarity

Rabinowitz (1981) pointed out that rarity means different things to different biologists. From a botanical perspective she suggested that rarity has three meanings—a species can be numerically rare throughout its range (numerical rarity), it can occur in a rare habitat or habitats (habitat rarity), and it can be geographically rare with a restricted range (geographical rarity). Species may be rare along one, two, or all three of her axes, leading to her 'seven forms of rarity'. Her classification works perfectly well with respect to the subterranean fauna.

There is little doubt that the majority of the subterranean fauna is geographically rare. In an analysis of stygobionts in Belgium, France, Italy, Portugal, Slovenia, and Spain (Michel *et al.* 2009), part of the PASCALIS project (Chapter 8), it was found that 464 of 1059 stygobionts (44 per cent) occurring in caves and interstitial habitats were limited to a single 12 min by 12 min (12′ x 12′) cell approximately 400 km² in area. In a study of the calcretes of the Pilbara region of Western Australia, Eberhard *et al.* (2009) suggested that nearly the entire fauna is comprised of short range endemics (SREs), species with ranges less than 10 000 km² (Harvey 2002). In an analysis of the obligate cave fauna of the United States, Culver *et al.* (2000) found that 463 out of 673 troglobionts (69 per cent) and 131 out of 300 stygobionts (44 per cent) were limited to a single county, with the average size of counties with stygobiotic or troglobiotic species being approximately 10 000 km². In the case of the US troglobiotic fauna east of the Mississippi River, 211 out of 467 species (45 per cent) are known from a single cave. One of the hallmarks of cave organisms is their endemism.

All of these levels of endemism are much higher than that recorded for any surface habitat. Endemism seems higher in troglobionts than stygobionts, and troglobionts in general have smaller ranges (Lamoreaux 2004), but to date no pattern has been reported with respect to trophic position or body size (Culver *et al.* 2009).

The pattern with respect to habitat rarity is mixed. In the main karst areas (Fig. 1.1) and lava tube areas, caves are quite common, reaching up to 5 per km² in Slovenia and in lava flows in Hawaii (Culver and Pipan 2014), and it is areas such as these that harbour most of the stygobionts and troglobionts found in caves. There are a few stygobionts and troglobionts found in isolated areas. An example of this is the troglobiotic beetle *Choleva septentrionis holsatica*, endemic to Segeberger Höhle, a cave in an isolated rock salt dome in northern Germany (Ipsen 2000). Some cave habitats, especially the cave hygropetric (Sket 2004a), are probably quite rare, so hygropetric specialists, such as the beetle *Nauticiella stygivaga*, are rare because of habitat rarity. As a habitat, epikarst is probably more common than caves, being more or less continuous in karst areas except in the tropics and glaciated areas (Williams 2008). Interstitial habitats, especially fluvial aquifers, are common, or at least as common as surface streams and rivers. Calcrete aquifers, while confined to very arid regions, are quite common in these regions. That most superficial of shallow subterranean habitats, the hypotelminorheic, is certainly not a common habitat but its geographical distribution is not known. In the environs of Washington, DC, where it has been best studied, it is localized to a few bands of habitat. All in all, most stygobionts and troglobionts are not rare because of habitat rarity, but there are some exceptions.

The question of numerical rarity of stygobionts and troglobionts is a very interesting one. A considerable number of troglobionts and stygobionts from caves are known from only a handful of specimens, in some cases a single specimen. The obvious rarity of these species may be more apparent than real because it is likely that the primary habitats of many of these species are not caves but habitats such as epikarst and phreatic water, and that they are accidentals in caves. Other populations of stygobionts and troglobionts are known to be quite large. The best known example is that of the Baget ecosystem in France (see Chapter 9), where the number of individuals of stygobiotic copepod species that washed out of the system was in the thousands (Rouch 1970).

There are two general methods available to estimate population size that bring some clarity to the range of observations about population size. One of these is mark–recapture, based on recapturing in a second sample individuals that were marked in the first sample. Population size (X) and its standard error can be estimated as follows (Begon 1979):

$$X = an/r$$
$$\text{SE}\left(X\right) = \left[a^2 n\left(n-r\right)/r^3 + 1\right]^{0.5}$$

where a is the number marked in the first sample, n is the number of individuals in the second sample, and r is the number of marked individuals in the second sample. Such studies with subterranean animals are technically difficult and standard errors are often very large because of relatively small sample size (Knapp and Fong 1999; Fong 2003) but nonetheless very informative. These estimates of the size of the local population and the geographical extent of the population being estimated are dependent on the extent of dispersal and mixing of the individuals marked. The small number of these estimates available (Culver 1982) indicate population sizes between 100 (the crayfish *Aviticambarus sheltie* in Shelta Cave, Alabama, USA) and 9000 (the crayfish *Orconectes inermis inermis* in Pless Cave, Indiana, USA). Those on the small end of the estimates range are certainly numerically rare. Populations can be quite dense—Porter and Hobbs (1997) estimated the density of the amphipod *Crangonyx indianensis* as $777 \pm 108/\text{m}^2$ in Dillion Cave, Indiana.

Population geneticists use a different measure of population size, 'effective population size', basically the size of a randomly mating population that would result in the same levels of heterozygote frequency and same levels of genetic variation as the observed population. A population with low genetic variation or low heterozygosity has a smaller effective population size. Except for a few fish populations, there is little evidence of reduced genetic variability in stygobionts and troglobionts (Sbordoni *et al.* 2000, 2012). Buhay and Crandall (2005), using mitochondrial DNA sequences, calculated an effective population size for several stygobiotic species of crayfish of between 20 000 and 80 000. Sbordoni *et al.* (2012) attribute the large estimates of effective population sizes of subterranean species in general to selection for heterozygotes rather than population size per se. In general it appears that subterranean populations, as with surface-dwelling populations, can be either large or small and some unknown fraction of stygobionts and troglobionts are numerically rare.

Thus, the majority of stygobionts and troglobionts are geographically rare, many are likely to be numerically rare, and a few are in rare habitats. This rarity has important conservation implications. Geographically rare species are subject to catastrophic losses as the result of relatively minor and frequent environmental insults, if for no other reason than their geographically restricted range. Numerically rare species are more likely to go extinct than common species because of genetic inbreeding, demographic stochasticity (such as the appearance of a single-sexed population in a generation), and environmental stochasticity (minor environmental insults). Taken together, conservation biologists call these phenomena the extinction vortex (Groom *et al.* 2005).

10.3 Other biological risk factors

Because many stygobionts and troglobionts, as well as bats, have relatively low reproductive rates (although lifetime rates may be high due to increased longevity—see Chapter 6), their rate of population growth following an environmental insult will be low, resulting in a smaller population for a longer period of time relative to surface-dwelling populations. This results in increased extinction risk because they are in the extinction vortex longer.

Stygobionts and troglobionts may also be especially sensitive to some kinds of environmental fluctuations. For example, troglobionts are often especially sensitive to changes in relative humidity as a result of exoskeleton thinning (Howarth 1980, 1983). Stygobionts may be more or less sensitive than surface relatives to heavy metals (Notenboom *et al.* 1994), but similarly to their surface counterparts, stygobionts, especially interstitial species, frequently must cope with heavy metal contamination (Vesper 2012). Stygobionts and troglobionts appear to be especially sensitive to non-subterranean competitors and predators that can occur in subterranean sites as a result of pollution events, especially organic pollution of streams (Sket 1977; Chapter 5).

Cave-inhabiting bats have special biological risk factors associated with cave use. For species that concentrate in large numbers in a small number

Table 10.1 Biotic factors of the subterranean fauna that increase its extinction risk.

Biotic factor	Applicability
Geographical rarity	Most stygobionts and troglobionts, especially troglobionts; some hibernating bats
Numerical rarity	Some stygobionts and troglobionts, especially vertebrates
Habitat rarity	Cave hygropetric, a few cave areas, hypotelminorheic?
Low reproductive rate	Bats and most stygobionts and troglobionts
Sensitivity to environmental flucations	Most stygobionts and troglobionts
Sensitivity to surface-dwelling competitors and predators	Most stygobionts and troglobionts
Strong clustering	Bats
Sensitivity to arousal from hibernation	Bats

of caves, such as *Myotis grisescens* in the southeastern United States, the very fact of their concentration makes them vulnerable to any changes that occur in that cave. Hibernating bats are also sensitive to arousal from hibernation, an energetically expensive proposition, and after repeated arousals they may not have enough fat reserves to survive the winter (Kunz and Fenton 2003). For species that use caves as nursery roosts, with females gathering in large numbers to raise their young (usually one per year), disturbance at this critical time can have a damaging effect on breeding success and survival of the population. These are some of the reasons that white nose syndrome (Moore and Kunz 2012) has had such a devastating effect on the North American bat fauna. The biotic factors that increase the vulnerability of the subterranean fauna are summarized in Table 10.1.

10.4 Threats to the subterranean fauna

Threats to subterranean fauna are about as diverse as threats to surface-dwelling species, especially since most environmental disasters in surface habitats are environmental disasters for subsurface habitats as well. And there are, as we shall see, disasters limited to the subsurface. Jones *et al.* (2003) divide threats into three overarching categories:

- Alteration of the physical habitat.
- Water quality and quantity.
- Direct changes to the subterranean fauna.

To this list the threats of global climate change to the subterranean fauna must be added (Mammola *et al.* 2017).

Threats to subterranean faunas are present throughout the world, and there are regional differences. Threats in developed and developing countries may be different. There is more detailed information available about threats to subterranean habitats, especially caves, from the United States than elsewhere, and many of the examples we cite are from the United States. Whether there are actually more immediate threats in the United States is not clear. Certainly monitoring of subterranean habitats, including both caves and interstitial aquifers, is not as extensive in developing countries. However, it may also be that, relative to other developed countries, there are more environmental threats in the United States. This is not as implausible as it appears, because population growth rates in the United States are higher than most developed countries, especially Europe. Finally, the US Endangered Species Act may be more thoroughly implemented with respect to the subterranean fauna than legislation elsewhere, such as the European Union Habitats Directive.

10.4.1 Alteration of the physical habitat

Quarrying of limestone, especially for the making of cement, is the ultimate kind of threat because it completely removes the habitat. Worldwide, the area of fastest growth in limestone quarrying is southeast Asia. Annually, 1.75×10^7 metric tons of limestone are quarried from southeast Asia and it is growing at a rate of 6 per cent per year, higher than any other area of the world (Clements *et al.* 2006). The limestone karst regions of Indonesia, Thailand, and Vietnam cover an area of 400 000 km² and are 'arks of biodiversity' (Clements *et al.* 2006), both because they are biodiversity hotspots and because they are relatively untouched by agricultural and forestry practices because of the rugged terrain, such as tower karst and karst pinnacles (Ford and Williams 2007). Both the surface and subsurface biota are very incompletely known, and this is one of these situations where species disappear before they are even discovered, as can also happen with cryptic species in well-studied areas (Niemiller *et al.* 2013a). Because of the island-like nature of these karst outcrops (see Chapter 7), endemism of not only the subsurface species, but also the surface-dwelling species, is high. Vermeulen and Whitten (1999) report six land snails endemic to the Sarang karst, an area of only 0.2 km²; and 50 endemic to the Subis karst, a 'large' area of 15 km², both in Borneo. Clements *et al.* (2006) also point out that the subsurface and surface of karst areas in southeast Asia is understudied, even relative to other habitats in the region. Quarrying is not the result of entirely local factors. Much of the cement produced is exported, and funding for the development of cement plants in southeast Asia is international.

Quarrying is a threat to cave and karst faunas elsewhere as well. Hamilton-Smith and Eberhard (2000) point out quarrying is a threat in much of Australia, including the highly diverse Cape Range area of northwest Australia (Humphreys 2004). Jones *et al.* (2003) report that Zink Cave,

Indiana, was almost completely quarried away, resulting in the extirpation of the stygobiotic fish *Amblyopsis spelaea* from the cave. A major bat hibernaculum in eastern North America is Hellhole in West Virginia (Dasher 2001), with 45 per cent of the known individuals of the US federally endangered bat *Corynorhinus townsendii virginianus*[1] and a hibernating population of another endangered bat, *Myotis sodalis*. An extensive limestone quarry is adjacent to the cave and negotiations concerning where quarrying is allowed continue between the owner of the quarry and various government agencies. Access to the cave is controlled by the quarry, making both negotiations and monitoring especially difficult.

In karst areas, road construction both reveals new caves, such as Inner Space Caverns in Texas, USA (Elliott 2000), and destroys caves or portions of caves (Knez and Slabe 2016). The details of the location of highways through karst regions can make a big difference in terms of environmental impacts. For example, small changes in routing of a highway in southwestern Virginia, USA, protected Young–Fugate Cave, a cave with a hibernating colony of the federally listed *M. grisescens* (Hubbard and Balfour 1993; Tercafs 2001).

Especially in the past, caves themselves were directly mined, especially for guano from large bat colonies, and in North America for saltpetre, an ingredient of gunpowder (Hubbard 2012). Devastation of the fauna in these cases must have been extreme. Saltpetre mining was generally carried out in drier portions of the cave, which have few troglobionts because of low humidity.

Another type of site alteration is the development of a cave for tourist visitation. The commercial caves, such as Postojnska jama and Mammoth Cave, both of which are biodiversity hotspots (see Chapter 9) have over 1 million visitors per year. Commercialization of a cave requires physical alteration of natural passages and installation of lights, with the concomitant development of a lampenflora, the growth of plants associated with electric lighting (Aley 2004). Speleobiologists have frequently warned against excessive commercialization of caves because of these changes. In caves such as Postojnska jama and Mammoth Cave, a relatively small percentage of the total known passage has been altered and lit for tourist visitation. Those sections of caves that are commercialized, are, as a rule, depauperate with respect to stygobionts and troglobionts.

Physical alteration of a cave entrance, either by filling it in, enlarging it, or putting a gate on it, can have an impact on the fauna, especially the terrestrial fauna. Either filling in or gating can alter the movement of animals in and out of the cave, an important source of organic carbon for many troglobionts (Chapter 2). Enlarging an entrance or creating an artificial entrance can increase air flow and drying, reducing relative humidity. However, the major impact of alteration of entrances is on bat populations. Historically,

---

[1] Formerly *Plecotus townsendii virginianus*

Mammoth Cave in Kentucky was an important bat hibernation site but after the Historic Entrance was modified to block incursions of cold winter air, bats abandoned the cave (Elliott 2000). Before the effect of gates on bats were fully understood and before gate design was perfected, gating of caves, ostensibly to protect bats, probably caused much of the decline of *M. sodalis* from the 1960s to the 1980s (MacGregor 1993). Improperly designed gates can cause bats to abandon a roost, and if the spaces in the gate are too close together, bats are forced to crawl through the gate. In this situation, the cave entrance becomes a magnet for bat predators, such as snakes and feral cats. Cave gates that are 'bat-friendly' have horizontally stiffened angle-irons at 15 cm, with vertical supports at least 1.2 m apart (Fig. 10.1) (Elliott 2000). Some bats, such as *M. grisescens*, do not tolerate complete gates and more complex gate designs are needed (Elliott 2012; Hildreth-Werker and Werker 2006). Some bat species, such as *Miniopterus schreibersii* in Europe, avoid caves with gates, especially during the breeding season (Mitchell-Jones *et al.* 2007). In such cases, other systems to prevent unauthorized access, such as fences around the entrance, are needed.

10.4.2 Water quality and quantity

Interstitial aquifers, especially shallow fluvial aquifers, are subject to most of the same environmental threats as rivers and streams. There are few places

Fig. 10.1   Gate at the entrance to Fisher Cave, Missouri, USA, designed to allow unimpeded access for bats. Photo by H. Hobbs III, with permission. See Plate 21.

left in the world where human activity has not had a negative impact on water quality. In surface habitats, species living in rivers are typically more at risk of extinction than species in other habitats, based on data for the US fauna (Master *et al.* 1998). The threats to groundwater are summarized in Table 10.2. Several environmental drivers are of critical importance. One is agriculture and the nearly universal overapplication of fertilizers and pesticides that accompanies it. For example, nitrate levels in groundwater throughout Europe continue to rise, and due to overuse of fertilizers and pesticides they are frequently found in shallow groundwater (Notenboom 2001). Another important driver is water extraction for agriculture, industry, and urban activities, especially the extensive use of irrigation in agriculture. Groundwater levels have fallen in many areas, more than 30 m in some cases

Table 10.2 Summary of drivers and associated pressures on groundwater.

Drivers	Pressures
Above-ground activities	
Climate and natural processes	Influence of water discharge and recharge; input of organic and inorganic substances.
Urban activities and infrastructure	Seeping of oil products from fuel storage tanks; heavy metals, salts, PAHs (polyaromatic hydrocarbons; leakage of sewage systems. Growth of water demand.
Tourism	Additional water demand, waste and sewage in sensitive areas.
Industry	Spillage of chemicals, seeping of oil products from fuel storage tanks. Local groundwater withdrawal.
Agriculture	Leaching of persistent pesticides and metabolic products; and of nitrate and metals from fertilizers and manure. Groundwater withdrawal for irrigation, lowering water table.
Waste treatment (illegal, improper)	Leaching of pollutants from waste-disposal sites.
Below-ground activities	
Mining	Lowering groundwater levels and changing flows; changes in physicochemical conditions; pollution with mining spoils (heavy metals).
Water extraction	Overexploitation; saltwater intrusion; upwelling of mineral rich groundwater; decline of groundwater levels; changes in physicochemical conditions.
Waste injection and underground storage	Direct introduction of contaminants; changes in physicochemical conditions.
Heat and cold storage	Changes in physicochemical conditions.
Infrastructure building below ground	Lowering groundwater levels and changing flows; changes in physicochemical conditions.
Surface water infiltration	Nutrient enrichment of groundwater; infiltration of pollutants.
Gas and CO_2 storage	Changes in physicochemical conditions.

Source: Modified from Notenboom (2001).

(Danielopol *et al.* 2003). This results in changed and reduced connections between surface and subsurface with the concomitant change in nutrients (see Chapter 4).

Even relatively mild alterations to aquifer recharge affect the fauna. In a shallow groundwater aquifer in the city of Lyon, France, Datry *et al.* (2005) found that artificial recharge of some sites with storm water changed the species composition at those sites, although this was probably the result of the increase in organic carbon rather than in recharge per se.

One very interesting and instructive example of recharge effects is the Trebišnjica river system in southern Bosnia & Herzegovina (Čučković 1983; Minanović 1990). The Trebišnjica had a surface water course of between 35 and 90 km, depending on water flow, and it sinks in Popovo Polje. Poljes are spring-fed large karst depressions with flat floors that are commonly covered by river sediments. During the heavy rains, especially in the fall, a temporary lake forms in the bottom of the polje due to flooding of the river. Along the sides of Popovo Polje are many caves, including Vjetrenica, one of the most diverse caves in the world (see Chapter 8). Many of these caves were periodically flooded and with the floods came organic material necessary for the survival of many stygobiotic species (see Chapters 2 and 4). To provide hydroelectric power and to control flooding, the government of what was at that time Yugoslavia instituted construction of a dam and a channelization of the Trebišnjica (Fig. 10.2). With the completion of construction in 1979, Trebišnjica was no longer a sinking ('lost') river (Čučković 1983), and the surrounding caves were starved of nutrients. A fish species, *Paraphoxinus ghetaldi*, endemic to the polje and especially common in ecotones between surface and subsurface water, is possibly extinct. The population of the salamander *Proteus anguinus* in the polje was perhaps the largest one anywhere and it is now decimated. Populations of other unique species such as the Dinaric cave clam *Congeria kusceri*, the only stygobiotic cave clam in the world, and the polychaete worm *Marifugia cavatica* have likewise been decimated as a consequence of damming of the river and the resulting hydrological changes.

Dam construction can have direct negative effects on caves. When the New Melones Reservoir on the Stanislaus River, California, USA, was constructed in the late 1970s, about 30 caves were inundated, including McClean's Cave, one of the two known localities at the time for the troglobiotic harvestman, *Banksula melones* (Elliott *et al.* 2017). The population was transplanted to a nearby mine, which was stocked similar to a terrarium with cave soil, rocks, and rotting wood (Elliott 1981), and the population was still in the mine 20 years later (Elliott 2000). Transplanting to an artificial site was preferable to transplanting to another cave since the mine had no natural community that would have been disrupted. One positive outcome of this ecological disaster was that a thorough inventory of nearby caves revealed 16 other caves where *B. melones* was found.

Fig. 10.2   Channelized Trebišnija watercourse in Popovo polje, Bosnia & Herzegovina in 2005. Photograph by M. Zagmajster, with permission. See Plate 22.

Excessive groundwater use, especially from karst aquifers, can also have a major negative impact on fauna. The Devil's Hole pupfish, *Cyprinodon diabolis*, is known from a single sinkhole in Nevada, USA, where it lives in the ecotone between surface and groundwater. On the US Endangered Species list, it was reduced in numbers due to excessive pumping, and now groundwater extraction in the region is limited because of the fish. Similarly in the Edwards Aquifer (see Chapter 9), species living in springs (ecotones between surface and groundwater) are threatened, including two beetles, one fish, a salamander, and a species of wild rice (Elliott 2000).

The impact of organic pollution can be drastic and catastrophic, as the next examples show. All subterranean habitats, especially SSHs, are at risk from spills of toxic materials, typically from trucks on roads, but also from railroads and leaking storage tanks, especially associated with petrol stations. The number of underground storage tanks is staggering. In the mid-1980s, at the start of the USA Leaking Underground Storage Tank (LUST) Trust Fund, there were more than two million storage tanks in the USA, and by 2018, there had been over 540 000 toxic releases (U.S. EPA 2018). These spills either quickly enter SSHs or are buried in SSHs to begin with. Most storage materials are nonaqueous phase liquids (NAPLs), which includes fuels, solvents, and insulators (Loop 2012). NAPLs have, by definition, limited solubility in water and hence are difficult to remove from the subsurface especially because groundwater extraction and treatment ('pump and treat') is very inefficient (Herman *et al.* 2001). Even the limited solubility of many NAPLs results in groundwater contamination that exceeds the maximum concentration level set by the U.S. Environmental Protection Agency (Loop 2012). NAPLs include hydrocarbons, vinyl chloride, PCE, PCBs, and many insecticides and herbicides such as atrazine. NAPLs in general come to reside in the shallow subsurface, and are moved out of this zone by precipitation, but are often held for extensive periods of time in the soil, epikarst, and regolith, and continue to contaminate the aquifer below. NAPLs can be retained in pores (epikarst and MSS), and in the sediments (soil and hypotelminorheic). Natural degradation (attenuation) is faster for hydrocarbons than chlorinated compounds such as chlorinated solvents and PCBs, but for a large spill may take many decades.

The organic pollution of Hidden River (Horse) Cave System, Kentucky, USA, is in some ways the equivalent of the hydrological alteration of Popovo polje—they were both catastrophic and instructive. A large cave near Mammoth Cave, with a large, attractive entrance in the small town of Horse Cave, Hidden River Cave was commercialized in 1916. It had a rich stygobiotic and troglobiotic fauna, including fish and crayfish (Elliott 2000). Increasing contamination of the cave water from indiscriminate sewage disposal from Cave City and Horse Cave, as well as wastes from a nearby creamery, led to the closing of the cave's tourist operation in 1943. By the 1960s, the stench from the cave made it unpleasant to even walk by it. The stygobiotic fauna, at the least, was extirpated from the cave. By the late 1980s, the major sources of pollution had stopped and the cave stream began to recover. By about 1995, the original animal community had recolonized the formerly polluted cave stream from unpolluted upstream reaches of the caves (Lewis 1996). In Thompson Cedar Cave in Virginia, another cave with serious organic pollution (in this case from sawdust waste), the fauna recovered about 15 years after the original sources of pollution had been removed, having recolonized from unaffected upstream reaches (Culver *et al.* 1992). In all of these cases of organic pollution, the decline in water quality was accompanied by invasions of organisms typical of polluted waters, such as tubificid worms. Competition and predation from these invading species may actually be a bigger threat to the stygobionts than water quality itself (Sket 1977).

10.4.3 Direct changes to the subterranean fauna

A fungal disease, white nose syndrome, has wreaked havoc on the hibernating North American bat fauna. First discovered in the winter of 2006–2007 in Howe Caverns, a commercial cave in New York, bats were observed prematurely emerging from hibernation, flying erratically, and dying. The cause of death was determined to be the fungus *Pseudogymnoascus destructans*[2].

---

[2] Formerly *Geomyces destructans*

Although estimates vary, six million bats have died from the disease since 2007. Nine species have developed WNS, including seven in the genus *Myotis*. Declines in affected bat populations in the northeast United States have averaged 73 per cent (Frick *et al.* 2010). Characteristics of the disease include (Moore and Kunz 2012):

- Fungal infections and tissue damage.
- Depleted fat reserves.
- Atypical winter behaviour.
- Changes in immune response during hibernation.
- Wing damage.

Since its discovery it has spread extensively throughout the eastern half of the US, and cases have been reported from Washington state (www. whitenosesyndrome.org). The transmission of WNS by bats is well documented but transmission by humans is possible and even likely. To minimize this problem, decontamination procedures for cavers have been implemented. The response to WNS has been controversial. If some individuals have developed immunity, or the disease has declined in virulence, then more aggressive intervention is not needed. If on the other hand, there is no immunity or decline in virulence, then possible solutions such as captive breeding, immunization, and application of anti-fungal chemicals become more relevant.

The most direct human impact on cave faunas is that caused by human visitation. The most egregious example is that of deliberate destruction of bats. Elliott (2000) provides several examples, including a case where four men were convicting of shooting and crushing to death the endangered Indiana bat (*M. sodalis*) in Thornhill Cave, Kentucky. For many people, bats are vermin, nothing more than carriers of disease such as rabies and histoplasmosis, even though the threat of these diseases is very slight (Woloszyn 1998). There is no doubt that large concentrations of bats are unhealthy to be around if for no other reason than the concentration of ammonia resulting from urine in the air, but there is no reason, except for bat biologists, to be around large concentrations of bats. The combination of fear and loathing of bats has made bat protection difficult, and makes gates the primary tool of protection (see section 10.7). More benign visits to caves with hibernating bats may also have a negative impact, because the activities of cavers may awaken bats from hibernation, causing significant energy expenditure. Repeated arousals from hibernation may result in mortality, as is the case with WNS. Cave visitation may also have some negative impact on stygobiotic and troglobiotic species, primarily by compaction of terrestrial habitat and disruption of stream habitats when stream walking is necessary.

Overcollecting, especially of vertebrates, may well have negative effects, especially on single-site endemics (Elliot 2000). Most speleobiologists have had the experience of collecting from a site, then returning to find as many animals as before, or returning to find very few. Prudence dictates that collecting be kept to the minimum necessary.

In Europe, the problem is exacerbated by a collectors' market for some cave animals, such as beetles (Simečić 2017). A potentially specially damaging collecting technique is pitfall trapping. Pitfall traps baited with cheese or rotting meat can attract hundreds of troglobionts in a day or two.

10.4.4 Global warming

A final human activity may, in the coming years, dwarf all other impacts. This is global warming. Since subterranean habitats in general are less variable than surface habitats, organisms in subterranean habitats will experience less in the way of temperature extremes. However, the rise in average temperature will increase average temperatures of subterranean habitats since their temperature approximates the mean annual temperature. Since any organism, surface or subterranean, is rarely adapted to temperatures it never encounters, rising temperatures may result in lethal conditions for some stygobionts and troglobionts. Since stygobionts and troglobionts have very limited dispersal capability, if temperatures are rising relatively rapidly, species may go extinct. Mammola *et al.* (2017) modelled the impact of expected change in temperature on *Troglophantes* spiders in the Italian Alps, and predicted that most of the highly endemic species would go extinct.

10.5   Site selection

To protect subterranean species, their location needs to be known. The problem of sampling completeness was considered in Chapter 8, and it is as relevant here. Thorough inventories are critical if appropriate decisions about protection are to be made. Several useful inventories are available on the World Wide Web. The International Union for the Conservation of Nature (IUCN) maintains a list of species at risk that includes subterranean species (www.iucnredlist.org) and the US Fish and Wildlife Service (www.usfws.gov/endangered/) maintains a list of threatened and rare species. Both these lists are limited in that not all subterranean species, or even a significant fraction of them, have been evaluated. The non-governmental US conservation organization NatureServe (www.natureserve.org) maintains a more complete list for the United States of species of concern, which includes most subterranean species because of their small ranges.

The selection of sites for protection on a regional basis is a complex problem, one that has different solutions, depending on the criteria used. Possible criteria that can be used in reserve design include (Izquierdo *et al.* 2001; Michel *et al.*, 2009):

1. maximizing the number of species in the reserve network;
2. maximizing the number of endemic species in the reserve network;
3. maximizing the number of species in each reserve; and
4. minimizing the number of reserves while keeping total area constant.

More complex site selection criteria are possible. Borges *et al.* (2012) include a variety of other criteria, such as threat level, in a sophisticated analysis of site selection for lava tubes in the Azores. Phelps *et al.* (2016) used surface features of the landscape to predict high diversity bat caves in Bohol Island, Philippines. Using principal component analysis, they found that the third eigenvector, loading on low surface disturbance, was a good predictor of bat species richness. Christman *et al.* (2016) showed that it was possible to predict the presence or absence of different taxonomic groups of troglobionts and stygobionts using surface features. However, the predictors were complex and varied from group to group, limiting its utility in conservation planning.

Michel *et al.* (2009) have investigated the optimal reserve design for stygobionts for a large part of Europe—Belgium, France, Italy, Portugal, Slovenia, and Spain. They placed a 12′ × 12′ grid of 4675 cells over the study area and assigned 10 183 records of 1059 stygobiotic species to these cells. Of the 4675 cells, 1280 (27.4 per cent) had one or more stygobionts. It is characteristic of data on subterranean fauna that a relatively small fraction of the area actually has stygobionts or troglobionts. Using a much larger (and irregular) grid size of US counties, Culver *et al.* (2000) found that only 16.5 per cent of 3112 counties had a stygobiont or troglobiont, but they did not include stygobionts from interstitial aquifers, unlike Michel *et al.* (2009).

Michel and colleagues used three different criteria to determine a network of 128 cells (10 per cent of the cells with stygobionts and 2.7 per cent of the total cells). One preserve design was determined by finding the 128 cells that had the most stygobionts (the species richness criterion); the second preserve design was determined by finding the 128 cells that had the most single-cell endemic stygobionts (the endemics criterion), and the third preserve design was determined by finding the 128 cells that, in aggregate, had the most stygobionts represented at least once (the complementarity criterion, Fig. 10.3). The complementarity criterion included the most species, and nearly as many endemics as the endemic criterion (Table 10.3). The complementarity preserve design was more fragmented than the other two, and included a broader geographical spread. For example, a hotspot of stygobiotic richness occurs in several grid cells in southwest Slovenia. Nearly all of these grid cells are included in the richness and endemism design, but fewer are included in the complementarity design. The complementarity design includes a cell in Sicily, a region of only moderate richness, and the other two do not (Fig. 10.3). A preserve of only 2.7 per cent of the total area of the six countries could include nearly 80 per cent of all known stygobionts. Culver *et al.* (2000) and Izquierdo *et al.* (2001) also found that reserves designed around complementarity principles required relatively little land area in the United States and Canary Islands, respectively.

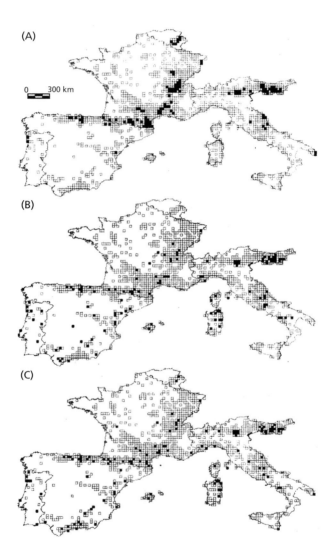

Fig. 10.3 Distribution of 128 cells selected by three methods to represent the groundwater fauna of Europe: A) richness hotspots; B) endemism hotspots; C) complementarity areas. See Table 10.3 for information on completeness of representation. From Michel *et al.* (2009). Used with permission of Blackwell Publishing.

Table 10.3 Percent of stygobionts and percent of single-cell endemics included in preserve designs with 128 2′ × 2′ cells.

Selection criteria	Percent of total species	Percent of endemics
Species richness	65.1	48.9
Endemism	72.4	71.1
Complementarity	79.7	67.7

There were 1059 stygobionts and 464 single-cell endemics.
Source: Data from Michel *et al.* (2009).

10.6 Protection strategies

There are two international agreements that have been used to protect subterranean sites. The Convention on Wetlands is an intergovernmental treaty, commonly known as the Ramsar Convention. It is a global treaty on conservation and sustainable use of wetlands as natural resources. Technical and policy guidelines are available to assist countries in protecting their Ramsar wetlands (Beltram 2004). The number of karst wetlands has been growing steadily. As of February 2018 there were 130 sites listed under the wetland type 'karst and other subterranean hydrological systems' (https://rsis.ramsar.org). Several of the sites are rich in stygobionts and troglobionts, including Škocjanske jame in Slovenia, and the Baradala/Domica transboundary cave system in Hungary and Slovakia.

A second international list of importance for the protection of subterranean habitats is the World Heritage Site list, developed under the auspices of UNESCO. Natural heritage sites are nominated by the countries involved, who also agree to continue to protect their integrity and provide access to all peoples. In return UNESCO provides support for the restoration and protection of sites (Hamilton-Smith 2004). Among the sites are large parks with significant karst features, such as Halong Bay in Vietnam, and Rocky Mountain Parks in Canada, as well as cave-focused sites, such as Carlsbad Caverns National Park, New Mexico, USA. Among the sites with high subterranean biodiversity on the list are Gunung Mulu National Park in Malaysia, Škocjanske jame in Slovenia, Mammoth Cave in Kentucky, and Durmitor National Park in Montenegro.

At the national level, there are a variety of designations offering varying levels of protection. One of the most interesting examples of this is the Danube Floodplain National Park in Austria, created to protect the highly diverse interstitial aquifer, the Lobau wetlands (Danielopol and Pospisil 2001; Table 9.3). Non-governmental organizations, especially The Nature Conservancy in the United States, have also been active in protecting subterranean sites, either by outright purchase or purchase of development rights. Small groups of individuals interested in cave protection have also been active in the United States and elsewhere, in some cases buying caves outright, as evidenced by the many small cave conservancies that have sprung up in the United States.

While most efforts at protection have been focused on the protection of sites, such as the European Union Habitats Directive, the US Endangered Species Act focuses on individual species rather than sites or communities. Because of the high levels of endemism, nearly the entire subterranean fauna is at risk. In practice, threats are unevenly distributed taxonomically and geographically, and because of the lengthy process of petitioning to have a species listed and the difficulty in adding species to the list, it is in areas of imminent threat where most of the species are listed. Of the five bat species and subspecies listed from the continental United States, four are ones that hibernate in caves. Species in springs, usually not stygobionts, are well represented on the list. For example, six species of pupfish (*Cyprinodon*), all of which occur in desert springs and sinkholes in the southwestern United States, are on the list. This reflects their vulnerability to excessive groundwater extraction and the consequent drop in the water table. Of the 30 stygobionts and troglobionts on the list, 15, mostly troglobionts, are from the state of Texas. Stygobionts tend to be found in deeper caves and especially in wells that tap into the Edwards Aquifer (see Chapter 9). This is not because endemism levels are higher in Texas, but rather because the major cave region—the Balcones Escarpment/Edwards Aquifer—faces much higher development pressure than other US karst areas. This development pressure comes from the growing urban areas of San Antonio and Austin, and the area in between. Thus, three groups of subterranean species in the United States are at high risk—species in springs (especially in arid regions), bats hibernating in caves, and troglobionts in Texas.

European countries often have a variety of legal protections of subterranean habitats and fauna (Tercafs 2001). One the leading countries in this regard is Slovenia, especially the 2004 Zakon o ohranjanju narave (Nature Conservation Act), which parallels the European Union Habitats Directive and the 2004 Zakon o varstvu podzemnih jam (Cave Protection Act), which is unique to Slovenia and specific for cave and cave fauna protection. Protections are in place for both species and habitat types, with strict controls on access and collecting. Additional protections for especially important sites, such as Postojna–Planina Cave System (PPCS), are in place at the local level as well.

10.7 Preserve design

As humans enter caves through an entrance, much of the focus of protection efforts for cave fauna has been on the protection of the entrance of the cave. There are numerous examples of caves thought to be protected when the entire protection strategy consists of a gate to control human access, a gate that often alters the movement of animals (and other organic carbon) in and out of the cave. A gate is often critical to providing protection for hibernating and maternity colonies of bats, provided it is of the appropriate construction (Elliott 2012; Fig. 10.1) and one that allows the movement of organic carbon. A gate provides no protection against habitat destruction or alterations in water quality or quantity.

A more appropriate starting place for preserve design is to first consider what the species and communities are that are to be protected. It is often the case that more than one species or community is of conservation interest. In fluvial aquifers, the benthic community as well as the interstitial community is often in need of protection. At least in the United States, the fauna of freshwater streams and rivers is often at risk (Master *et al.* 1998), and protection efforts should involve both. In caves with important bat populations, there are often important populations of troglobionts and stygobionts. In addition, karst areas often harbour rare plants, ones adapted to the thin, acidic soils and exposed rock outcrops present in karst. For example, the discovery of a new species of clover endemic to a threatened karst area in Virginia made for a stronger case for land acquisition and changed the design of the preserve.

Subterranean species and communities of conservation interest fall into four broad categories, categories that dictate different preserve designs. The first category is that of streams, both surface and subsurface, and the associated terrestrial biota along the banks and flood plain. This includes fluvial interstitial aquifers, cave streams, terrestrial riparian communities along cave streams, and even the hypotelminorheic, albeit at a smaller scale (Fig. 1.15). The area of concern is the upstream part of the drainage basin. In a karst basin, this includes the drainage area of surface streams that sink into the karst (Fig. 4.1). It is rare indeed when an entire basin is completely protected but the basin must be the focus of protection.

The second category is one where the direction of movement of both water and organic carbon is vertical, and includes some interstitial aquifers, epikarst (Fig. 1.7), MSS (Fig. 1.16), and lava tubes (Fig. 1.12). Here the protection focus is the immediate area around the site, such as sink holes above a cave with an epikarst fauna of conservation interest.

The third category is that of deep aquifers, accessed either by wells or caves. For this category, the area of concern is often more nebulous, and often quite large. For example, the US threatened isopod species *Antrolana lira* occurs in groundwater along a linear distance of more than 200 km in the Shenandoah Valley of Virginia and West Virginia (Holsinger *et al.* 1994). Protection of this species requires protection of the regional aquifer. On the other hand, flow rates in deep aquifers are often in the range of centimetres per day and so contaminant plumes move very slowly (Heath 1982).

The fourth category brings us back to cave entrances. Some of the terrestrial cave fauna is dependent on nutrients that come into the system from the entrance. The cricket–beetle interactions discussed in Chapter 5 are an example of such a system. Bats and other species that periodically enter and exit caves are also dependent on the entrance. Once these animals leave the cave, they need a place to forage. The terrestrial fauna of Robber Baron Cave in Texas is disappearing not because the cave is not protected but because there is not an adequate foraging area for crickets that form the base of the food web. In the case of crickets, Taylor *et al.* (2005) found that they used a foraging radius of 100 m around a cave entrance. For species dependent on the carcasses of pack rats, racoons and the like (Table 2.5), they would require a sufficient area for these mammals to be present as well.

10.8 Summary

A critical factor in the biology of the subterranean fauna and one that increases the risk of extinction is its geographical rarity. Endemism even at the scale of a single site is common. Some but not all stygobionts and troglobionts are also numerically rare. Subterranean organisms are also at increased risk of extinction because of low reproductive rates, increased sensitivity to environmental stress, and in the case of bats, because of their propensity to cluster in large numbers in a few caves. Threats to the subterranean fauna are of four general kinds—alteration of the physical habitat (such as quarrying, especially in southeast Asia), changes in water quality and quantity (a global problem), direct changes to the subterranean fauna (such as the effects of human visitation on cave faunas), and global warming. The selection of sites for preservation requires detailed inventory data, but available evidence suggests that a majority of species can be protected at least at one site and that a relatively small percentage of total land area is required. A variety of mechanisms are available for site protection, including listing as a Ramsar wetland and as a UNESCO World Heritage Site. Design of a preserve is highly dependent on the nature of the fauna being protected.

Glossary

Abiotic Non-living.

Accumulation curve The graph of the number of species and the number of samples, typically created by repeated random samples of 1, 2, 3...collections.

Acetogenic Chemoautotrophic process in which acetate is produced in the acetyl-Coenzyme A pathway of carbon fixation.

Adaptation A process of genetic change resulting in improvement of a character with reference to a specific function or feature that has become prevalent in a population because of a selective advantage.

Adaptive radiation Evolutionary divergence of members of a single phylogenetic line into a variety of different adaptive forms.

Allochthonous Originating from outside the habitat.

Allopatric A population or species occupying a geographical region different from that of another population or species.

Alluvial Sediment deposited by a flowing river.

Ambient Surrounding, as in ambient temperature.

Amphibiont Aquatic species or population whose life cycle requires both surface water and groundwater habitats.

Amplexus The grasping and, in *Gammarus*, carrying of females by males prior to fertilization.

Anaerobic Environment devoid of oxygen.

Anchialine (or anchihaline) Subterranean habitats with more or less extensive connections to the sea, and showing noticeable marine as well as terrestrial influences.

Anophthalmy Without an eye.

Aphotic The absence of light.

Apoptosis Programmed cell death.

Aquifer A groundwater reservoir, usually a rock of high permeability capable of delivering water to a well.

Archaea One of the three domains of life. They are similar to Bacteria in lacking a cell nucleus and similar to Eukaryota in the method of genetic transcription and translation.

Area cladogram A diagram that hypothesizes the historical relationships of the areas concerned based on the cladistic relationships of the disjunct taxa living in those areas.

Aridification The drying out of a region; arid regions are strictly defined as areas receiving less than 10 cm of rainfall per year.

Ash-free dry mass The dry mass of organic matter less the weight after oxidizing (ashing) in a high temperature muffle furnace.

Autotroph Organism converting inorganic carbon to organic carbon; 'self-feeding'.

Basalt A dark grey igneous rock containing microscopic crystals. Basalt forms by rapid solidification of lava at or near the land surface.

Biodiversity Any measure of taxonomic or genetic diversity of organisms.

Biofilm A coating on rocks and other surfaces composed of microorganisms, extracellular polysaccharides, other materials that the organisms produce, and particles trapped or precipitated within the biofilm.

Biological species concept Under this concept a species is a group of potentially interbreeding populations that are reproductively isolated from all other populations.

Biospeleology The branch of speleology dealing with subterranean organisms and their habitats.

Bou–Rouch pump A special hand pump designed to collect water samples from shallow interstitial aquifers.

Calcicole Plant that grows best in calcareous soil.

Calcrete Carbonate deposits that form in the soil or in the vicinity of the water table as a result of the evaporation of soil water or groundwater respectively.

Caprock An impermeable rock layer, such as sandstone, covering soluble carbonates and restricting erosion.

Carotenoids Organic pigments produced by plants—xanthophylls and carotenes; with yellow, orange, and red colours.

Catchment The area from which precipitation is collected and exits at a stream or spring; a natural drainage area.

Cave Underground void, generally of a size sufficient for people to enter, and with an aphotic zone.

Chao2 A formula to estimate total species richness based on the number of species occurring in one and two samples.

Chemoautotrophs Organisms deriving nourishment from chemical reactions of inorganic substances, as in sulfur bacteria.

Clade An evolutionary branch in a phylogeny; a group of species with a common ancestor.

Cladistic Pertaining to branching patterns; a cladistic classification classifies organisms on the basis of the historical sequences by which they diverged from common ancestors.

Complementarity The procedure in site selection where sites are picked in the order of the largest number of new species that are added.

Congeneric Species in the same genus.

Conglomerate Composed of the fragments of pre-existing rocks cemented together.

Connate Water trapped in sedimentary rock during its deposition.

Convergence The independent evolution of similar traits in two or more genetically distinct taxa by different genetic and developmental pathways.

Coprophage Organism feeding on dung.

CPOM Coarse particulate organic matter, >1 mm in diameter.

Cryptic species Species within a genus that are morphologically so similar that they cannot be visually distinguished but are genetically distinct.

Cyanobacteria Photosynthetic bacteria, sometimes (incorrectly) called blue-green algae.

Demography The study of the population growth characteristics of populations.

Detritivore Organism feeding on detritus.

Dissolution The process by which a rock or mineral dissolves (usually in water).

DNAPL Dense non-aqueous phase liquids—ones that sink in groundwater.

DOC Dissolved organic carbon.

Dolina Simple closed circular depression with subterranean drainage, and commonly funnel-shaped.

DOM Dissolved organic matter.

Ecosystem The set of biotic and abiotic components in a given environment.

Ecosystem engineers Organisms that by their activities alter their habitat for future generations.

Ecosystem services The goods and services provided to the human population by the natural world that would otherwise have to be provided in another way.

Ecotone The boundary between two communities or habitats.

eDNA DNA released by the organism into the environment.

Effective population size A term in population genetics theory to denote the size of a randomly mating population that would have the same level of inbreeding as that observed in the study population. Effective population size can be affected by many factors, such as population size and mating patterns.

Endemic Pertaining to a taxon that is restricted to the geographical area specified.

Epikarst The highly porous uppermost zone.

Epilithic Growing on rock.

Epiphreatic Lowest level of unsaturated zone, immediately above ground (phreatic) water.

Eutroglophile Facultative permanent resident of subterranean habitats; called troglophiles by some authors.

Eutrophic Pertaining to a large amount of available organic matter or nutrient enrichment.

Evaporites Rocks formed primarily by evaporation of surface water in arid regions, most commonly in lagoons and closed basins. Evaporites include gypsum, anhydrite, and rock salt.

Evapotranspiration The combination of evaporation from water surfaces and transpiration from plants.

Exaptation Adaptation for one function serving for another function.

Exothermic Chemical reaction characterized by the production of heat.

Fitness The ability of an organism with a particular genotype to leave offspring in the next or succeeding generations as compared to that of organisms with other genotypes.

Fluvial Of or pertaining to rivers.

Flux A flow of energy or matter.

Flysch Sequence of interbedded shales and sandstones deposited contemporaneously with mountain building.

FPOM Fine particulate organic matter, between 0.45 μm and 1000 μm (1 mm) in diameter.

Furcula The 'spring-tail' of Collembola that allows them to jump.

Gastrovascular cavity Internal extracellular cavity of some invertebrates, lined by the gastrodermis.

Ghyben–Herzberg lens Fresh water lens on top of salt water; its depth below sea level is approximately 40 times the height of the water table above sea level.

Gibbs free energy The chemical potential that is minimized when a system reaches equilibrium at constant pressure and temperature. As such, it is a convenient criterion of spontaneity for processes with constant pressure and temperature.

Gneiss A metamorphic rock, composed like granite, of quartz, feldspar, or orthoclase, but distinguished from it by its laminated structure.

Gypsum A rock or mineral composed of hydrated calcium sulfate ($CaSO_4 \cdot 2H_2O$) that is deposited mainly in areas where water evaporates.

Habitat The locality in which animal (or plant) lives.

Haplotype A mitochondrial genotype, which is haploid.

Heritability The proportion of variance among individuals for a trait that is attributable to differences in genotype.

Heterozygosity In genetics the frequency of individuals that carry two different genes at the same locus.

Hibernaculum A hibernation site.

Hibernation A state of inactivity and metabolic depression in animals, characterized by lower body temperature, slower breathing, and lower metabolic rate.

Holarctic In the geographical distribution of plants and animals, ones with an extratropical northern distribution. Includes both the Palearctic and Nearctic regions.

Homeotherm An animal that maintains a fairly constant body temperature.

Homoplasy A character shared by a set of species but not present in their common ancestor.

Hotspot An area of relatively high number of species or high number of endemics.

Hydrogenotrophic Bacteria and Archaea that utilize hydrogen as an electron donor in chemoautotrophy.

Hygropetric A steep or vertical rocky surface, covered by a thin layer of moving water.

Hyporheic Interstitial spaces within the sediments of a stream bed; a transition zone between surface water and groundwater.

Hypotelminorheic A persistent wet spot, a kind of perched aquifer; fed by subsurface water in a slight depression in an area of low to moderate slope; rich in organic matter; underlain by a clay layer typically 5 to 50 cm beneath the surface; with a drainage area typically of less than 10 000 m^2; and with a characteristic dark colour derived from decaying leaves which are usually not skeletonized.

Intermittent lake A lake in a karst region that sporadically (intermittently) fills and drains. Also known as a turlough.

Interstitial Spaces between particles.

Karren field Area of superficial solution features in bare bedrock (generally at the land surface).

Karst Landscape in soluble rock where solution rather than erosion is the primary geomorphic agent, typically with caves, sinkholes, and springs.

Karst basin A drainage basin in karst which contributes water to given point on a stream or to a spring.

Lateral line system In fishes, an extra-optic sensory system, including neuromasts, that detects motion.

Lava tube Tubular caves within lava flows.

Lineage A branch of a phylogenetic tree that is both complete and has a single origin.

Littoral The marginal zone of the sea, and in fresh water, the shallow zone that may contain rooted plants.

Meiofauna Assemblage of animals that pass through a 500 μm sieve but are retained by a 40 μm sieve.

Melanoblast A cell, or a precursor to a cell, that produces melanin pigment.

Mesocavern Cavities smaller than caves, between 0.1 and 20 cm in diameter.

Messenger RNA RNA that translates the DNA triplet code into amino acids.

Messinian salinity crisis Time during the Miocene (5.5 to 6.5 million years ago) when the Mediterranean area was landlocked and almost completely dried out, leaving a series of hypersaline lakes.

Methanogenesis Process where Archaea oxidize hydrogen and reduce CO_2 to methane in chemoautotrophy.

Methanotrophy Utilization of methane as a carbon source by Bacteria.

Milieu souterrain superficiel (MSS) Interconnected cracks and crevices in scree slopes and similar habitats.

Monophyletic Having arisen from one ancestral form; in the strictest sense, from one initial population.

Neo-Darwinism Dating from the 1930s, the reconciliation of Darwin's theory of evolution with the facts of genetics.

Neo-Lamarckism Evolutionary theory, largely developed in the late nineteenth century in North America, that emphasized the evolution of acquired characters through use and disuse.

Neoteny Retardation of somatic development, so that sexual maturity is attained in an organism retaining juvenile characters.

Neuromast Sensory receptor, part of the acoustico-lateralis system of aquatic chordates, to detect vibration and movement in the water.

Neutral mutation A genetic mutation that has no advantage or disadvantage to the organism.

Niche The total requirements of a population or species for resources and physical conditions.

Nutrient spiralling Nutrients in streams cycling between abiotic and biotic components while continuously or periodically moving downstream.

Oligotrophic Pertaining to a low amount of available organic matter or nutrients.

Optical vesicle Precursor to eye in developing eye.

Orthogenesis Evolution towards a 'perfect form', determined by factors internal to the organism.

Paedomorphic Precocious sexual maturity in an organism that is still at a morphologically juvenile stage.

Paleodrainage A drainage basin that has subsequently been altered.

Pangaea The supercontinent of the Permian that was composed of essentially all the present continents and major continental islands.

Panhoehoe Volcanic rock with smooth ropy surface formed from the solidification of fluid lavas.

Parapatric Pertaining to species or populations that have contiguous but non-overlapping geographical distributions.

PASCALIS Protocol for the ASsessment and Conservation of Aquatic Life In the Subsurface, which involved seven research groups from six European countries (France, Spain, Portugal, Italy, Slovenia, and Belgium). The main objectives of this programme were to demonstrate the major distribution patterns of subterranean aquatic biodiversity in Europe and to develop operational tools for its assessment and conservation.

Percolating water Water moving vertically from epikarst through the unsaturated zone.

Permeability The ability of rock or soil to permit water or other fluids to pass through.

Photoautotroph Organism that obtains metabolic energy from light by a photochemical process.

Phreatic Below the groundwater table; below the unsaturated zone.

Phreatobiological net Net designed to sample deep groundwater, usually through bores and wells. Also called the Cvetkov net.

Phylogenetic species concept Under this concept a species is a monophyletic group of populations that share a derived (synapomorphic) character.

Phylogeny The genealogy of a group of taxa such as species.

Phylogeography The study of the historical processes that may be responsible for the contemporary geographical distributions of individuals.

Phytophagous Feeding on plants.

Pleiotropy Pertaining to a gene that has more than one phenotypic effect.

POC Particulate organic carbon.

Polje A large spring-fed karst depression with a flat floor commonly covered by river sediment.

Pre-adaptation Possession by an organism of the necessary properties to permit a shift into a new niche or habitat. A structure is pre-adapted if it can assume a new function before it becomes modified itself.

Predictability The ability to predict (forecast) the values of environmental parameters by date or season.

Productivity Biomass produced by autotrophs (primary productivity) or produced by heterotrophs (secondary productivity).

Prokaryote Organisms lacking membrane-bound intracellular organelles, including nuclei.

Quartzite A rock consisting of quartz, the crystalline form of silicon dioxide (SiO_2).

Ramiform Branch-like.

RAPD Randomly Amplified Polymorphic DNA.

Recharge That part of precipitation or surface water that penetrates the Earth's surface and eventually reaches the water table.

Redox A reversible reaction in which one compound is oxidized and another reduced.

Refractory When referring to organic carbon, difficult to metabolize, such as cellulose.

Refugium An area in which climate has remained relatively unchanged while areas surrounding it have changed markedly, and which has served as a refuge for species requiring the particular conditions.

Regressive evolution The loss of morphological and/or behavioural characters that accompanies isolation in caves.

Relic The last survivors of an ancient radiation.

Relict Population of organisms separated from a parent population by some vicarious event.

Resample The statistical procedure of choosing k samples at random out of total of n samples, typically repeatedly. Used for accumulation curves.

Resurgence Spring where a stream, which has a course on the surface higher up, reappears at the surface.

Reverse evolution The change of a character state to a state similar in appearance to an ancestral state, encompassing patterns associated with both reversion and regression.

Riffle In a stream course, an area of shallower, faster-moving water often associated with whitewater. They alternate with pools.

Riparian Pertaining to the banks of a river or stream.

Schiner–Racovitza system Classification of subterranean animals on ecological grounds into troglobionts, troglophiles, and trogloxenes.

Schist A crystalline rock whose component minerals are arranged in a more or less parallel manner.

Seep A small spring where water oozes out of the ground. Often associated with hypotelminorheic habitats.

Sister taxa The two taxa that are most closely (and therefore most recently) related.

Spatial correlation (spatial autocorrelation) When a variable is correlated with itself in other locations. Analogous to temporal autocorrelation.

Spatial subsidy When resources from one system (e.g., surface) are transferred to another adjoining system (e.g., caves).

Speleobiology The branch of biology dealing with subterranean organisms and their habitats.

Speleothem A mineral deposit in a cave; popularly known as formations.

Standard deviation In statistics, the square root of the variance (the mean squared deviation of observations from the mean).

Standard error In statistics, the standard error of the mean is the standard deviation divided by the square root of sample size.

Stochasticity Random variation, as in demographic stochasticity.

Stygobiont Obligate, permanent resident of aquatic subterranean habitats.

Stygofauna Fauna inhabiting the various types of groundwater.

Subtroglophile Obligate or facultative resident of subterranean habitats but associated with surface habitats for some part of its life cycle. Called either trogloxenes or troglophiles by some authors.

Swallet (swallow hole) Hole into which a stream flows.

Sympatric (sympatry) Individuals living in the same local community, close enough to interact.

Teneral Adult insect recently emerged from a pupa and with a soft exoskeleton.

Torpid State of suspended activity, dormant.

Troglobiont Obligate, permanent resident of terrestrial subterranean habitats; used by some authors for aquatic species as well (see stygobiont).

Troglomorphic Pertaining to morphological and behavioural characters that are convergent in subterranean populations.

Troglophile See eutroglophile and subtroglophile.

Trogloxene Species appearing sporadically in subterranean habitats; called accidentals by some authors.

Trophic The nutritional structure of a community, e.g., primary producer, herbivore, carnivore.

Turbulent Fluid flow that contains eddies that allow mixing between adjacent flow paths.

Vadose The zone above the water table in which water moves by gravity and capillarity. Water does not fill all the openings and does not build up pressures greater than atmospheric.

Vicariance Speciation as a result of range disruption, typically the result of some non-biological process.

Würm glacier The last major Pleistocene glaciation in Europe, approximately equivalent to the Wisconsin glaciation in North America.

Literature Cited

Aley, T. (2004). Tourist caves: algae and lampenflora. In J. Gunn, ed. *Encyclopedia of caves and karst science*, pp. 733–4. Fitzroy Dearborn, New York.

Al-fares, W., Bakalowicz, M., Guerin, R.T., and Dukhan, M. (2002). Analysis of the karst aquifer structure of the Lamalou area (Herault, France) with ground penetrating radar. *Journal of Applied Geophysics*, **51**, 97–106.

Aljančič, G. (2008). Jamski laboratorij Tular in človeška ribica. V spomin na očeta, Marka Aljančiča (1933–2007). *Proteus*, **70**, 246–58.

Aljančič, M., Bulog, B., Kranjc, A., Josipovič, D., Sket, B., and Skoberne, P. (1993). *Proteus: the mysterious ruler of the karst darkness*. Vitrum, Ljubljana, Slovenia.

Allred, K. (2012). Kazumura Cave, Hawaii. In W.B. White and D.C. Culver, eds. *Encyclopedia of caves. Second edition*, pp. 438–42. Elsevier/Academic Press, Amsterdam, The Netherlands.

Arechavaleta, M., Sala, L.L., and Oromí, P. (1999). La fauna invertebrada de la Cueva de Felipe Reventón (Icod de los Vinos, Tenerife, Islas Canarias). *Viraea*, **27**, 229–44.

Armbruster, J.W., Niemiller, M.L., and Hart, P.B. (2016). Morphological evolution of the cave-, spring-, and swampfishes of the Amblyopsidae (Percopsiformes). *Copeia*, **104**, 763–77.

Arnedo, M.A., Oromí, P., Múrria, C., Macías-Hernández, M., and Ribera, C. (2007). The dark side of an island radiation: systematics and evolution of troglobiotic spiders of the genus *Dysdera* Latreille (Araneae: Dysderidae) in the Canary Islands. *Invertebrate Systematics*, **21**, 623–60.

Arntzen, J.W., and Sket, B. (1997). Morphometric analysis of black and white European cave salamanders, *Proteus anguinus*. *Journal of Zoology (London)*, **241**, 699–707.

Ashmole, N.P., and Ashmole, M.J. (2000). Fallout of dispersing arthropods supporting invertebrate communities in barren volcanic habitats. In H. Wilkens, D.C. Culver, and W.F. Humphreys, eds. *Subterranean ecosystems*, pp. 269–86. Elsevier Press, Amsterdam, The Netherlands.

Aspiras, A.C., Rohner, N., Martineau, B., Borowsky, R.L., and Tabin, C.J. (2015). Melanocortin 4 receptor mutations contribute to the adaptation of cavefish to nutrient-poor conditions. *Proceedings of the National Academy of Sciences*, **112**, 9668–73.

Audra, P., and Maire, R. (2004). Nakanai caves, Papua New Guinea, In J. Gunn, ed. *Encyclopedia of caves and karst science*, pp. 538–40. Fitzroy Dearborn, New York.

Audra, P., and Palmer, A.N. (2015). Research frontiers in speleogenesis. Dominant processes, hydrogeological conditions and resulting cave patterns. *Acta Carsologica*, **44**, 315–48.

Auler, A.S. (2012). Quartzite caves of South America. In W.B. White and D.C. Culver, eds. *Encyclopedia of caves. Second edition*, pp. 635–9. Elsevier/Academic Press, Amsterdam, The Netherlands.

Auler, A.S., and Piló, L.B. (2015). Caves and mining in Brazil: the dilemma of cave preservation within a mining context. In B. Andreo, F. Carrasco, J.J. Durán, P. Jiménez, and J.W. LaMoreaux, eds. *Environmental earth sciences*, pp. 487–96. Springer, Berlin.

Auler, A.S., Piló, L.B., Senko, J.M., Sasowsky, I.D., and Barton, H.A. (2014). Hypogene cave patterns in iron ore caves: convergence of forms or processes? In A. Klimchouk, I.D. Sasowsky, J. Mylroie, S.A. Engel, A.S. Engel, eds. *Hypogene caves morphologies*, pp. 15–9. Karst Waters Institute Special Publication 18, Leesburg, Virginia.

Balogová, M., Jelić, D., Kyselová, M., and Uhrin, M. (2017). Subterranean systems provide a suitable overwintering habitat for *Salamandra salamandra*. *International Journal of Speleology*, **46**, 321–9.

Ban, F, Tan, M. Cai, B., and Pan, G. (2006). Variations in dissolved organic carbon of cave drip waters in Shihua Cave, Beijing. In B.P. Onac, T. Tămaş, S. Constantin, and A. Perşoiu, eds. *Archives of Climate Change in Karst*, pp. 11–12. Karst Waters Institute Special Publication 10, Charles Town, West Virginia.

Banta, A.M. (1907). The fauna of Mayfield's Cave. *Carnegie Institution of Washington Publications*, **67**, 1–114.

Barbour, R.W., and W.H. Davis. (1969). *Bats of America*. University of Kentucky Press, Lexington, Kentucky.

Bardgett, R. (2005). *The biology of soil. A community and ecosystem approach*. Oxford University Press, Oxford, UK.

Bareth, C., and J. Pages. (1994). Diplura. In C Juberthie and V Decu, eds. *Encylopaedia biospeologica, tome II*, pp. 277–83. Société de Biospéologie, Moulis, France.

Barr, T.C. (1960). A synopsis of cave beetles of the genus *Pseudanophthalmus* of the Mitchell Plain in southern Indiana (Coleoptera, Carabidae). *American Midland Naturalist*, **63**, 307–20.

Barr, T.C. (1967). Ecological studies in the Mammoth Cave system of Kentucky. I. The biota. *International Journal of Speleology*, **3**, 147–204.

Barr, T.C. (1968). Cave ecology and the evolution of troglobites. *Evolutionary Biology*, **2**, 35–102.

Barr, T.C. (1979). The taxonomy, distribution, and affinities of *Neaphaenops*, with notes on associated species of *Pseudanophthalmus*. *American Museum Novitates*, no. 2682, 20 pp.

Barr, T. C. (2004). A Classification and Checklist of the Genus *Pseudanophtalmus* Jeannel (Coleoptera: Carabidae: Trechinae). *Virginia Museum of Natural History Scientific Publication Series, No. 11*, 52 pp. Martinsville, Virginia.

Barton, H.A. (2015). Starving artists: bacterial oligotrophic heterotrophy in caves. In A.S. Engel, ed. *Microbial life of cave systems*, pp. 79–104. De Gruyter, Berlin, Germany.

Begon, M. (1979). *Investigating animal abundance: capture–recapture for biologists*. University Park Press, Baltimore, Maryland.

Behrmann-Godel, J., Nolte, A.W., Kreiselmaier, J., Berka, R., and Freyhof, J. (2017). The first European cave fish. *Current Biology*, **27**, R257–8.

Beltram, G. (2004). Ramsar sites—wetlands of international importance. In J. Gunn, ed. *Encyclopedia of caves and karst science*, pp. 277–83. Fitzroy Dearborn, New York.

Bhullar, K., Waglechner, N., Pawlowski, A., Koteva, K., Banks, E.D., Johnston, M.D., Barton, H.A., and Wright, G.D. (2012). Antibiotic resistance is prevalent in an isolated cave microbiome. *PloS ONE*, **7**, e34953.

Bichuette, M.E., and Trajano, E. (2004). Three new subterranean species of *Ituglanis* from central Brazil (Siluriformes: Trichomycteridae). *Ichthyological Explorations of Freshwaters*, **15**, 243–56.

Bilandžija, H., Ćetković, H., and Jeffery, W.R. (2012). Evolution of albinism in cave planthoppers by a convergent defect in the first step of melanin biosynthesis. *Evolution and Development*, **14**, 196–203.

Bilandžija, H., Lazlo, M., Porter, M.L., and Fong, D.W. (2017). Melanization in response to wounding is ancestral in arthropods and conserved in albino cave species. *Scientific Reports*, **7**, 17148.

Bilandžija, H., Ma, L., Parkhurst, A., and Jeffery, W.R. (2013). A potential benefit of albinism in *Astyanax* cavefish: downregulation of the *oca2* gene increases tyrosine and catecholamine levels as an alternative to melanin synthesis. *PLoS ONE*, **8**, e80823.

Bole, J., and Velkovrh, F. (1986). Mollusca from continental subterranean aquatic habitats. In L. Botosaneanu, ed. *Stygofauna mundi*, pp. 177–206. E.J. Brill, Leiden, The Netherlands.

Borges, P.A., Cardoso, P., Amorim, I.R., Pereira, F.E., Constância, J.P., Nunes, J.C., Barcelos, P., Costa, P., Gabriel, R., and Dapkevicius, M.L. (2012). Volcanic caves: priorities for conserving the Azorean endemic troglobiont species. *International Journal of Speleology*, **41**, 101–12.

Borowsky, R. (2016). Regressive evolution: testing hypotheses of selection and drift. In A.C. Keene, M. Yoshizawa, and S.E. McGaugh, eds. *The biology and evolution of Mexican cavefish*, pp. 93–110. Academic/Elsevier, Amersterdam, The Netherlands.

Bosák, P. (2002). Karst processes from the beginning to the end: how can they be dated? In F. Gabrovšek, ed. *Evolution of karst: from prekarst to cessation*, pp. 191–223. Založba ZRC, Ljubljana, Slovenia.

Boston, P. (2004). Biofilms. In J. Gunn, ed. *Encyclopedia of caves and karst science*, pp. 145–7. Fitzroy Dearborn, New York.

Boston, P.J., Spilde, M.N., Northup, D.E., Melim, L.A., Soroka, D.A., Kleina, L.G., Lavoie, K.H., Hose, L.D., Mallory, L.M., Dahm, C.N., Crossey, L.J., Scheble, R.T. (2001). Cave biosignature suites: microbes, minerals and Mars. *Astrobiology*, **1**, 25–55.

Botosaneanu, L., ed. (1986) *Stygofauna mundi*. E.J. Brill, Leiden, The Netherlands.

Botosaneanu, L., ed. (1998). *Studies in crenobiology. The biology of springs and spring-brooks*. Backhuys Publishers, Leiden, The Netherlands.

Bou C. (1974). Recherches sur les eaux souterraines -25- Les méthodes de récolte dans les eaux souterraines interstitielles. *Annales de Spéléologie*, **29**, 611–9.

Bou, C., and Rouch, R. (1967). Un nouveau champ de recherches sur la faune aquatique souterraine. *Compte Rendus de l'Académie des Sciences de Paris*, **265**, 369–70.

Boutin, C., and Coineau, N. (2000). Evolutionary rates and phylogenetic age in some stygobiontic species. In H. Wilkens, D.C. Culver, and W.F. Humphreys, eds. *Subterranean ecosystems*, pp. 433–51. Elsevier Press, Amsterdam, The Netherlands.

Brandon, R.N. (1990). *Adaptation and environment*. Princeton University Press, Princeton, New Jersey.

Bregović, P., and Zagmajster, M. (2016). Understanding hotspots within a global hotspot—identifying the drivers of regional species richness patterns in terrestrial subterranean habitats. *Insect Conservation and Diversity*, **9**, 268–81.

Broadley, R.A., and Stringer, I.A.N. (2001) Prey attraction by larvae of the New Zealand glowworm, *Arachnocampa luminosa* (Diptera: Mycetophilidae). *Invertebrate Biology*, **120**, 170–7.

Brunet, A.K., and Medellín, R.A. (2001). The species–area relationship in bat assemblages in tropical caves. *Journal of Mammalogy*, **82**, 1114–22.

Buhay, J.E., and Crandall, K.E. (2005). Subterranean phylogeography of freshwater crayfishes shows extensive gene flow and surprisingly large population sizes. *Molecular Ecology*, **14**, 4259–73.

Bulog, B. (2004). Amphibia: *Proteus*. In J. Gunn, ed. *Encyclopedia of caves and karst science*, pp. 62–4. Fitzroy Dearborn, New York.

Buss, S., Cai, Z., Cardenas, B., Fleckenstein, J., Hannah, D., Heppell, K., Hulme, P., Ibrahim, T., Kaeser, D., Krause, S., Lawler, D. Lerner, D., Mant, J., Malcolm, I., Old, G., Parkin, G., Pickup, R., Pinay, G., Porter, J., Rhodes, G., Ritchie, H., Riley, J., Robertson, A., Sear, D., Shields, B., Smith, J., Tellam, J., and Wood, P. (2009). *The hyporheic handbook*. Environmental Agency, Bristol, UK.

Caccone, A. Milinkovitch, M.C., Sbordoni, V., and Powell, J.R. (1997). Mitochondrial rates and biogeography in European newts (genus *Euproctus*). *Systematic Biology*, **46**, 126–44.

Camp, C.D., and Jensen, J.B. (2007). Use of twilight zones of caves by plethodontid salamanders. *Copeia*, **2007**, 594–604.

Carlini, D.B., and Fong, D.W. (2017). The transcriptomes of cave and surface populations of *Gammarus minus* (Crustacea: Amphipoda) provide evidence for positive selection on cave downregulated transcripts. *PLoS ONE*, **12**, e0186173.

Carlini, D.B., Manning, J., Sullivan, P.G., and Fong D.W. (2009). Molecular genetic variation and population structure in morphologically differentiated cave and surface populations of the freshwater amphipod *Gammarus minus*. *Molecular Ecology*, **18**, 1932–45.

Carlini, D.B., Satish, S., and Fong, D.W. (2013). Parallel reduction in expression, but no loss of functional constraint, in two opsin paralogs within cave populations of *Gammarus minus* (Crustacea: Amphipoda). *BMC Evolutionary Biology*, **13**, 89.

Carmichael, S.K., and Bräuer, S.L. (2015). Microbial diversity and manganese cycling: a review of manganese-oxidizing microbial cave communities. In A.S. Engel, ed. *Microbial life of cave systems*, pp. 137–60. De Gruyter, Berlin, Germany.

Cartwright, R.A., Schwartz, R.S., Merry, A.L., and Howell, M.M. (2017). The importance of selection in the evolution of blindness in cavefish. *BMC Evolutionary Biology*, **17**, 45.

Caumartin, V. (1963). Review of the microbiology of underground environments. *Bulletin of the National Speleological Society*, **25**, 1–14.

Chao, A. (1984). Non-parametric estimation of the number of classes in a population. *Scandanavian Journal of Statistics*, **11**, 783–91.

Chapman, P. (1982). The origin of troglobites. *Proceedings of the University of Bristol Spelaeological Society*, **16**, 133–41.

Chapman, P. (1993). *Caves and cave life*. Harper Collins, London.

Chappuis, P.A. (1925). Sur les Copépodes et les Syncarides des eaux souterraines de Cluj et des Monts Bihar. *Bulletin de Sociéte Scientifique Cluj*, **2**, 157–82.

Chen, Y.X., Bai, J., Fang, D., Jiang, W., Qiu, Y., Jiang, W., Yuan, H., Bian, C., Lu, J., He, S., Pan, X., Zhang, Y., Wang, X. You, X., Wang, Y., Sun, Y., Mao, D., Liu, Y., Fan, G., Zhang, H., Chen, X., Zhang, X., Zheng, L., Wang, J., Chen, J., Ruan, Z., Li, J., Yu, H., Peng, C., Ma, X., Xu, J., He, Y., Xu, Z., Xu, P., Wang, P., Yang, H., Wang, J., Whitten, T., Xu, X., and Shi, Q. (2016). The *Sinocyclocheilus* cavefish genome provides insights into cave adaptation. *BMC Biology*, **14**, DOI 10.1186/s12915-015-0223-4.

Christiansen, K.A. (1961). Convergence and parallelism in cave Entomobryinae. *Evolution*, **15**, 288–301.

Christiansen, K.A. (1962). Proposition pour la classification des animaux caverni-coles. *Spelunca*, **2**, 75–8.

Christiansen, K.A. (1965). Behavior and form in the evolution of cave Collembola. *Evolution*, **19**, 529–37.

Christiansen, K.A., and Culver D.C. (1969). Geographical variation and evolution in *Pseudosinella violenta*. *Evolution*, **23**, 602–21.

Christiansen, K.A., and Culver, D.C. (1987). Biogeography and the distribution of cave Collembola. *Evolution*, **19**, 529–37.

Christman, M.C., and Culver, D.C. (2001). The relationship between cave biodiver-sity and available habitat. *Journal of Biogeography*, **28**, 367–80.

Christman, M.C., Culver, D.C., Madden, M., and White, D. (2005). Patterns of endemism of the eastern North American cave fauna. *Journal of Biogeography*, **32**, 1441–52.

Christman M.C., Doctor, D.H., Niemiller, M.L., Weary, D.J., Young, J.A., Zigler, K.S., and Culver, D.C. (2016). Predicting the occurrence of cave-inhabiting fauna based on features of the Earth surface environment. *PLoS ONE*, **11**, e0160408.

Christman, M.C., and Zagmajster, M. (2012). Mapping subterranean biodiversity. In W.B. White and D.C. Culver, eds. *Encyclopedia of caves. Second edition*, pp. 474–81. Elsevier/Academic Press, Amsterdam, The Netherlands.

Cigna, A.A. (2002). Modern trend[s] in cave monitoring. *Acta Carsologica*, **31**, 35–54.

Clements, R., Sodhi, N.S., Schilthuizen, M., and Ng, P.K.L. (2006). Limestone karsts of southeast Asia: imperiled arks of biodiversity. *Bioscience*, **56**, 733–42.

Coiffait, H. (1958). Les coléoptères du sol. *Vie et Milieu Supplement*, **7**, 1–204.

Coineau, N. (1998). Syncarida. In C. Juberthie and V. Decu, eds. *Encylopaedia bios-peologica, tome II*, pp. 863–76. Société de Biospéologie, Moulis, France.

Coineau, N. (2000). Adaptations to interstitial groundwater life. In H. Wilkens, D.C. Culver, and W.F. Humphreys, eds. *Subterranean ecosystems*, pp. 189–210. Elsevier Press, Amsterdam, The Netherlands.

Coineau, N., and Camacho, A. (2004). Crustacea: Syncarida. In J. Gunn, ed. *Encyclopedia of caves and karst science*, pp. 268–70. Fitzroy Dearborn, New York.

Coineau, N., Henry, J.-P., Magniez, G., and Negoescu, I. (1994). Isopoda aquatica. In C Juberthie and V Decu, eds. *Encylopaedia biospeologica, tome I*, pp. 123–40. Société de Biospéologie, Moulis, France.

Cokendolpher, J. C., and Polyak, V. J. (1996). Biology of the caves at Sinkhole Flat, Eddy County, New Mexico. *Journal of Caves and Karst Studies*, **58**, 181–92.

Colwell, R.K. (2016). *EstimateS: Statistical estimation of species richness and shared species from samples. Version 9*. User's Guide and application published at http://purl.oclc.org/estimates.

Colwell, R.K., Chao, A., Gotelli, N.J., Lin, S.-Y., Mao, C.X., Chazdon, R.L., and Longino, J.T. (2012). Models and estimators linking individual-based and sample-based rarefaction, extrapolation, and comparison of assemblages. *Journal of Plant Ecology*, **5**, 3–21.

Colwell, R.K., Mao, C.X., and Chang, J. (2004). Interpolating, extrapolating, and comparing incidence-based species accumulation curves. *Ecology*, **85**, 2717–27.

Cooper, S.J.B., Bradbury, J.H., Saint, K.M., Leys, R., Austin, A.D., and Humphreys, W.F. (2007). Subterranean archipelago in the Australian arid zone: mitochondrial

DNA phylogeography of amphipods from central Western Australia. *Molecular Ecology*, **16**, 1533–44.

Cooper, S.J.B., S. Hinze, S., Leys, R., Watts, C.H.S., and Humphreys, W.F. (2002). Islands under the desert: molecular systematics and evolutionary origins of stygobitic water beetles (Coleoptera: Dytiscidae) from central Western Australia. *Invertebrate Systematics*, **16**, 589–98.

Crandall, K.A., and Hillis, D.M. (1997). Rhodopsin evolution in the dark. *Nature*, **387**, 667–8.

Creuzé des Châtelliers, M., Juget, J., Lafont, M., and Martin, P. (2009). Subterranean aquatic Oligochaeta. *Freshwater Biology*, **54**, 678–90.

Crouau-Roy, B., Crouau, V., and Ferre, C. (1992). Dynamic and temporal structure of the troglobitic beetle *Speonomus hydrophilus* (Coleoptera: Bathysciinae). *Ecography*, **15**, 12–18.

Culver, D.C. (1970). Analysis of simple cave communities. I. Caves as islands. *Evolution*, **24**, 463–74.

Culver, D.C. (1973). Competition in spatially heterogeneous systems: an analysis of simple cave communities. *Ecology*, **54**, 102–10.

Culver, D.C. (1975). The interaction of predation and competition in cave stream communities. *International Journal of Speleology*, **7**, 229–45.

Culver, D.C. (1976). The evolution of aquatic cave communities. *American Naturalist*, **110**, 949–57.

Culver, D.C. (1982). *Cave life*. Harvard University Press, Cambridge, Massachusetts.

Culver, D.C. (1987). Eye morphometrics of cave and spring populations of *Gammarus minus* (Amphipoda: Gammaridae). *Journal of Crustacean Biology*, **7**, 136–47.

Culver, D.C. (1994). Species interactions. In J. Gibert, D.L. Danielopol, and J.A. Stanford, eds. *Groundwater ecology*, pp. 271–85. Academic Press, San Diego, California.

Culver, D.C. (2012a). Molluscs. In W.B. White and D.C. Culver, eds. *Encyclopedia of caves. Second edition*, pp. 512–7. Elsevier/Academic Press, Amsterdam, The Netherlands.

Culver, D.C. (2012b). Species interactions. In W.B. White and D.C. Culver, eds. *Encyclopedia of caves. Second edition*, pp. 743–8. Elsevier/Academic Press, Amsterdam, The Netherlands.

Culver, D.C. (2012c). Life history evolution. In W.B. White and D.C. Culver, eds. *Encyclopedia of caves. Second edition*, pp. 465–8. Elsevier/Academic Press, Amsterdam, The Netherlands.

Culver, D.C., Christman, M.C., Elliott, W.R., Hobbs III, H.H., and Reddell, J.R. (2003). The North American obligate cave fauna: regional patterns. *Biodiversity and Conservation*, **12**, 441–68.

Culver, D.C., Christman, M.C., Sket, B., and Trontelj, P. (2004b). Sampling adequacy in an extreme environment: species richness patterns in Slovenian caves. *Biodiversity and Conservation*, **13**, 1209–29.

Culver, D.C., Deharveng, L., Bedos, A., Lewis, J.J., Madden, M., Reddell, J.R., Sket, B., Trontelj, P., and White, D. (2006b). The mid-latitude biodiversity ridge in terrestrial cave fauna. *Ecography*, **29**, 120–8.

Culver, D.C., and Ehlinger, T.J. (1982). Determinants of size of two subterranean isopods *Caecidotea cannulus* and *Caecidotea holsingeri* (Isopoda: Asellidae). *Polskie Archirum Hydrobiologii*, **29**, 463–70.

Culver, D.C., Fong, D.W., and Jernigan, R.W. (1991). Species interactions in cave stream communities: experimental results and microdistribution effects. *American Midland Naturalist*, **126**, 364–79.

Culver, D.C., and Hobbs III, H.H. (2017). Biodiversity of Mammoth Cave. In H.H. Hobbs III, R.A. Olson, R.G. Winkler, and D.C. Culver, eds. *Mammoth Cave. A Human and Natural History*, pp. 227–34. Springer, Cham, Switzerland.

Culver, D.C., Hobbs III, H.H., Christman, M.C., and Master, L.L. (1999). Distribution map of caves and cave animals of the United States. *Journal of Cave and Karst Studies*, **61**, 139–40.

Culver, D.C., and Holsinger, J.R. (1992). How many species of troglobites are there? *Bulletin of the National Speleological Society*, **54**, 79–80.

Culver, D.C., Holsinger, J.R., and Feller, D.J. (2012). The fauna of seepage springs and other subterranean habitats in the mid-Atlantic Piedmont and Coastal Plain. *Northeastern Naturalist*, **19**, (Monograph 9), 1–42.

Culver, D.C., Jones, W.K., Fong, D.W., and Kane, T.C. (1994). Organ Cave karst basin. In J. Gibert, D.L. Danielopol, and J. Stanford, eds. *Groundwater Ecology*, pp. 451–73. Academic Press, San Diego, California.

Culver, D.C., Jones, W.K., and Holsinger, J.R. (1992). Biological and hydrological investigation of the Cedars, Lee County, Virginia, an ecologically significant and threatened karst area. In J.A. Stanford and J.J. Simons, eds. *Proceedings of the first international conference on groundwater ecology*, pp. 281–90. American Water Resources Association, Bethesda, Maryland.

Culver, D.C., Kane, T.C., and Fong, D.W. (1995). *Adaptation and natural selection in caves*. Harvard University Press, Cambridge, Massachusetts.

Culver, D.C., Master, L.L., Christman, M.C., and Hobbs III, H.H. (2000). Obligate cave fauna of the 48 contiguous United States. *Conservation Biology*, **14**, 386–401.

Culver, D.C., and Pipan, T. (2008a). Superficial subterranean habitats—gateway to the subterranean realm? *Cave and Karst Science*, **35**, 5–12.

Culver, D.C., and Pipan, T. (2008b). Caves as islands. In R Gillespie, ed. *Encyclopedia of islands*, pp. 150–3. University of California Press, Berkeley, California.

Culver, D.C., and Pipan, T. (2011). Redefining the extent of the aquatic subterranean biotope—shallow subterranean habitats. *Ecohydrology*, **4**, 721–30.

Culver, D.C., and Pipan, T. (2013). Subterranean ecosystems. In S.A. Levin, ed. *Encyclopedia of biodiversity. Second edition, volume 7*, pp. 49–62. Elsevier, Amsterdam, The Netherlands.

Culver, D.C., and Pipan, T. (2014). *Shallow subterranean habitats. Ecology, evolution, and conservation*. Oxford University Press, Oxford.

Culver, D.C., Pipan, T., and Gottstein, S. (2006a). Hypotelminorheic—a unique freshwater habitat. *Subterranean Biology*, **4**, 1–8.

Culver, D.C., Pipan, T., and Schneider, K. (2009). Vicariance, dispersal and scale in the aquatic subterranean fauna of karst regions. *Freshwater Biology*, **54**, 918–29.

Culver, D.C., and Shear, W.A. (2012). Myriapods. In W.B. White and D.C. Culver, eds. *Encyclopedia of caves. Second edition*, pp. 438–41. Elsevier/Academic Press, Amsterdam, The Netherlands.

Culver, D.C., and Sket, B. (2000). Hotspots of subterranean biodiversity in caves and wells. *Journal of Cave and Karst Studies*, **62**, 11–7.

Curl, R. (1966). Caves as a measure of karst. *Journal of Geology*, **74**, 798–830.

Cvetkov, L. (1968). Un filet phréatobiologique. *Bulletin de l'Institut de Zoologie et Musee de Academie Bulgare Sciences*, **27**, 215–8.

Čučković, S. (1983). The influence of the change in the water-course regime of the Trebišnjica water-system on the fauna of underground karst regions. *Naš Krš* (Sarajevo), **9**, 129–42.

Daily, G.C., ed. (1997). *Nature's services. Societal dependence on natural ecosystems.* Island Press, Washington, DC.

Danielopol, D.L. (1981). Distribution of ostracods in the groundwater of the north western coast of Euboea (Greece). *International Journal of Speleology*, **11**, 91–104.

Danielopol, D.L., Griebler, C., Gunatilka, A., and Notenboom, J. (2003). Present state and future prospects for groundwater ecosystems. *Environmental Conservation*, **30**, 104–30.

Danielopol, D.L., Marmonier, P., Boulton, A.J., and Bonaduce, G. (1994). World subterranean ostracod biogeography: dispersal or vicariance. *Hydrobiologia*, **287**, 119–29.

Danielopol, D.L., and Pospisil, P. (2001). Hidden biodiversity in the groundwater of the Danube Flood Plain National Park, Austria. *Biodiversity and Conservation*, **10**, 1711–21.

Danielopol, D.L., Pospisil, P., and Dreher, J. (2001). Structure and functioning of groundwater ecosystems in a Danube wetland at Vienna. In C. Griebler, D.L. Danielopol, J. Gibert, H.P. Nachtnebel, and J. Notenboom, eds. *Groundwater ecology. A tool for management of water resources*, pp. 121–42. European Communities, Luxembourg.

Danielopol, D.L., Pospisil, P., Dreher, J., Mösslacher, F., Torreiter, P., Geiger-Kaiser, M., and Gunatilaka, A. (2000). A groundwater ecosystem in the Danube wetlands at Wien (Austria). In H. Wilkens, D.C. Culver, and W.F. Humphreys, eds. *Subterranean ecosystems*, pp. 481–511. Elsevier Press, Amsterdam, The Netherlands.

Danielopol, D.L., and Rouch, R. (2012). Invasion, active versus passive. In W.B. White and D.C. Culver, eds. *Encyclopedia of caves. Second edition*, pp. 404–9. Elsevier/Academic Press, Amsterdam, The Netherlands.

Darwin, C. (1859). *On the origin of species by means of natural selection, or the preservation of favoured races in the struggle for life. First edition.* John Murray, London.

Dasher, G.R. (2001). *The caves and karst of Pendleton County.* West Virginia Speleological Survey Bulletin 15. Barrackville, West Virginia.

Datry, T., Malard, F., and Gibert, J. (2005). Response of invertebrate assemblages to increased groundwater recharge rates in a phreatic aquifer. *Journal of the North American Benthological Society*, **24**, 461–77.

Day, M., and Mueller, B. (2005). Aves (Birds). In J. Gunn, ed. *Encyclopedia of caves and karst science*, pp. 130–1. Fitzroy Dearborn, New York.

Decu, V. (1981). Quelques aspects de la biospéologiques cubano-roumaines à Cuba. In *Résultats des expéditions biospéologiques Cubano-Roumaines à Cuba. Vol 3*, pp. 9–16. Editura Academiei Republicii Socialiste România, Bucharest, Romania.

Decu, V., and Juberthie, C. (1998). Coléoptères (généralités et synthèse). In C. Juberthie and V. Decu, eds. *Encylopaedia biospeologica, tome II*, pp. 1025–30. Société de Biospéologie, Moulis, France.

Decu, V., and Juberthie, C. (2004). Insecta: Coleoptera. In J. Gunn, ed. *Encyclopedia of caves and karst science*, pp. 447–51. Fitzroy Dearborn, New York.

Deharveng, L., and Bedos, A. (2000). The cave fauna of southeast Asia. Origin, evolution, and ecology. In H. Wilkens, D.C. Culver, and W.F. Humphreys, eds. *Subterranean ecosystems*, pp. 603–32. Elsevier Press, Amsterdam, The Netherlands.

Deharveng, L., and Bedos, A. (2004). Salukkan Kallang, Indonesia: biospeleology. In J. Gunn, ed. *Encyclopedia of caves and karst science*, pp. 631–3. Fitzroy Dearborn, New York.

Deharveng, L., and Bedos, A. (2012). Diversity patterns in the tropics. In W.B. White and D.C. Culver, eds. *Encyclopedia of caves. Second edition*, pp. 238–50. Elsevier/ Academic Press, Amsterdam, The Netherlands.

Deharveng, L., and Bedos, A. (2019). Diversity patterns in the tropics. In W.B. White, D.C. Culver, and T. Pipan, eds. *Encyclopedia of caves. Third edition*. Elsevier/ Academic Press, Amsterdam, The Netherlands.

Deharveng, L., Gibert, J., and Culver, D.C. (2012). Diversity patterns in Europe. In W.B. White and D.C. Culver, eds. *Encyclopedia of caves. Second edition*, pp. 219–28. Elsevier/Academic Press, Amsterdam, The Netherlands.

Deharveng, L., Stoch, F., Gibert, J., Bedos, A., Galassi, D., Zagmajster, M., Brancelj, A., Camacho, A., Fiers, F., Martin, P., Giani, N., Magniez, G., and Marmonier, P. (2009). Groundwater biodiversity in Europe. *Freshwater Biology*, **54**, 709–26.

Deharveng, L., and Thibaud, M. (1989). Acquisitions récentes sur les Insectes Collemboles cavernicoles d'Europe. *Mémoires de Biospéologie*, **16**, 145–61.

Deleurance-Glaçon, M. (1963). Recherches sur les coleopteres troglobies da la sous-familie die Bathysciinae. *Annales de Sciences Naturelles, Zoologie*, **5**, 1–172.

Desai, M.S., Assig, K., and Dattagupta, S. (2013). Nitrogen fixation in distinct microbial niches within a chemoautotrophically-driven cave ecosystem. *ISME Journal*, **7**, 2411–23.

Desutter-Grandcolas, L., and Grandcolas, P. (1996). The evolution toward troglobitic life: a phylogenetic reappraisal of climatic relict and local habitat shift hypotheses. *Mémoires de Biospéologie*, **23**, 57–63.

Dickson, G.W., and Kirk Jr, P.W. (1976). Distribution of heterotrophic microorganisms in relation to detritivores in Virginia caves (with supplemental bibliography on cave mycology and microbiology). In B.C. Parker and M.K. Roane, eds. *The distribution history of the biota of the southern Appalachians. Part IV. Algae and Fungi*, pp. 205–26. University Press of Virginia, Charlottesville.

Di Russo, C., and Sbordoni, V. (1998). Gryllacridoidea. In C. Juberthie and V. Decu, eds. *Encyclopaedia Biospeologica, Tome II*, pp. 976–88. Société de Biospéologie, Moulis, France.

Dobat, K. (1998). Flore (Lichens, Bryophytes, Pteridophytes, Spermatophyes). In C. Juberthie and V. Decu, eds. *Encyclopaedia Biospeologica, Tome II*, pp. 1311–24. Société de Biospéologie, Moulis, France.

Doledec, S., Chessel, D., and Gimaret-Carpentier, C. (2000). Niche separation in community analysis: a new method. *Ecology*, **81**, 2914–27.

Dole-Olivier, M.-J., Castellarini, F., Coineau, N., Galassi, D.M.P., Martin, P., Mori, N., Valdecasas, A., and Gibert, J. (2009a). Towards an optimal sampling strategy to assess groundwater biodiversity: comparison across six regions of Europe. *Freshwater Biology*, **54**, 777–96.

Dole-Olivier, M.-J., Malard, F., Martin, D., Lefébure, T., and Gibert, J. (2009b). Relationships between environmental gradients and stygobiotic biodiversity in ground waters. Species requirements at a regional scale. *Freshwater Biology*, **54**, 797–813.

Dole-Olivier, M.-J., and Marmonier, P. (1992). Patch distribution of interstitial communities: prevailing factors. *Freshwater Biology*, **27**, 177–91.

Dole-Olivier, M.-J., Marmonier, P., Creuzé des Châtelliers, M., and Martin, D. (1994). Interstitial fauna associated with the alluvial floodplain of the Rhône River (France). In J. Gibert, D.L. Danielopol, and J.A. Stanford, eds. *Groundwater ecology*, pp. 313–46. Academic Press, San Diego, California.

Dreybrodt, W., and Gabrovšek, F. (2002). Basic processes and mechanisms governing the evolution of karst. In F. Gabrovšek, ed. *Evolution of karst from prekarst to cessation*, pp. 115–54. Založba ZRC, Ljubljana, Slovenia.

Dreybrodt, W., Gabrovšek, F., and Romanov, D. (2005). *Processes of speleogenesis: a modeling approach*. Založba ZRC, Ljubljana, Slovenia.

Dudich, E. (1930). Az Aggteleki-barlang állatvilágának élelemforrásai. *Állattani Közlemények*, **27**, 62–85.

Dudich, E. (1932–1933). Die speläobiologische Station zu Postumia und ihre Bedeutung für die Höhlenkunde. *Speläologische Jahrbuch*, **13–14**, 51–71.

Dumnicka, E. (2012). Worms. In W.B. White and D.C. Culver, eds. *Encyclopedia of caves. Second edition*, pp. 910–6. Elsevier/Academic Press, Amsterdam, The Netherlands.

Eberhard, S.M., Halse, S.A., Williams, M.R., Scanlon, M.D., Cocking, J.S., and Barron, H.J. (2009). Exploring the relationship between sampling efficiency and short range endemism for groundwater fauna in the Pilbara region, Western Australia. *Freshwater Biology*, **54**, 885–901.

Egemeier, S. J. (1981). Cavern development by thermal waters. *Bulletin of the National Speleological Society*, **43**, 31–51.

Eigenmann, C.H. (1909). *Cave vertebrates of America. A study in degenerative evolution*. Carnegie Institution of Washington, Washington, D.C.

Elliott, W.R. (1981). Damming up the caves. *Caving International*, **10**, 38–41.

Elliott, W.R. (2000). Conservation of the North American cave and karst biota. In H. Wilkens, D.C. Culver, and W.F. Humphreys, eds. *Subterranean ecosystems*, pp. 665–89. Elsevier Press, Amsterdam, The Netherlands.

Elliott, W.R. (2012). Protecting caves and cave life. In W.B. White and D.C. Culver, eds. *Encyclopedia of caves. Second edition*, pp. 624–33. Elsevier/Academic Press, Amsterdam, The Netherlands.

Elliott, W.R., Reddell, J.R., Rudolph, D.C., Graening, G.O., Briggs, T.L., Ubick, D., Aalbu, R.L., Krejca, J., and Taylor, S.J. (2017). The cave fauna of California. *Proceedings of the California Academy of Sciences*, **64**, 1–311.

Emblanch, C., Blavoux, B., Puig, J., and Mudry, J. (1998). Dissolved organic carbon of infiltration with the autogenic karst hydrosystem. *Geophysical Research Letters*, **25**, 1459–62.

Eme, D., Malard, F., Konecny-Dupré, L, Lefébure, T., and Douady, C.J. (2013). Bayesian phylogeographic inferences reveal contrasted colonization dynamics among European groundwater isopods. *Molecular Ecology*, **22**, 5865–99. doi 10.1111/mec 12520.

Eme, D., Zagmajster, M., Fišer, C., Galassi, D., Marmonier, P., Stoch, F., Cornu, J.F., Oberdorff, T., and Malard, F. (2015). Multi-causality and spatial non-stationarity in the determinants of groundwater crustacean diversity in Europe. *Ecography*, **38**, 531–40.

Engel, A.S. (2007). Observations on the biodiversity of sulfidic karst habitats. *Journal of Cave and Karst Studies*, **69**, 187–206.

Engel, A.S. (2012a). Chemoautotrophy. In W.B. White and D.C. Culver, eds. *Encyclopedia of caves. Second edition*, pp. 125–34. Elsevier/Academic Press, Amsterdam, The Netherlands.

Engel, A.S. (2012b). Microbes. In W.B. White and D.C. Culver, eds. *Encyclopedia of caves. Second edition*, pp. 490–9. Elsevier, Amsterdam, The Netherlands.

Engel, A.S., ed. (2016). *Microbial life of cave systems*. De Gruyter, Berlin.

Engel, A.S., Lee, N., Porter, M.L., Stern, L.A., Bennett, P.C., and Wagner, M. (2003). Filamentous '*Epsilonproteobacteria*' dominate microbial mats from sulfidic cave springs. *Applied and Environmental Microbiology*, **69**, 5503–11.

Engel, A.S., Porter, M.L., Kinkle, B.K., and Kane, T.C. (2001). Ecological assessment and geological significance of microbial communities from Cesspool Cave, Virginia. *Geomicrobiology Journal*, **18**, 259–74.

Engel, A.S., Stern, L.A., and Bennett, P.C. (2004). Microbial contributions to cave formation: new insights into sulfuric acid speleogenesis. *Geology*, **32**, 369–72.

Esmaeili-Rineh, S., Sari, A., Delić, T., Moškrič, A., and Fišer, C. (2015). Molecular phylogeny of the subterranean genus *Niphargus* (Crustacea: Amphipoda) in the Middle East: a comparison with European Niphargids. *Zoological Journal of the Linnean Society*, **175**, 812–26.

Espinasa, L., and Espinasa, M. (2016). Hydrogeology of caves in the Sierra del Abra region. In A.C. Keene, M. Yoshizawa, and S.E. McGaugh, eds. *The biology and evolution of Mexican cavefish*, pp. 41–58. Academic/Elsevier, Amersterdam, The Netherlands.

Evans, A.M. (1982). The Hart's tongue fern—an endangered plant in cave entrances. In R.C. Wilson and J.J. Lewis, eds. *National Cave Management Symposium proceedings, Carlsbad, New Mexico 1978 and Mammoth Cave, Kentucky 1980*, pp. 143–45. Pygmy Dwarf Press, Oregon City, Oregon.

Fagan, W.F., Lutscher, F., and Schneider, K. (2007). Population and community consequences of spatial subsidies derived from central-place foraging. *American Naturalist*, **170**, 902–15.

Faille, A., Bourdeau, C., Belles, X., and Fresneda, J. (2015). Allopatric speciation illustrated: The hypogean genus *Geotrechus* Jeannel, 1919 (Coleoptera: Carabidae: Trechini), with description of four new species from the Eastern Pyrenees (Spain). *Arthropod Systematics and Phylogeny*, **73**, 439–55.

Faille, A., Ribera, I., Deharveng, L., Bourdeau, C., Garnery, L., Queinnec, E., and Deuve, T. (2010). A molecular phylogeny shows the single origin of the Pyrenean subterranean Trechini ground beetles (Coleoptera: Carabidae). *Molecular Phylogenetics and Evolution*, **54**, 97–105.

Falconer, D.S., and McKay, T.F.C. (1996). *Introduction to quantitative genetics. Fourth edition*. Benjamin Cummings, San Francisco, California.

Fanenbruck, M., Herzsch, S., and Wägele, J.W. (2004). The brain of Remipedia (Crustacea) and an alternative hypothesis on their phylogenetic relationships. *Proceedings of the National Academy of Sciences*, **101**, 3868–73.

Fenolio, D.B. (2016). *Life in the dark: Illuminating biodiversity in the shadowy haunts of planet Earth*. Johns Hopkins University Press, Baltimore, Maryland.

Fenolio, D.B., Graening, G.O., Collier, B.A., and Stout, J.F. (2006). Coprophagy in a cave-adapted salamander; the importance of bat guano examined through nutritional and stable isotope analysis. *Proceedings of the Royal Society B: Biological Sciences*, **273**, 439–43.

Ferreira D. (2005). *Biodiversité aquatique souterraine de France: base de données, patrons de distribution et implications en termes de conservation*. Ph.D. Thesis, Lyon 1 University, Lyon, France.

Ferreira, D., Malard, F., Dole-Olivier, M.J., and Gibert, J. (2007). Obligate groundwater fauna of France: diversity patterns and conservation implications. *Biodiversity and Conservation*, **16**, 567–96.

Ferreira, R.L., Prous, X., and Martins, R.P. (2007). Structure of bat guano communities in a dry Brazilian cave. *Tropical Zoology*, **20**, 55–74.

Ficetola, G.F., Miaud, C., Pompanon, F., and Taberlet, P. (2008). Species detection using environmental DNA from water samples. *Biology Letters*, **4**, 423–5.

Fišer, C. (2012). *Niphargus*: a model system for evolution and ecology. In W.B. White and D.C. Culver, eds. *Encyclopedia of caves. Second edition*, pp. 555–64. Elsevier/ Academic Press, Amsterdam, The Netherlands.

Fišer, C., Blejec, A., and Trontelj, P. (2012). Niche-based mechanisms operating within extreme habitats: a case study of subterranean amphipod communities. *Biology Letters*, **8**, 578–81.

Fišer, C., Keber, R., Kereži, V., Moškrič, A., Palandančić, A., Petkovska, H., Potočnik, H., and Sket, B. (2007). Coexistence of species of two amphipod genera: *Niphargus timavi* (Niphargidae) and *Gammarus fossarum* (Gammaridae). *Journal of Natural History*, **41**, 2641–51.

Fišer, C, Trontelj, P., and Sket, B. (2006). Phylogenetic analysis of the *Niphargus orcinus* complex (Crustacea: Amphipoda: Niphargidae) with description of new taxa. *Journal of Natural History*, **40**, 2265–315.

Fišer, Ž., Novak, L., Luštrik, R., and Fišer, C. (2016). Light triggers habitat choice of eyeless subterranean but not of eyed surface amphipods. *Science Natura*, **103**, 7.

Fong, D.W. (1989). Morphological evolution of the amphipod *Gammarus minus* in caves: quantitative analysis. *American Midland Naturalist*, **121**, 361–78.

Fong, D.W. (2003). Intermittent pools at headwaters of subterranean drainage basins as sampling sites for epikarst fauna. In W.K. Jones, D.C. Culver, and J.S. Herman, eds. *Epikarst. Proceedings of the symposium held October 1 through 4, 2003, Sheperdstown, West Virginia, USA*, pp. 114–8. Karst Waters Institute Special Publication 9, Charles Town, West Virginia.

Fong, D.W., and Culver, D.C. (1994). Fine-scale biogeographic differences in the Crustacean fauna of a cave stream. *Hydrobiologia*, **287**, 29–37.

Fong, D.W., Kane, T.C., and Culver, D.C. (1995). Vestigialization and loss of nonfunctional characters. *Annual Review of Ecology and Systematics*, **26**, 249–68.

Ford, D., and Williams, P. (2007). *Karst hydrogeology and geomorphology.* John Wiley & Sons, New York.

Frederickson, J.K., Garland, T.R., Hicks, R.J., Thomas, J.M., Li, S.W., and McFadden, S.M. (1989). Lithotrophic and heterotrophic bacteria in deep subsurface sediments and their relation to sediment properties. *Geomicrobiology Journal*, **7**, 53–66.

Frick, W.F., Pollock, J.F., Hicks, A.C., Langwig, K.E., Reynolds, B.S., and Turner, G.G. (2010). An emerging disease causes regional population collapse of a common North American bat species. *Science*, **329**, 679–82.

Friedrich, M., Chen, R., Daines, B., Bao, R., Caravas, J., Rai, P. K., Zagmajster, M., and Peck, S. B. (2011). Phototransduction and clock gene expression in the troglobiont beetle *Ptomaphagus hirtus* of Mammoth Cave. *Journal of Experimental Biology*, **214**, 3532–41.

Galassi, D.M., Huys, R., and Reid, J.W. (2009). Diversity, ecology, and evolution of groundwater copepods. *Freshwater Biology*, **54**, 691–708.

Gallaõ, J.E., and Bichuette, M.E. (2018), Brazilian obligatory subterranean fauna and threats to the hypogean environment. *ZooKeys*, **746**, 1–23.

Gautier, F., Clauzon, G., Suc, J.P., Cravatte, J., and Violanti, D. (1994). Age and duration of the Messinian salinity crisis. *Compte Rendu de Academie de Sciences, Paris* (IIA), **318**, 1103–9.

Gerić, B., Pipan, T., and Mulec, J. (2004). Diversity of culturable bacteria and meio-fauna in the epikarst of Škocjanske jame caves (Slovenia). *Acta Carsologica*, **33**, 301–9.

Gers, C. (1992). *Ecologie et biologie des Arthropodes terrestres du milieu souterrain superficiel fonctionnement et ecologie evolutive*. Ph.D. Dissertation, Université Paul Sabatier de Toulouse, France.

Gers, C. (1998). Diversity of energy fluxes and interactions between arthropod communities, from soil to cave. *Acta Oecologia*, **19**, 205–13.

Gibert J. (1986). Ecologie d'un systeme karstique jurassien. Hydrogéologie, dérive animale, transits de matières, dynamique de la population de *Niphargus* (Crustacé Amphipode). *Mémoires de Biospéologie*, **13**, 1–379.

Gibert, J. (1991). Groundwater systems and their boundaries: conceptual framework and prospects in groundwater ecology. *Verhaltlungen der Internationalen Vereinigung für Theoretische und Angewandte Limnologie*, **24**, 1605–8.

Gibert, J., ed. (2005). *World subterranean biodiversity. Proceedings of an international symposium held on 8–10 December in Villeurbanne, France*. Equipe Hydrobiologie et Ecologie Souterraines, Université Claude Bernard I, Villeurbanne, France.

Gibert, J., Danielopol, D.L., and Stanford, J.A., eds. (1994). *Groundwater ecology*. Academic Press, San Diego, California.

Gibert, J., and Deharveng, L. (2002). Subterranean ecosystems: a truncated functional biodiversity. *Bioscience*, **52**, 473–81.

Gibert, J., Dole-Olivier, M.J., Marmonier, P., and Vervier, P. (1990). Surface water/groundwater ecotones. In R.J. Naiman and H. Décamps, eds. *Ecology and management of aquatic-terrestrial ecotones*, pp. 199–225. Parthenon Publishing, Carnforth, United Kingdom.

Ginet, R., and David, J. (1963). Présence de *Niphargus* (Amphipode Gammaridae) dans certaines eaux épigées des forêts de la Dombes (départment de l'Ain, France). *Vie et Milieu*, **14**, 299–310.

Glazier, D.S., Horne, M.T., and Lehman, M.E. (1992). Abundance, body composition, and reproductive output of *Gammarus minus* (Crustacea: Amphipoda) in ten cold springs differing in pH and ionic content. *Freshwater Biology*, **28**, 149–63.

Gnaspini, P., and Trajano, E. (2000). Guano communities in tropical caves. In H. Wilkens, D.C. Culver, and W.F. Humphreys, eds. *Subterranean ercosystems*, pp. 251–68. Elsevier Press, Amsterdam, The Netherlands.

Goricki, Š. (2006). Filogeogratska in morfološka populacij močerila (*Proteus anguinus*). Ph.D. Dissertation, University of Ljubljana, Ljubljana, Slovenia.

Goricki, Š., Niemiller, M.L., and Fenolio, D.B. (2012). Salamanders. In W.B. White and D.C. Culver, eds. *Encyclopedia of caves. Second edition*, pp. 665–76. Elsevier/Academic Press, Amsterdam, The Netherlands.

Goricki, Š., Stanković, D., Snoj, A., Kuntner, M., Jeffery, W.R., Trontelj, P., Pavićević, M., Grizelj, Z., Năpăruş-Aljančič, M., and Aljančič, G. (2017). Environmental DNA in subterranean biology: range extension and taxonomic implications for *Proteus*. *Scientific Reports*, **7**, 45054.

Goricki, Š., and Trontelj, P. (2006). Structure and evolution of the mitochondrial control region and flanking sequences in the European cave salamander *Proteus anguinus*. *Gene*, **378**, 31–41.

Gould, S.J., and Lewontin, R.C. (1979). The spandrels of San Marcos and the Panglossian paradigm: a critique of the adaptationist programme. *Proceedings of the Royal Society, Series B*, **205**, 581–98.

Gourbault, N. (1994). Turbellaria, Tricladida. In C. Juberthie and V. Decu, eds. *Encyclopaedia Biospeologica, Tome I*, pp. 41–4. Société Internationale de Biospéologie, Moulis, France.

Graening, G.O., and Brown, A.V. (2003). Ecosystem dynamics and pollution effects in an Ozark cave stream. *Journal of the American Water Resources Association*, **39**, 1497–505.

Graham, R.E. (1968). The twilight moth, *Triphosa haesitata* (Lepidoptera: Geometridae) from California and Nevada caves. *Caves and Karst*, **10**, 41–8.

Grant, P.R. (1986). *Ecology and evolution of Darwin's finches*. Princeton University Press, Princeton, New Jersey.

Griebler, C., and Avramov, M. (2015). Groundwater ecosystem services: a review. *Freshwater Science*, **34**, 355–67.

Griffith, D.M., and Poulson, T.L. (1993). Mechanisms and consequences of intraspecific competition in a carabid cave beetle. *Ecology*, **74**, 1373–83.

Groom, M.J., Meffe, G.K., and Carroll, C.R. (2005). *Principles of conservation biology. Third edition*. Sinauer Associates, Sunderland, Massachusetts.

Guéorguiev, V.B. (1977). *La faune troglobie terrestre de la péninsule Balkanique. Origine, formation et zoogéographie*. Editions de l'Academie Bulgare des Sciences, Sofija, Bulgaria.

Gulden, R. (2017). http://www.caverbob.com/wdeep.htm. Accessed October 15, 2017.

Guzik, M.T., Austin, A.D., Cooper, S.J.B., Harvey, M.S., Humphreys, W.F., Bradford, T., Eberhard, S.M., King, R.A., Leys, R., Muirhead, K.A., and Tomlinson, M. (2011). Is the Australian subterranean fauna uniquely diverse? *Invertebrate Systematics*, **24**, 407–18.

Guzik, M.T., Cooper, S.J.B., Humphreys, W.F., and Austin, A.D. (2009). Fine-scale comparative phylogeography of a sympatric sister species triplet of subterranean diving beetles from a single calcrete aquifer in Western Australia. *Molecular Ecology*, **18**, 3683–98.

Hamilton-Smith, E. (2004). World heritage sites. In J. Gunn, ed. *Encyclopedia of caves and karst science*, pp. 777–9. Fitzroy Dearborn, New York.

Hamilton-Smith, E., and Eberhard, S. (2000). Conservation of cave communities in Australia. In H. Wilkens, D.C. Culver, and W.F. Humphreys, eds. *Subterranean ecosystems*, pp. 647–64. Elsevier Press, Amsterdam, The Netherlands.

Harvey, M.S. (2002). Short range endemism among the Australian fauna: some examples from non-marine environments. *Invertebrate Systematics*, **16**, 555–70.

Harvey, M.S., Shear, W.A., and Hoch, H. (2000). Onychophora, Arachnida, Myriapods, and Insecta. In H. Wilkens, D.C. Culver, and W.F. Humphreys, eds. *Subterranean ecosystems*, pp. 79–94. Elsevier Press, Amsterdam, The Netherlands.

Hawes, R.S. (1939). The flood factor in the ecology of caves. *Journal of Animal Ecology*, **8**, 1–5.

Heath, R.C. (1982). *Basic ground-water hydrology*. U.S. Geological Survey Water Supply Paper No. 2220.

Helf, K.L. (2003). *Foraging ecology of the cave cricket* Hadenoecus subterraneus: *effects of climate, ontogeny and predation*. Ph.D. Dissertation, University of Illinois at Chicago, Chicago, Illinois.

Helf, K.L., and Olson, R.A. (2017). Subsurface aquatic ecology of Mammoth Cave, pp. 209–26. In H.H. Hobbs III, R.A. Olson, R.G. Winkler, and D.C. Culver, eds. *Mammoth Cave. A Human and Natural History*. Springer, Cham, Switzerland.

Herman, J.W., Culver, D.C., and Salzman, J. (2001). Groundwater ecosystems and the service of water purification. *Stanford Environmental Law Journal*, **20**, 479–95.

Hershler, R., and Longley, G. (1986). Phreatic hydrobiids (Gastropoda: Prosobranchia) from the Edwards (Balcones Fault Zone) aquifer region, south-central Texas. *Malacologia*, **27**, 127–72.

Hildreth-Werker, V., and Werker, J. (2006). *Cave Conservation and Restoration*. National Speleological Society, Huntsville. Alabama.

Hinaux, H., Poulain, J., Da Silva, C., Noirot, C., Jeffery, J.R., Casane, D., and Rétoux, S. (2013). *De novo* sequencing of *Astyanax mexicanus* surface fish and Pachon cavefish transcriptomes reveals enrichment of mutations in cavefish putative eye genes. *PLoS ONE*, **8**, doi: 10.371/journal.pone.0053553.

Hobbs III, H.H. (1975). Distribution of Indiana cavernicolous crayfishes and eco-commensal ostracods. *International Journal of Speleology*, **7**, 273–302.

Hobbs III, H.H. (2012). Crustacea. In W.B. White and D.C. Culver, eds. *Encyclopedia of caves. Second edition*, pp. 177–94. Elsevier/Academic Press, Amsterdam, The Netherlands.

Hobbs III, H.H., and Lawyer, R. (2002). A preliminary population study of the cave cricket, *Hadenoecus cumberlandicus* Hubbell and Norton, from a cave in Carter County, Kentucky. *Journal of Cave and Karst Studies*, **65**, 174.

Hoch, H. (1994). Homoptera (Auchenorrhyncha: Fulgoroidea). In C. Juberthie and V. Decu, eds. *Encyclopaedia Biospeologica, Tome I*, pp. 307–25. Société Internationale de Biospéologie, Moulis, France.

Hoch, H. (2000). Acoustic communication in darkness. In H. Wilkens, D.C. Culver, and W.F. Humphreys, eds. *Subterranean ecosystems*, pp. 211–9. Elsevier Press, Amsterdam, The Netherlands.

Hoch, H., Oromí, P., and Arechavaleta, M. (1999). *Nisia subfogo* sp. n., a new cave-dwelling planthopper from the Cape Verde Islands (Hemiptera: Fulgoromorpha: Meenoplidae). *Revista de la Academia Canaria de Ciencias*, **11**, 189–99.

Hoenemann, M., Neiber, M.T., Humphreys, W.F., Iliffe, T.M., Li, D., Schram, F.R., and Koenemann, S. (2013). Phylogenetic analysis and systematic revision of Remipedia (Nectiopoda) from Bayesian analysis of molecular data. *Journal of Crustacean Biology*, **33**, 603–19.

Holsinger, J.R. (1994). Amphipoda. In C. Juberthie and V. Decu, eds. *Encyclopaedia Biospeologica, Tome I*, pp. 147–64. Société Internationale de Biospéologie, Moulis, France.

Holsinger, J.R. (2000). Ecological derivation, colonization, and speciation. In H. Wilkens, D.C. Culver, and W.F. Humphreys, eds. *Subterranean ecosystems*, pp. 399–432. Elsevier Press, Amsterdam, The Netherlands.

Holsinger, J.R. (2004). Crustacea: Amphipoda. In J. Gunn, ed. *Encyclopedia of caves and karst science*, pp. 258–9. Fitzroy Dearborn, New York.

Holsinger, J.R. (2009). Three new species of the subterranean amphipod crustacean genus *Stygobromus* (Crangonyctidae) from the District of Columbia. In S.M. Roble and J.C. Mitchell, eds. *A lifetime of contributions to myriapodology and the natural history of Virginia: a Festschrift in honor of Richard L. Hoffman's 80th birthday*, pp. 261–76. Virginia Museum of Natural History Special Publication No. 16, Martinsville, Virginia.

Holsinger, J.R. (2012). Vicariance and dispersalist biogeography. In W.B. White and D.C. Culver, eds. *Encyclopedia of caves. Second edition*, pp. 849–58. Elsevier/Academic Press, Amsterdam, The Netherlands.

Holsinger, J.R., and Culver, D.C. (1970). Morphological variation in *Gammarus minus* Say (Amphipoda, Gammaridae), with emphasis on subterranean forms. *Postilla* No. 146, 24 pp.

Holsinger, J.R., and Culver, D.C. (1988). The invertebrate cave fauna of Virginia and a part of east Tennessee: zoogeography and ecology. *Brimleyana* No. 14, 162 pp.

Holsinger, J.R., Hubbard, Jr, D.A., and Bowman, T.E. (1994). Biogeographic and ecological implications of newly discovered populations of the stygobiont isopod crustacean *Antrolana lira* Bowman (Cirolanidae). *Journal of Natural History*, **28**, 1047–58.

Holsinger, J. R., and Longley, G. (1980). The subterranean amphipod crustacean fauna of an artesian well in Texas. *Smithsonian Contributions to Zoology*, **308**, 1–59.

Holthuis, L.B. (1986). Decapoda. In L. Botosaneanu, ed. *Stygofauna mundi*, pp. 589–615. E.J. Brill, Leiden, The Netherlands.

Hose, L.D., Palmer, A.N., Palmer, M.V., Northup, D.E., Boston, P.J., and DuChene, H.J. (2000). Microbiology and geochemistry in a hydrogen-sulphide rich karst environment. *Chemical Geology*, **169**, 399–423.

Hose, L.D., and Rosales-Lagarde, L. (2017). Sulfur-rich caves of southern Tabasco, Mexico. In A. Klimchouk, A.N. Palmer, J. De Waele, A.S. Auler, and P. Audra, eds. *Hypogene karst regions and caves of the world*, pp. 803–16. Springer, Cham, Switzerland.

Howarth, F. G. (1972). Cavernicoles in lava tubes on the island of Hawaii. *Science*, **175**, 325–6.

Howarth, F.G. (1980). The zoogeography of specialized cave animals: a bioclimatic model. *Evolution*, **28**, 365–89.

Howarth, F.G. (1983). Ecology of cave arthropods. *Annual Review of Ecology and Systematics*, **28**, 365–89.

Howarth, F.G. (1987). The evolution of non-relictual tropical troglobites. *International Journal of Speleology*, **16**, 1–16.

Howarth, F.G., and Hoch, H. (2012). Adaptive shifts. In W.B. White and D.C. Culver, eds. *Encyclopedia of caves. Second edition*, pp. 9–17. Elsevier/Academic Press, Amsterdam, The Netherlands.

Hubbard, D.A. (2012). Saltpetre mining. In W.B. White and D.C. Culver, eds. *Encyclopedia of caves. Second edition*, pp. 676–79. Elsevier/Academic Press, Amsterdam, The Netherlands.

Hubbard, D.A., and Balfour, W. (1993). An investigation of engineering and environmental conerns relating to proposed highway construction in a karst terrane. *Environmental Geology*, **22**, 326–9.

Hubbell, T.H., and Norton, R.M. (1978). The systematics and biology of the cave-crickets of the North American tribe Hadenoecini (Orthoptera Saltatoria: Ensifer: Rhaphidophoridae: Dolichopodinae). *Miscellaneous Publications of the Museum of Zoology, University of Michigan*, No. 156.

Hubbs, C.L., and Innes, W.T. (1936). The first known blind fish of the family Characidae: a new genus from Mexico. *Occasional Papers of the Museum of Zoology of the University of Michigan. Vol. 342*. 8 pp.

Humphreys, W.F. (2000). Relict faunas and their derivation. In H. Wilkens, D.C. Culver, and W.F. Humphreys, eds. *Subterranean ecosystems*, pp. 417–32. Elsevier Press, Amsterdam, The Netherlands.

Humphreys, W.F. (2001). Groundwater calcrete aquifers in the Australian arid zone: the context to an unfolding plethora of stygal biodiversity. *Records of the Western Australian Museum Supplement*, **64**, 63–83.

Humphreys, W.F. (2004). Cape Range, Australia: biospeleology. In J. Gunn, ed., *Encyclopedia of caves and karst science*, pp. 181–3. Fitzroy Dearborn, New York.

Humphreys, W.F., and Freinberg, M.N. (1995). Food of the blind cave fishes of northwestern Australia. *Records of Western Australian Museum*, **17**, 29–33.

Humphreys, W.F., Watts, C.H.S., Cooper, S.J.B., and Leijs, R. (2009). Groundwater estuaries of salt lakes: buried pools of endemic biodiversity on the western plateau, Australia. *Hydrobiologia*, **626**, 79–95.

Huntsman B.M., Venarsky M.P., Benstead J.P., and Huryn A.D. (2011). Effects of organic matter availability on the life history and production of a top vertebrate predator (Plethodontidae: *Gyrinophilus palleucus*) in two cave streams. *Freshwater Biology*, **56**, 1746–60.

Hüppop, K. (2000). How do cave animals cope with the food scarcity in caves? In H. Wilkens, D.C. Culver, and W.F. Humphreys, eds. *Subterranean ecosystems*, pp. 159–88. Elsevier Press, Amsterdam, The Netherlands.

Hutchins, B.T. (2017). The conservation status of Texas groundwater invertebrates. *Biodiversity and Conservation*, **27**, 475–501.

Hutchins, B.T., Engel, A.S., Nowlin, W.H., and Schwartz, B.F. (2016). Chemolithoautotrophy supports macroinvertebrate food webs and affects diversity and stability in groundwater communities. *Ecology*, **97**, 1530–42.

Hutchins, B.T., Fong, D.W., and Carlini, D.B., (2010). Genetic population structure of the Madison Cave isopod, *Antrolana lira* (Cymothoida: Cirolanidae) in the Shenandoah Valley of the eastern United States. *Journal of Crustacean Biology*, **30**, 312–22.

Hutchins, B.T., Schwartz, B.F., and Engel, A.S. (2013). Environmental controls on organic matter production and transport across surface-subsurface and geochemical boundaries in the Edwards Aquifer, Texas, USA. *Acta Carsologica*, **42**, 245–59.

Hutchinson, G.E. (1958). Concluding remarks. *Cold Spring Harbor Symposia on Quantitative Biology*, **22**, 425–7.

Hynes, H.B.N. (1983). Groundwater and stream ecology. *Hydrobiologia*, **100**, 93–9.

Ipsen, A. (2000). The Segeberger Höhle—a phylogenetically young cave ecosystem in northern Germany. In H. Wilkens, D.C. Culver, and W.F. Humphreys, eds. *Subterranean ecosystems*, pp. 569–79. Elsevier Press, Amsterdam, The Netherlands.

Izquierdo, I., Martin, J.L., Zurita, N. and Medina, A.L. (2001). Geo-referenced computer recordings as an instrument for protecting cave-dwelling species of Tenerife (Canary Islands). In D.C. Culver, L. Deharveng, J. Gibert, and I. Sasowsky, eds. *Mapping subterranean biodiversity. Cartographie de la biodiversité souterraine*, pp. 45–8. Karst Waters Institute Special Publication 6, Charles Town, West Virginia.

Jaffé, R., Prous, X., Calux, A., Gastauer, M., Nicacio, G., Zampaulo, R., Souza-Filho, P.W., Oliveira, G., Brandi, I.V., and Siqueira, J.O. (2018). Conserving relics from ancient underground worlds: assessing the influence of cave and landscape features on obligate iron cave dwellers from the Eastern Amazon. *PeerJ*, **6**, e4531.

Jaffé, R., Prous, X., Zampaulo, R., Giannini, T., Imperatriz-Fonseca, V., Maurity, C., Oliveira, G., Brandi, I., and Siqueira, J. (2016). Reconciling mining with the conservation of cave biodiversity: a quantitative baseline to help establish conservation priorities. *PLoS ONE*, **11**, e0168348

Jasinska, E., and Knott, B. (2000). Root-driven faunas in cave waters. In H. Wilkens, D.C. Culver, and W.F. Humphreys, eds. *Subterranean ecosystems*, pp. 287–307. Elsevier Press, Amsterdam, The Netherlands.

Jeannel, R. (1943). *Les fossiles vivants des cavernes*. Gallimard, Paris, France.

Jeffery, W.R. (2001). Cavefish as a model system in evolutionary developmental biology. *Developmental Biology*, **231**, 1–12.

Jeffery, W.R. (2005a). Adaptive evolution of eye degeneration in the Mexican blind cavefish. *Journal of Heredity*, **96**, 185–96.

Jeffery, W.R. (2005b). Evolution of eye degeneration in cavefish: the return of pleiotropy. *Subterranean Biology*, **3**, 1–11.

Jeffery, W.R. (2006). Regressive evolution of pigmentation in the cavefish *Astyanax*. *Israel Journal of Ecology and Evolution*, **52**, 405–22.

Jeffery, W.R. (2009). Regressive evolution in *Astyanax* cavefish. *Annual Review of Genetics*, **43**, 25–47.

Jeffery, W.R., Ma, L., Parkhurst, A., and Bilandžija, H. (2016). Pigment regression and albinism in *Astyanax* cavefish. In A.C. Keene, M. Yoshizawa, and S.E. McGaugh, eds. *The biology and evolution of Mexican cavefish*, pp. 155–174. Academic/Elsevier, Amersterdam, The Netherlands.

Jeffery, W.R., and Martasian, D.P. (1998). Evolution of eye regression in the cavefish *Astyanax*. Apoptosis and the *Pax6* gene. *American Zoologist*, **38**, 685–96.

Jones, C.G., Lawton, J.H., and Shachak, M. (1994). Organisms as ecosystem engineers. *Oikos*, **69**, 373–86.

Jones, R.D., Culver, D.C., and Kane, T.C. (1992). Are parallel morphologies of cave organisms the result of similar selection pressures? *Evolution*, **46**, 353–65.

Jones, W.K. (1997). *Karst hydrology atlas of West Virginia*. Karst Waters Institute Special Publication 4, Charles Town, West Virginia.

Jones, W.K., Hobbs III, H.H., Wicks, C.M., Currie, R.R., Hose, L.D., Kerbo, R.C., Goodbar, J.R., and Trout, J. (2003). *Recommendations and guidelines for managing caves on protected lands*. Karst Waters Institute Special Publication 8, Charles Town, West Virginia.

Juberthie, C. (1985). Cycle vital de *Telema tenella* dans la grotte-laboratoire de Moulis et strategies de reproduction chez les araignées cavernicoles. *Mémoires de Biospéologie*, **12**, 77–89.

Juberthie, C. (2000). The diversity of the karstic and pseudokarstic hypogean habitats in the world. In H. Wilkens, D.C. Culver, and W.F. Humphreys, eds. *Subterranean ecosystems*, pp. 17–39. Elsevier Press, Amsterdam, The Netherlands.

Juberthie, C. and Decu, V., eds. (1994–2001). *Encyclopaedia Biospeologica*. 3 vols. Société Internationale de Biospéologie, Moulis, France.

Juberthie, C., Delay, B., and Bouillon, M. (1980). Extension du milieu souterrain en zone non-calcaire: description d'un nouveau milieu et de son peuplement par les coleopteres troglobies. *Mémoires de Biospéologie*, **7**, 19–52.

Juberthie-Jupeau, L. (1988). Mating behaviour and barriers to hybridization in the cave beelte of the *Speonomus delarouzeei* complex. *International Journal of Speleology*, **17**, 51–64.

Juberthie-Jupeau, L. (1994). Symphyla. In C. Juberthie and V. Decu, eds. *Encyclopaedia Biospeologica, Tome I*, pp. 365–6. Société Internationale de Biospéologie, Moulis, France.

Kane, T.C., Culver, D.C., and Jones, R.T. (1992). Genetic structure of morphologically differentiated populations of the amphipod *Gammarus minus*. *Evolution*, **46**, 272–8.

Kane, T.C., Norton, R.M., and Poulson, T.L. (1975). The ecology of a predaceous troglobitic beetle, *Neaphaenops tellkampfi* (Coleoptera: Carabidae, Trechinae) I. Seasonality of food input and early life history stages. *International Journal of Speleology*, **7**, 45–54.

Kane, T.C., and Poulson, T.L. (1976). Foraging by cave beetles: spatial and temporal heterogeneity of prey. *Ecology*, **57**, 793–800.

Kane, T.C., and Ryan, T. (1983). Population ecology of carabid cave beetles. *Oecologia*, **60**, 46–55.

Kayo, R.T., Marmonier, P., Togouet, S.H.Z., Nola, M., and Piscart, C. (2012). An annotated checklist of freshwater stygobiotic crustaceans of Africa and Madagascar. *Crustaceana*, **85**, 1613–31.

Keene, A.C., Yoshizawa, M., and McGaugh, S.E., eds. (2016). *Biology and evolution of the Mexican cavefish*. Academic/Elsevier, Amsterdam, The Netherlands.

Kempe, S., Al-Malabeh, A., Döppes, D., Frehat, M., Henschel, H.V., and Rosendahl, W. (2006). Hyena caves in Jordan. *Scientific Annalas, School of Geology, Aristotle University of Tessalonki*, **98**, 57–68.

Klimchouk, A. (2007). *Hypogene speleogenesis: hydrogeological and morphogenetic perspective*. National Cave and Karst Research Institute Special Paper No. 1. Carlsbad, New Mexico.

Klimchouk, A. (2015). The karst paradigm: changes, trends and perspectives. *Acta Carsologica*, **44**, 289–314.

Klimchouk, A. (2017). Types and settings of hypogene karst. In A. Klimchouk, A.N. Palmer, J. De Waele, A.S. Auler, and P. Audra, eds. *Hypogene karst regions and caves of the world*, pp. 1–39. Springer, Cham, Switzerland.

Klimchouk, A., and V. Andreychouk. (2017). Gypsum karst in the southwest outskirts of the Eastern European Platform (western Ukraine): a type region of artesian transverse speleogenesis. In A. Klimchouk, A.N. Palmer, J. De Waele, A.S. Auler, and P. Audra, eds. *Hypogene karst regions and caves of the world*, pp. 363–85. Springer, Cham, Switzerland.

Knapp, S.M., and Fong, D.W. (1999). Estimates of population size of *Stygobromus emarginatus* (Amphipoda: Crangonyctidae) in a headwater stream in Organ Cave, West Virginia. *Journal of Cave and Karst Studies*, **61**, 3–6.

Knez, M., and Slabe, T., eds. (2016). *Cave exploration in Slovenia. Discovering over 350 new caves during motorway construction on classical karst*. Springer, Cham, Switzerland.

Knight, L.R.F.D. (2011). The aquatic macro-invertebrate fauna of Swildon's Hole, Mendip Hills, Somerset, UK. *Cave and Karst Science*, **38**, 81–92.

Konec, M., Prevorčnik, S., Sârbu, Ş.M., Verovnik, R., and Trontelj, P. (2015). Parallels between two geographically and ecologically disparate cave invasions by the same species, *Asellus aquaticus* (Isopoda, Crustacea). *Journal of Evolutionary Biologoy*, **28**, 864–75.

Kornobis, E., and Pálsson, S. (2013). The ITS region of groundwater amphipods: length, secondary structure and phylogenetic information content in Crangonyctoids and Niphargids. *Journal of Zoological Systematics and Evolutionary Research*, **51**, 19–28.

Kosswig, C. (1965). Génétique et évolution regréssive. *Revue des Questions Scientifiques*, **136**, 227–57.

Kosswig, C., and Kosswig, L. (1940). Die Variabilität bei *Asellus aquaticus* unter besonderer Berucksichtigung der Variabilität in isolierten unter- und oberirdischen Populationen. *Revue de Facultie des Sciences (Istanbul), ser. B*, **5**, 1–55.

Kowalko, J.E., Rohner, N., Linden, T.A., Rompani, S.B., Warren, W.C., Borowsky, R., Tabin, C.J., Jeffery, W.R., and Yoshizawa, M. (2013). Convergence in feeding posture occurs through different genetic loci in independently evolved cave populations of *Astyanax mexicanus*. *Proceedings of the National Academy of Sciences*, **110**, 16933–8.

Krause, S., Hannah, D.M., Fleckenstein, J.H., Heppell, C.M., Kaeser, D., Pickup, R., Pinnay, G., Robertson, A.L., and Wood, P.J. (2011). Inter-disciplinary perspectives on processes in the hyporheic zone. *Ecohydrology*, **4**, 481–99.

Kristjánsson, B.K., and Svavarsson, J. (2007). Subglacial refugia in Iceland enabled groundwater amphipods to survive glaciations. *The American Naturalist*, **170**, 292–6.

Krumholz, L.R. (2000). Microbial communities in the deep subsurface. *Hydrogeology Journal*, **8**, 4–10.

Kumaresan, D.W., Stephenson, J., Hillebrand-Volculescu, A., and Murrell, J.C. (2014). Microbiology of Movile Cave—a chemolithoautotrophic ecosystem. *Geomicrobiology Journal*, **31**, 186–93.

Kunz, T.H., Braun de Torrez, E., Bauer, D., Lobova, T., and Fleming, T.H. (2011). Ecosystem services provided by bats. *Annals of the New York Academy of Sciences*, **1223**, 1–38.

Kunz, T.H., and Fenton, M.B., eds. (2003). *Bat ecology*. University of Chicago Press, Chicago, Illinois.

Kunz, T.H., Murrary, S.W., and Fuller, N.W. (2012). Bats. In W.B. White and D.C. Culver, eds. *Encyclopedia of caves. Second edition*, pp. 45–54. Elsevier/Academic Press, Amsterdam, The Netherlands.

Kurtén, B. (1968). *Pleistocene mammals of Europe*. Weidenfeld and Nicholson, London.

Laiz, L., Groth, I., Gonzalez, I., and Saiz-Jimenez, C. (1999). Microbiological study of the dripping waters in Altamira cave (Santillana del Mar, Spain). *Journal of Microbiological Methods*, **36**, 129–38.

Lamarck, J.B. (1984). *Zoological philosophy: an exposition with regard to the natural history of animals* (H. Elliot, trans.). University of Chicago Press, Chicago, Illinois.

Lamoreaux, J. (2004). Stygobites are more wide-ranging than troglobites. *Journal of Cave and Karst Studies*, **66**, 18–19.

Lande, R. (1996). Statistics and partitioning of species diversity, and similarity among multiple communities. *Oikos*, **76**, 5–13.

Langecker, T.G., and Longley, G. (1993). Morphological adaptations of the Texas blind catfishes *Trogloglanis pattersoni* and *Satan eurystomus* (Siluriformes: Ictaluridae) to their underground environment. *Copeia*, **1993**, 976–86.

Lankester, E. (1925). The blindness of cave-animals. *Nature*, **116**, 745–6.

Lascu, C. (2004). Movile Cave. In J. Gunn, ed. *Encyclopedia of caves and karst science*, pp. 528–30. Fitzroy Dearborn, New York.

Latinne, A., Waengsothorn, S., Rojanadilok, P., Eiamampai, K., Sribuarod, K., and Michaux, J. R. (2012). Combined mitochondrial and nuclear markers revealed a deep vicariant history for *Leopoldamys neilli*, a cave-dwelling rodent of Thailand. *PLoS ONE*, **7**, e47670.

Lavoie, K.H., Helf, K.L., and Poulson, T.L. (2007). The biology and ecology of North American cave crickets. *Journal of Cave and Karst Studies*, **69**, 114–34.

Lavoie, K.H., Winter, A.S., Read, K.J.H., Hughes, E.M., Spilde, M.N., and Northup, D.E. (2017). Comparison of bacterial communities from lava cave microbial mats to overlying surface soils from Lava Beds National Monument, USA. *PLoS ONE*, **12**, e0169339.

Lechleitner, F.A., Dittmar, T., Baldini, J.U.L., Prufer, K.M., and Eglinton, T.I. (2017). Molecular signatures of dissolved organic matter in a tropical karst system, *Organic Geochemistry*, **113**, 141–9.

Lefébure, T., Douady, C.J., Gouy, M., and Gibert, J. (2006a). Relationship between morphology, taxonomy, and molecular divergence with Crustacea: proposal of a molecular threshold to help species definition. *Molecular Phylogeny and Evolution*, **40**, 435–47.

Lefébure, T., Douady, C.J., Gouy, M., Trontelj, P., Briolay, J., and Gibert, J. (2006b). Phylogeography of a subterranean amphipod reveals cryptic diversity and dynamic evolution in extreme environments. *Molecular Ecology*, **15**, 1797–806.

Leijs, R., van Nes, E.H., Watts, C.H., Cooper, S.J.B., Humphreys, W.F., and Hogendoorn, K. (2012). Evolution of blind beetles in isolated aquifers: a test of alternative modes of speciation. *PLoS ONE*, **7**, e34260.

Lewis, J.J. (1996). Bioinventory as a management tool. In G.T. Rea, ed. *Proceedings of the 1995 Cave Management Symposium, Spring Mill State Park, Mitchell, Indiana*, pp. 228–36. Indiana Karst Conservancy, Indianapolis, Indiana.

Leys, R., Watts, C.H.S., Cooper, S.J.B., and Humphreys, W.F. (2003). Evolution of subterranean diving beetles (Coleoptera: Dytiscidae: Hydroporini: Bidessini) in the arid zone of Australia. *Evolution*, **57**, 2819–34.

Lomolino, M.V., and Heaney, L.R., eds. (2004). *Frontiers of biogeography. New directions in the geography of nature.* Sinauer Associates, Sunderland, Massachusetts.

Longley, G. (1981). The Edwards Aquifer: the most groundwater ecosystem? *International Journal of Speleology*, **11**, 123–8.

Longley, G. (2004). Edwards Aquifer, United States: Biospeleology. In J. Gunn, ed. *Encyclopedia of caves and karst science*, pp. 315–6. Fitzroy Dearborn, New York.

Loop, C.M. (2012). Contamination of cave waters by nonaqueous phase liquids. In W.B. White and D.C. Culver, eds. *Encyclopedia of caves. Second edition*, pp. 166–72. Elsevier/Academic Press, Amsterdam, The Netherlands.

López, H., and Oromí, P. (2010). A pitfall trap for sampling the mesovoid shallow substratum (MSS) fauna. *Speleobiology Notes*, **2**, 7–11.

López-Rodríguez, M.J., and Tierno de Figueroa, J.M. (2012). Life in the dark: On the biology of the cavernicolous stonefly *Protonemura gevi* (Insecta, Plecoptera). *The American Naturalist*, **180**, 684–91.

Losos, J.B., and Ricklefs, R.E., eds. (2010) *The theory of island biogeography revisited.* Princeton University Press, Princeton, New Jersey.

Lučić, I., and Sket, B. (2003). *Vjetrenica. Pogled u dušu zemlje.* Savez speleologa Bosne i Hercegovine and Hrvatsko biospeleološko društvo. Zagreb, Croatia.

Ludwig, W. (1942). Zur evoutorischen Erklärung der Höhlentiermerkmale durch Allelimination. *Biologisches Zentralblatt*, **62**, 447–82.

Luštrik, R., Turjak, M., Kralj-Fišer, S., and Fišer, C. (2011). Coexistence of surface and cave amphipods in an ecotone environment. *Contributions of Zoology*, **80**, 133–41.

Ma, L., and Zhao, Y. (2012). Cavefish of China. In W.B. White and D.C. Culver, eds. *Encyclopedia of caves. Second edition*, pp. 107–25. Elsevier/Academic Press, Amsterdam, The Netherlands.

MacArthur, R.H., and R. Levins. 1967. The limiting similarity, convergence, and divergence of coexisting species. *American Naturalist*, **101**, 377–85.

MacArthur, R.H., and Wilson, E.O. (1967). *The theory of island biogeography.* Princeton University Press, Princeton, New Jersey.

MacAvoy, S.E., Braciszewski, A., Tengi, E., and Fong, D.W. (2016). Trophic plasticity among spring vs. cave populations of *Gammarus minus. Ecological Research*, **31**, 589–95.

MacGregor, J. (1993). Responses of winter population of the federal endangered Indiana bat (*Myotis sodalis*) to cave gating in Kentucky. In D.L. Foster, ed. *Proceedings of the National Cave Management Symposium*, pp. 364–79. American Cave Conservation Association, Horse Cave, Kentucky.

Malard, F., ed. (2003). *Sampling manual for the assessment of regional groundwater biodiversity.* PASCALIS (Protocols for the Assessment and Conservation of Aquatic Life in the Subsurface), Lyon, France.

Malard, F., Boutin, C., Camacho, A.I., Ferreira, D., Michel, G., Sket, B., and Stoch, F. (2009). Diversity patterns of stygobiotic crustaceans across multiple spatial scales in western Europe. *Freshwater Biology*, **54**, 756–76.

Malard, F., Reygrobellet, J.-L., Laurent, R., and Mathieu, J. (1997). Developments in sampling the fauna of deep water-table aquifers. *Archiv für Hydrobiologie*, **138**, 401–32.

Malard, F., Tockner, K., Dole-Olivier, M.J., and Ward, J.V. (2002). A landscape perspective of surface-subsurface hydrological exchanges in river corridors. *Freshwater Biology*, **47**, 621–40.

Malard, F., Ward, J.V., and Robinson, C.T. (2000). An expanded perspective of the hyporheic zone. *Verhaltlungen der Internationalen Vereinigung für Theoretische und Angewandte Limnologie*, **27**, 431–7.

Mammola, S., Giachino, P.M., Piano, E., Jones, A., Barberis, M., Badino, G., and Isaia, M. (2016). Ecology and sampling techniques of an understudied subterranean habitat: the Milieu Souterrain Superficiel (MSS). *The Science of Nature*, **103**, 88.

Mammola, S., Goodacre, S.L., and Isaia, M. (2017). Climate change may drive cave spiders to extinction. *Ecography*, **41**, 233–43.

Mann, A.W., and Horwitz, R.C. (1979). Groundwater calcrete deposits in Australia: some observations from Western Australia. *Journal of the Geological Society of Australia*, **26**, 293–303.

Marmonier, P., Creuzé des Châtelliers, M., Dole-Olivier, M.-J., Plénet, S., and Gibert, J. (2000). Rhône groundwater systems. In H. Wilkens, D.C. Culver, and W.F. Humphreys, eds. *Subterranean ecosystems*, pp. 513–31. Elsevier Press, Amsterdam, The Netherlands.

Marsh, T.G. (1969). *Ecological and behavioral studies of the cave beetle* Darlingtonea kentuckensis. Ph.D. Dissertation, University of Kentucky, Lexington, Kentucky.

Martens, K. (2004). Crustacea: Ostracoda. In J. Gunn, ed. *Encyclopedia of caves and karst science*, pp. 267–8. Fitzroy Dearborn, New York.

Martin P., De Broyer, C., Fiers, F., Michel, G., Sablon, R., and Wouters, K. (2009). Biodiversity of Belgian groundwater fauna in relation to environmental conditions. *Freshwater Biology*, **54**, 814–29.

Martínez-Ansemil, E., Creuzé des Châtelliers, M., Martin, P., and Sambugar, B. (2012). The Parvidrilidae–a diversified groundwater family: description of six new species from southern Europe, and clues for its phylogenetic position within Clitellata (Annelida). *Zoological Journal of the Linnean Society*, **166**, 530–58.

Master, L.L., Flack, S.R., and Stein, B.A. (1998). *Rivers of life: critical watersheds for protecting freshwater biodiversity.* The Nature Conservancy, Arlington, Virginia.

Matjašič, J. (1958). Biologie und zoogeographie der europäischen Temnocephaliden. *Zoologischer Anzeiger*, **21**, 477–82.

Matjašič, J. (1994). Turbellaria, Temnocephala. In C. Juberthie and V. Decu, eds. *Encyclopaedia biospeologica, Tome I*, pp. 45–8. Société Internationale de Biospéologie, Moulis, France.

McAllister, C. T., and Bursey, C. R. (2004). Endoparasites of the dark-sided salamander, *Eurycea longicauda melanopleura*, and the cave salamander, *Eurycea lucifuga* (Caudata: Plethodontidae), from two caves in Arkansas, USA. *Comparative Parasitology*, **71**, 61–6.

McInerney, C., Maurice, L., Robertson, A., Knight, L.R.F.D., Arnscheidt, J., Venditti, C., Dooley, J., Mathers, T., Matthijs, S., Eriksson, K., Proudlove, G., and Hänfling, B. (2014). The ancient Britons: Groundwater fauna survived extreme climate changes over tens of millions of years across NW Europe. *Molecular Ecology*, **23**, 1153–66.

Mejía-Ortíz, L.M., Pipan, T., Culver, D.C., and Sprouse, P. (2018). The blurred line between photic and aphotic environments: a large Mexican cave with almost no dark zone. *International Journal of Speleology*, **37**, 69–80.

Meleg, I., Zakšek, V., Fišer, C., Kelemen, B.S., and Moldovan, O.T. (2013). Can environment predict cryptic diversity? The case of *Niphargus* inhabiting western Carpathian groundwater. *PLoS ONE*, **8**, e76760.

Meng, F., Braasch, I., Phillips, J.B., Lin, X., Titus, T., Zhang, C., and Postlethwait, J.H. (2013). Evolution of the eye transcriptome under constant darkness in *Sinocyclocheilus* cavefish. *Molecular Biology and Evolution*, **30**,1527–43.

Meštrov, M. (1962). Un nouveau milieu aquatique souterrain: le biotope hypotelminorheique. *Compte Rendus Academie des Sciences, Paris*, **254**, 2677–9.

Michel, G., Malard, F., Deharveng, L., Di Lorenzo, T., Sket, B., and De Broyer, C. (2009). Reserve selection for conserving groundwater biodiversity in Europe. *Freshwater Biology*, **54**, 861–76.

Minanović, P. (1990). Influence of construction on hydrogeological and environmental conditions in the karst region, eastern Herzegovina, Yugoslavia. *Environmental Geology*, **15**, 5–11.

Mitchell, R.W. (1968). Food and feeding habits of the troglobitic carabid beetle *Rhadine subterranea*. *International Journal of Speleology*, **3**, 249–70.

Mitchell, R.W. (1969). A comparison of temperate and tropical cave communities. *The Southwestern Naturalist*, **14**, 73–88.

Mitchell, R.W., Russell, W.H., and Elliott, W.R. (1977). Mexican eyeless characin fishes, genus *Astyanax*: environment, distribution, and evolution. *Special Publications of the Museum of Texas Tech University*, **12**, 1–89.

Mitchell-Jones, A.J., Bihari, Z., Measing, M., and Rodriguez, L. (2007). *Protecting and managing underground sites for bats*. UNEP/EUROBATS Secretariat, Bonn, Germany.

Moore, M.S., and Kunz, T.H. (2012). White nose syndrome: a fungal disease of North American bats. In W.B. White and D.C. Culver, eds. *Encyclopedia of caves. Second edition*, pp. 904–810. Elsevier/Academic Press, Amsterdam, The Netherlands.

Morton, B., Velkovrh, F., and Sket, B. (1998). Biology and anatomy of the 'living fossil' *Congeria kusceri* (Bivalvia: Dreissenidae) from subterranean rivers and caves in the Dinaric karst of former Yugoslavia. *Journal of Zoology (London)*, **245**, 147–74.

Mulec J., Kosi, G., and Vrhovšek, D. (2007). Algae promote growth of stalagmites and stalactites in karst caves (Škocjanske jame, Slovenia). *Carbonates and Evaporites*, **22**, 6–10.

Mulec J., Kosi, G., and Vrhovšek, D. (2008). Characterization of cave aerophytic algal communities and effects of irradiance levels on production of pigments. *Journal of Cave and Karst Studies*, **70**, 3–12.

Musgrove, M., and Banner, J.L. (2004). Controls on the spatial and temporal variability of vadose dripwater chemistry: Edwards Aquifer, central Texas. *Geochimica et Cosmochimica Acta*, **68**, 1007–20.

Najvar, P.A., Fries, J.N., and Baccus, J.T. (2007). Fecundity of San Marcos salamanders in captivity. *Southwestern Naturalist*, **52**, 145–7.

Nelson, G.J., and Platnick, N. (1981). *Systematics and biogeography: cladistics and vicariance*. Columbia University Press, New York, New York.

Niemiller, M.L., Fitzpatrick, B.M., Shah, P., Schmitz, L., and Near, T. J. (2012b). Evidence for repeated loss of selective constraint in rhodopsin of amblyopsid cavefishes (Teleostei: Amblyopsidae). *Evolution*, **67**, 732–48.

Niemiller, M.L., Graening, G.O., Fenolio, D.B., Godwin, J.C., Cooley, J.R., Pearson, W.D., Fitzpatrick, B.M., and Near, T.J. (2013a). Doomed before they are described? The need for conservation assessments of cryptic species complexes using an amblyopsid cavefish (Amblyopsidae: Typhlichthys) as a case study. *Biodiversity and Conservation*, **22**, 1799–820.

Niemiller, M.L., McCandless, J.R., Reynolds, R.G., Caddle, J., Tillquist, C.R., Near, T.J., Pearson, W.D., and Fitzpatrick, B.M. (2013b). Effects of climatic and geological processes during the Pleistocene on the evolutionary history of the northern cavefish, *Amblyopsis spelaea* (Teleostei: Amblyopsidae). *Evolution*, **67**, 1011–25.

Niemiller, M.L., Near, T.J., and Fitzpatrick, B.M. (2012a). Delimiting species using multilocus data: diagnosing cryptic diversity in the southern cavefish, *Typhlichthys subterraneus* (Teleostei: Amblyopsidae). *Evolution*, **66**, 846–66.

Niemiller, M.L., Porter, M.L., Keany, J., Gilbert, H., Culver, D.C., Fong, D.W., Hobson, C.S., Kendall, K.D., and Taylor, S.J. (2017). Evaluation of eDNA for groundwater invertebrate detection and monitoring: a case study with endangered *Stygobromus* (Amphipoda: Crangonyctidae). *Conservation Genetics Methods*, doi:10.1007/s12686-017-0785-2.

Niemiller, M.L., and Zigler, K.S. (2013). Patterns of cave biodiversity and endemism in the Appalachians and Interior Plateau of Tennessee, U.S.A. *PLoS ONE*, **8**, e64177.

Noltie, D.B., and Wicks, C.M. (2001). How hydrogeology has shaped the ecology of Missouri's Ozark cavefish, *Amblyopsis rosae*, and southern cavefish, *Typhlichthys subterraneus*: insights of the sightless from understanding the underground. *Environmental Biology of Fishes*, **62**, 171–94.

Northup, D.E., and Lavoie, K.H. (2001). Geomicrobiology of caves: a review. *Geomicrobiology Journal*, **18**, 199–222.

Northup, D.E., and Lavoie, K.H. (2004). Microbial processes in caves. In J. Gunn, ed. *Encyclopedia of caves and karst science*, pp. 505–9. Fitzroy Dearborn, New York.

Notenboom, J. (2001). Managing ecological risks of groundwater pollution. In C. Griebler, D.L. Danielopol, J. Gibert, H.P. Nachtnebel, and J. Notenboom, eds. *Groundwater ecology. A tool for management of water resources*, pp. 247–62. Director-General for Research, European Communities, Luxenbourg.

Notenboom, J., Plénet, S., and Turquin, M.-J. (1994). Groundwater contamination and its impact on groundwater animals and ecosystems. In J. Gibert, D.L. Danielopol, and J.A. Stanford, eds. *Groundwater ecology*, pp. 477–504. Academic Press, San Diego.

Novak, T., Janžekovič, F., and Lipovšček, S. (2013). Contribution of non-troglobiotic terrestrial invertebrates to carbon input in hypogean habitats. *Acta Carsologica*, **42**, 301–9.

Odum, E.P. (1953). *Fundamentals of ecology*. Saunders, Philadelphia, Pennsylvania.

Odum, H.T. (1957). Trophic structure and productivity of Silver Springs, Florida. *Ecological Monographs*, **27**, 55–112.

Olson, R. (2004). Mammoth Cave, United States: biospeleology. In J. Gunn, ed. *Encyclopedia of caves and karst science*, pp. 499–501. Fitzroy Dearborn, New York.

O'Meara, B.C. (2010). New heuristic methods for joint species delimitation and species tree inference. *Systematic Biology*, **59**, 59–73.

O'Quin, K., and McGaugh, S.E. (2016). Mapping the genetic basis of troglomorphy in *Astyanax:* How far we have come and where do we go from here? In A.C. Keene, M. Yoshizawa, and S.E. McGaugh, eds. *Biology and evolution of the Mexican cavefish*, pp. 111–36. Academic/Elsevier Press, Amsterdam.

Ornelas-Garcià, C.P., Dominguez-Dominguez, O., and Doadrio, I. (2008). Evolutionary history of the fish genus *Astyanax* Baird & Girard (1854) (Actinopterygii, Characidae) in Mesoamerica reveals multiple morphological homoplasies. *BMC Evolutionary Biology*, **8**, 340.

Oromí, P. (2004). Biospeleology in Macaronesia. *Association for Mexican Cave Studies Bulletin*, **19**, 98–104.

Oromí, P., and Martin, J.L. (1992). The Canary Islands subterranean fauna: characterization and composition. In A.I. Camacho, ed. *The natural history of biospeleology*, pp. 529–67. Museo Nacional de Ciencias Naturales, Madrid, Spain.

Ortuño, V.M., Gilgado, J.D., Jiménez-Valverde, A., Sendra, A., Pérez-Suárez, G., and Herrero-Borgoñón, J.J. (2013). The "alluvial mesovoid shallow substratum", a new subterranean habitat. *PloS ONE*, **8**, e76311.

Osborne, R.A.L. (2007). The world's oldest caves: how did they survive and what can they tell us? *Acta Carsologica*, **36**, 133–42.

Ozimec, R., and Lučić, I., 2009. The Vjetrenica cave (Bosnia & Herzegovina)–One of the world's most prominent biodiversity hotspots for cave-dwelling fauna. *Subterranean Biology*, **7**, 17–23.

Pabich, W.G., Aliela, I.V., and Hemond, H.F. (2001). Relationship between DOC concentrations and vadose zone thickness and depth below the water table in groundwater of Cape Cod. *Biogeochemistry*, **55**, 247–68.

Packard, A.S. (1888). The cave fauna of North America, with remarks on the anatomy of brain and the origin of the blind species. *Memoirs of the National Academy of Sciences (USA)*, **4**, 1–156.

Palmer, A.N. (2007). *Cave geology*. Cave Books, Dayton, Ohio.

Palmer, A.N. (2012). Passage growth and development. In W.B. White and D.C. Culver, eds. *Encyclopedia of caves. Second edition*, pp. 598–603. Elsevier/Academic Press, Amsterdam, The Netherlands.

Palmer, A.N. (2013). Sulfuric acid caves: Morphology and evolution. In A. Frumkin, ed. *Treatise on Geomorphology. Volume 6*, pp. 241–57. Elsevier/Academic Press, Amsterdam, The Netherlands.

Palmer, A.N. (2017). Geology of Mammoth Cave. In H.H. Hobbs, R.A. Olson, E.G. Winkler, and D.C. Culver, eds. *Mammoth Cave. A human and natural history*, pp. 97–109. Springer, Cham, Switzerland.

Pape, R.B. (2016). The importance of ants in cave ecology, with new records and behavioral observations of ants in Arizona caves. *International Journal of Speleology*, **45**, 185–205.

Parker, C.W., Wolf, J.A., Auler, A.S., Barton, H.A., and Senko, J.M. (2013). Microbial reducibility of Fe (III) phases associated with the genesis of iron ore caves in the Iron Quadrangle, Minas Gerais, Brazil. *Minerals*, **3**, 395–411.

Paterson, A., and Engel, A.S. (2015). Predicting bacterial diversity in caves associated with sulfuric acid speleogenesis. In A.S. Engel, ed. *Microbial life of cave systems*, pp. 193–214. De Gruyter, Berlin, Germany.

Peck, S.B. (1974). The food of the salamanders *Eurycea lucifuga* and *Plethodon glutinosus* in caves. *Bulletin of the National Speleological Society*, **36**, 7–10.

Peck, S.B. (1984). The distribution and evolution of cavernicolous *Ptomaphagus* beetles in the southeastern United States (Coleoptera: Leiodidae: Cholevinae) with new species and records. *Canadian Journal of Zoology*, **62**, 730–40.

Peck, S.B. (1986). Bacterial deposition of iron and manganese oxides in North American caves. *Bulletin of the National Speleological Society*, **48**, 26–30.

Peck, S. B., and Finston, T.L. (1993). Galapagos Islands troglobites: the questions of tropical troglobites, parapatric distributions with the eyed-sister-species, and their origin by parapatric speciation. *Mémoires de Biospéologie*, **20**, 19–37.

Phelps, K., Jose, R., Labonite, M., and Kingston, T. (2016). Correlates of cave-roosting bat diversity as an effective tool to identify priority caves. *Biological Conservation*, **201**, 201–9.

Pipan, T. (2005). *Epikarst—a promising habitat*. Založba ZRC, Ljubljana, Slovenia.

Pipan, T., Blejec, A., and Brancelj, A. (2006a). Multivariate analysis of copepod assemblages in epikarstic waters of some Slovenian caves. *Hydrobiologia*, **559**, 213–23.

Pipan, T., and Brancelj, A. (2001). Ratio of copepods (Crustacea: Copepoda) in fauna of percolation water in six karst caves in Slovenia. *Acta Carsologica*, **30**, 257–65.

Pipan T., Christman M.C., and Culver D.C. (2006b). Dynamics of epikarst communities: microgeographic pattern and environmental determinants of epikarst copepods in Organ Cave, West Virginia. *American Midland Naturalist*, **156**, 75–87.

Pipan, T., and Culver, D.C. (2005). Estimating biodiversity in the epikarstic zone of a West Virginia cave. *Journal of Cave and Karst Studies*, **67**, 103–9.

Pipan, T., and Culver, D.C. (2007a). Regional species richness in an obligate subterranean dwelling fauna—epikarst copepods. *Journal of Biogeography*, **34**, 854–61.

Pipan, T., and Culver, D.C. (2007b). Copepod distribution as an indicator of epikarst system connectivity. *Hydrogeology Journal*, **15**, 817–22.

Pipan, T., Culver, D.C., Papi, F., and Kozel, P. (2018). Partitioning diversity in subterranean invertebrates: The epikarst fauna of Slovenia. *PLoS ONE*, **13**, e0185991.

Pipan, T., Holt, N., and Culver, D.C. (2010). How to protect a diverse, poorly-known, inaccessible fauna: identification of source and sink habitats in the epikarst. *Aquatic Conservation: Marine and Freshwater Ecosystems*, **20**, 748–55.

Pipan, T., López H., Oromí, P., Polak, S., and Culver, D.C. (2011). Temperature variation and the presence of troglobionts in shallow subterranean habitats. *Journal of Natural History*, **45**, 253–73.

Pipan, T., Navodnik, V., Janžekovič, F., and Novak, T. (2008). First studies on the fauna of percolation water in Huda Luknja, a cave in the isolated karst in northeast Slovenia. *Acta Carsologica*, **37**, 33–43.

Polak, S., Delić, T., Kostanjšek, R., and Trontelj, P. (2016). Molecular phylogeny of the cave beetle genus *Hadesia* (Coleoptera: Leiodidae: Cholevinae: Leptodirini), with a description of a new species from Montenegro. *Arthropod Systematics and Phylogeny*, **74**, 241–54.

Polis, G. A., and Hurd, S. D. (1996). Linking marine and terrestrial food webs: allochthonous input from the ocean supports high secondary productivity on small islands and coastal land communities. *American Naturalist*, **147**, 396–423.

Por, F.D. (2007). Ophel: a groundwater biome based on chemoautotrophic resources. The global significance of the Ayyalon Cave finds, Israel. *Hydrobiologia*, **592**, 1–10.

Por, F.D., Dimentman C., Frumkin, A., and Naaman I. (2013). Animal life in the chemoautotrophic ecosystem of the hypogenic groundwater cave of Ayyalon (Israel): a summing up. *Natural Science*, **5**, 7–13.

Porter, M.L., and Crandall, K.A. (2003). Lost along the way: the significance of evolution in reverse. *Trends in Ecology and Evolution*, **18**, 541–7.

Porter, M.L., Dittmar, K., and Pérez-Losada, M. (2007). How long does evolution of the troglomorphic form take? Estimating divergence times in *Astyanax mexicanus*. *Acta Carologica*, **36**, 173–82.

Porter, M.L., Engel, A.S., Kane, T.C., and Kinkle, B.K. (2009). Productivity-diversity relationships from chemoautotrophically based sulfidic karst systems. *International Journal of Speleology*, **38**, 27–40.

Porter, M.L., and Hobbs III, H.H. (1997). Population studies of an undescribed species of *Crangonyx* in Dillion Cave, Orange County, Indiana USA (Crustacea: Amphipoda: Crangonyctidae). In I.D. Sasowsky, D.W. Fong, and E.L. White, eds. *Conservation and protection of the biota of karst*. Karst Waters Institute Special Publication 3, Charles Town, West Virginia.

Pospisil, P. (1994). The groundwater fauna of a Danube aquifer in the wetland Lobau at Vienna, Austria. In H. Wilkens, D.C. Culver, and W.F. Humphreys, eds. *Subterranean ecosystems*, pp. 347–66. Elsevier Press, Amsterdam, The Netherlands.

Poulson, T.L. (1963). Cave adaptation in amblyopsid fishes. *American Midland Naturalist*, **70**, 257–90.

Poulson, T.L. (1969). Population size, density and regulation in cave fishes. *Proceedings of the Fourth International Congress of Speleology, Yugoslavia*, **4–5**, 189–92.

Poulson, T.L. (1992). The Mammoth Cave ecosystem. In A.I. Camacho, ed. *The Natural History of Biospeleology*, pp. 569–611. Museo Nacional de Ciencias Naaturales Monografias, Madrid, Spain.

Poulson, T.L. (2012). Food sources. In W.B. White and D.C. Culver, eds. *Encyclopedia of caves. Second edition*, pp. 321–34. Elsevier/Academic Press, Amsterdam, The Netherlands.

Poulson, T.L. (2017a). Terrestrial cave ecology of the Mammoth Cave region. In H.H. Hobbs, R.A. Olson, E.G. Winkler, and D.C. Culver, eds. *Mammoth Cave. A human and natural history*, pp. 199–207. Springer, Cham, Switzerland.

Poulson, T.L. (2017b). Book review: 'Evolution in the dark. Darwin's loss without selection.' *Journal of Cave and Karst Studies*, **79**, 135–7.

Poulson, T.L., and White, W.B. (1969). The cave environment. *Science*, **165**, 971–81.

Prendini, L., Francke, O.F., and Vignoli, V. (2010). Troglomorphism, trichobothriotaxy and typhlochactid phylogeny (Scorpiones, Chactoidea): more evidence that troglobitism is not an evolutionary dead-end. *Cladistics*, **26**, 117–42.

Prevorčnik, S., Blejec, A., and Sket, B. (2004). Racial differentiation in *Asellus aquaticus* (L.) (Crustacea: Isopoda: Asellidae). *Archiv für Hydrobiologie*, **160**, 193–214.

Protas, M.C., Conrad, M., Gross, J.B., Tabin, C., and Borowsky, R. (2007). Regressive evolution in the Mexican cave tetra, *Astyanax mexicanus*. *Current Biology*, **17**, 452–4.

Protas, M.E., Hersey, C., Kochanek, D., Zhou, Y., Wilkens, H., Jeffery, W.R., Zon, L.I., Borowsky, R., and Tabin, C.J. (2006). Genetic analysis of cavefish reveals molecular convergence in the evolution of albinism. *Nature Genetics*, **38**, 107–11.

Protas, M., and Jeffery, W.R. (2013). Evolution and development in cave animals: from fish to crustaceans. *Wiley Interdisciplinary Reviews: Developmental Biology*, **1**, 823–45.

Proudlove, G.S. (2001). Subterranean biodiversity in the British Isles. In D.C. Culver, L. Deharveng, J. Gibert, and I. Sasowsky, eds. *Mapping subterranean biodiversity. Cartographie de la biodiversité souterraine*, pp. 56–8. Karst Waters Institute Special Publication 6, Charles Town, West Virginia.

Proudlove, G.S. (2006). *Subterranean fishes of the world*. International Society for Subterranean Biology, Moulis, France.

Proudlove, G.S. (2010). Biodiversity and distribution of subterranean fishes of the world. In E. Trajano, E.M. Bichuette, and B.G. Kapoor, eds. *Biology of subterranean fishes*, pp. 41–64. CRC Press, Boca Raton, Florida.

Rabinowitz, D. (1981). Seven forms of rarity. In H. Synge, ed. *Aspects of rare plant conservation*, pp. 205–17. Wiley, New York.

Rabinowitz, D., Cairns, S., and Dillon, T. (1986). Seven forms of rarity and their frequency in the flora of the British Isles. In M.E. Soulé, ed. *Conservation biology: the science of scarcity and diversity*, pp. 182–204. Sinauer Associates, Sunderland, Massachusetts.

Racoviță, E.G. (2006). Essay on biospeological problems (D.C. Culver and O.T. Moldovan, trans.). In O.T. Moldovan, ed. *Emil George Racovitza. Essay on biospeological problems—French, English, Romanian version*, pp. 127–83. Casa Cărți de Știință, Cluj-Napoca, Romania.

Rambla, M., and Juberthie, C. (1994). Opiliones. In C. Juberthie and V. Decu, eds. *Encyclopaedia Biospeologica, Tome I*, pp. 215–30. Société Internationale de Biospéologie, Moulis, France.

Rasquin, P. (1947). Progressive pigmentary reduction in fishes associated with cave environments. *Zoologica (New York)*, **32**, 35–42.

Ravbar, N. (2007). *The protection of karst waters. A comprehensive Slovene approach to vulnerability and contamination risk mapping*. Založba ZRC, Ljubljana, Slovenia.

Reddell, J.R. (2012). Spiders and related groups. In W.B. White and D.C. Culver, eds. *Encyclopedia of caves. Second edition*, pp. 786–97. Elsevier/Academic Press, Amsterdam, The Netherlands.

Reid, J.W. (2001). A human challenge: discovering and understanding continental copepod habitats. *Hydrobiologia*, **453/454**, 201–26.

Rétaux, S., and Casane, D. (2013). Evolution of eye development in the darkness of caves: adaptation, drift, or both? *EvoDevo*, **4**, 26.

Ribera, C., and Juberthie, C. (1994). Araneae. In C. Juberthie and V. Decu, eds. *Encyclopaedia Biospeologica, Tome I*, pp. 197–214. Société Internationale de Biospéologie, Moulis, France.

Riesch, R., Plath, M., and Schlupp, I. (2011). Speciation in caves: experimental evidence that permanent darkness promotes reproductive isolation. *Biology Letters*, **7**, 909–12.

Rivera, M.A.J., Howarth, F.G., Taiti, S., and Roderick, G.K. (2002). Evolution in Hawaiian cave-adapted isopods (Oniscidea: Philosciidae): vicariant speciation or adaptive shifts? *Molecular Phylogenetics and Evolution*, **25**, 1–9.

Rohner, N., Jarosz, D.F., Kowalko, J.E., Yoshizawa, M., Jeffery, W.R., Borowsky, R.L., Lindquist, S., and Tabin, C.J. (2013). Cryptic variation in morphological evolution: HSP90 as a capacitor for loss of eyes in cavefish. *Science*, **342**, 1372–5.

Romero, A. (2001). Scientists prefer them blind: the history of hypogean fish research. *Environmental Biology of Fishes*, **62**, 43–71.

Romero, A. (2004). Evolution of hypogean fauna. In J. Gunn, ed. *Encyclopedia of caves and karst science*, pp. 347–50. Fitzroy Dearborn, New York.

Romero, A. (2009). *Cave biology. Life in darkness*. Cambridge University Press, Cambridge, UK.

Rosa, G., and Penado, A. (2013). *Rana iberica* (Boulenger, 1879) goes underground: subterranean habitat usage and new insights on natural history. *Subterranean Biology*, **11**, 15–29.

Rouch, R. (1968). Contribution a la connaissance des Harpacticides hypogés (Crustacés-Copépodes). *Annales de Spéléologie*, **23**, 9–167.

Rouch, R. (1970). Recherches sur les eaux souterraines—12—Le système karstique du Baget. I. Le phénomène d''hémorragie' au niveau de l'exutoire principal. *Annales de Spéléologie*, **25**, 665–709.

Rouch, R. (1977). Considérations sur l'écosystème karstique. *Compte Rendu de Academie de Sciences, Paris*, **284**, 1101–3.

Rouch, R. (1986). Sur l'ecologie des eaux souterraines sur al karst. *Stygologia*, **2**, 352–98.

Rouch, R. (1991). Structure de peuplement des harpacticides dans le milieu hyorhéic d'un ruisseau des Pyrénées. *Annales de Limnologie*, **27**, 227–41.

Rouch, R. (1994). Copepoda. In C. Juberthie and V. Decu, eds. *Encyclopaedia Biospeologica, Tome I*, pp. 105–11. Société Internationale de Biospéologie, Moulis, France.

Rouch, R., Bakalowicz, M., Mangin, A., and D'Hulst, D. (1989). Sur les caractéristiques de sub-écoulement d'un ruisseau des Pyrénées. *Annales de Limnologie*, **25**, 3–16.

Rouch, R., and Danielopol, D.L. (1997). Species richness of microcrustacea in subterranean freshwater habitats. Comparative analysis and approximate evaluation. *International Revue Gesellshaft für Hydrobiologie*, **82**, 121–45.

Ruffo, S. (1957). La attuali conoscenze sulla fauna cavernicola della Regione Pugliese. *Memorie di biogeografica Adriatica*, **3**, 1–143.

Sârbu, Ş.M. (2000). Movile Cave: a chemoautotrophically based groundwater ecosystem. In H. Wilkens, D.C. Culver, and W.F. Humphreys, eds. *Subterranean ecosystems*, pp. 319–43. Elsevier Press, Amsterdam, The Netherlands.

Sârbu, Ş.M., Kane, T.C., and Culver, D.C. (1993). Genetic structure and morphological differentiation: *Gammarus minus* (Amhipoda: Gammaridae) in Virginia. *American Midland Naturalist*, **129**, 145–52.

Sârbu, Ş.M., Kane, T.C., and Kinkle, B.K. (1996). A chemoautotrophically based groundwater ecosystem. *Science*, **272**, 1953–5.

Sbordoni, V., Allegrucci, G., and Cesaroni, D. (2000). Population genetic structure: speciation and evolutionary rates in cave-dwelling organisms. In H. Wilkens, D.C. Culver, and W.F. Humphreys, eds. *Subterranean ecosystems*, pp. 453–78. Elsevier Press, Amsterdam, The Netherlands.

Sbordoni, V., Allegrucci, G., and Cesaroni, D. (2012). Population structure. In W.B. White and D.C. Culver, eds. *Encyclopedia of caves. Second edition*, pp. 608–18. Elsevier/Academic Press, Amsterdam, The Netherlands.

Schemmel, C. (1974). Genetische Untersuchungen zur Evolution des Geschmack-sapparates bei cavenicolen Fischen. *Zeitschrift für Zoologische Systematik und Evolutionforschung*, **12**, 169–215.

Schindel, G., Hoyt, J., and Johnson, S. (2004). Edwards Aquifer, United States. In J. Gunn, ed. *Encyclopedia of caves and karst science*, pp. 313–5. Fitzroy Dearborn, New York.

Schiner, J.R. (1854). Fauna der Adelsberger-, Luegger-, and Magdalenen Grotte. In A. Schmidt, ed. *Die Grotten und Höhlen von Adelsberg, Lueg, Planina, and Laas*, pp. 231–72. Braunmüller, Vienna, Austria.

Schminke, H.K. (1981). Perspectives in the study of the zoogeography of interstitial crustacean: Bathynellacea (Syncarida) and Parstenocarididae (Copepoda). *International Journal of Speleology*, **11**, 83–9.

Schneider, K., Christman, M.C., and Fagan, W.F. (2011). The influence of resource subsidies on cave invertebrates: results from an ecosystem-level manipulation experiment. *Ecology*, **92**, 765–76.

Schneider, K., and Culver, D.C. (2004). Estimating subterranean species richness using intensive sampling and rarefaction curves in a high density cave region in West Virginia. *Journal of Cave and Karst Studies*, **66**, 39–45.

Schneider, K., Kay, A.D., and Fagan, W.T. (2010). Adaptation to a limiting environment: the phosphorus content of terrestrial cave arthropods. *Ecological Research*, **25**, 565–77.

Sendra, A., Garay, P., Ortuño, V.M., Gilgado, J.D., Teruel, S., and Reboleira, A.S.P. (2014). Hypogenic versus epigenic subterranean ecosystem: lessons from eastern Iberian Peninsula. *International Journal of Speleology*, **43**, 253–64.

Shaw, T.R. (1999). *Proteus* for sale and for science in the 19th century. *Acta Carsologica*, **28**, 119–29.

Simečić, V. (2017). Poachers threaten Balkans' underground biodiversity. *Science*, **358**, 1116–7.

Simon, K.S., and Benfield, E.F. (2001). Leaf and wood breakdown in cave streams. *Journal of the North American Benthological Society*, **20**, 550–63.

Simon, K.S., and Benfield, E.F. (2002). Ammonium retention and whole-stream metabolism in cave streams. *Hydrobiologia*, **482**, 31–9.

Simon, K.S., Benfield E.F., and Macko, S.A. (2003). Food web structure and the role of epilithic films in cave streams. *Ecology*, **84**, 2395–406.

Simon, K.S., Pipan, T., and Culver, D.C. (2007a). A conceptual model of the flow and distribution of organic carbon in caves. *Journal of Cave and Karst Studies*, **69**, 279–84.

Simon, K.S., Pipan, T., and Culver, D.C. (2007b). Spatial and temporal heterogeneity in the flux of organic carbon in caves. In L. Ribeiro, T.Y. Stigter, A. Chambel, M.T. Condesso de Melo, J.P. Monteiro, and A. Medeiros, eds. *Groundwater and Ecosystems*, p. 367. International Association of Hydrogeologists, Lisbon, Portugal.

Simon, K.S., Pipan, T., Ohno, T., and Culver, D.C. (2010). Spatial and temporal patterns in abundance and character of dissolved organic matter in two karst aquifers. *Fundamental and Applied Limnology*, **177**, 81–92.

Sket, B. (1977). Gegenseitige Beeinflussung der Wasserpollution und des Höhlenm-ilieus. *Proceedings of the 6th International Congress of Speleology, Olomouc, ČSSR*, **4**, 253–62.

Sket, B. (1979). Jamska favna notranjskega trikotnika (Cerknica-Postojna-Planina) njena ogroženost in naravovarstveni pomen. *Varstvo Narave*, **12**, 45–59.

Sket, B. (1986). Ecology of the mixohaline hypogean fauna along the Yugoslav coasts. *Stygologia*, **2**, 317–38.

Sket, B. (1996). The ecology of the anchialine caves. *Trends in Ecology and Evolution*, **11**, 221–5.

Sket, B. (1999). The nature of biodiversity in subterranean waters and how it is endangered. *Biodiversity and Conservation*, **8**, 1319–38.

Sket, B. (2002). The evolution of karst versus the distribution and diversity of the hypogean fauna. In F. Gabrovšek, ed. *Evolution of karst: from prekarst to cessation*, pp. 225–32. Založba ZRC, Ljubljana, Slovenia.

Sket, B. (2004a). The cave hygropetric—a little known habitat and its inhabitants. *Archives für Hydrobiologie*, **160**, 413–25.

Sket, B. (2004b). Subterranean habitats. In J. Gunn, ed. *Encyclopedia of cave and karst science*, pp. 709–13. Fitzroy Dearborn, New York.

Sket, B. (2004c). Anchialine habitats. In J. Gunn, ed. *Encyclopedia of caves and karst science*, pp. 64–6. Fitzroy Dearborn, New York.

Sket, B. (2008). Can we agree on an ecological classification of subterranean animals. *Journal of Natural History*, **42**, 1549–63.

Sket, B. (2012). Anchialine caves and fauna. In W.B. White and D.C. Culver, eds. *Encyclopedia of caves. Second edition*, pp. 17–25. Elsevier/Academic Press, Amsterdam, The Netherlands.

Sket, B., and Arntzen, J.W. (1994). A black non-troglomorphic amphibian from the karst of Slovenia—*Proteus anguinus parkelj* (Urodela, Proteidae). *Bijdragen tot de Dierkunde*, **64**, 33–53.

Sket, B., Paragamian, K., and Trontelj, P. (2004). A census of the obligate subterranean fauna of the Balkan Peninsula. In H.W. Griffiths, B. Kryštufek, and J.M. Reed, eds. *Balkan biodiversity. Pattern and process in the European hotspot*, pp. 309–22. Kluwer Academic Publishers, Dordrecht, The Netherlands.

Sket, B., and Velkovrh, F. (1981). Postojnsko-Planinski jamski sistem kot model za preučevanje onesnaženja podzemeljskih voda. *Naše Jame*, **22**, 27–44.

Souza-Silva, M., Bernardi, L.F.O., Martins, R.P., and Ferreira R.L. (2012). Transport and consumption of organic detritus in a neotropical limestone cave. *Acta Carsologica*, **41**, 139–50.

Souza-Silva, M., and Ferreira, R.L. (2016). The first two hotspots of subterranean biodiversity in South America. *Subterranean Biology*, **19**, 1–21.

Souza-Silva, M., Martins, R.P., and Ferreira, R.L. (2011). Cave lithology determining the structure of the invertebrate communities in the Brazilian Atlantic Rain Forest. *Biodiversity and Conservation*, **20**, 1713–29.

Stahl, B.A., and Gross, J.B. (2017). A comparative transcriptomic analysis of development in two *Astyanax* cavefish populations. *Journal of Experimental Zoology*, **328B**, 515–32.

Stahl, B.A., Gross, J.B., Speiser, D.I., Oakley, T.H., Patel, N.H., Gould, D.B., and Protas, M.E. (2015). A transcriptomic analysis of cave, surface, and hybrid isopod crustaceans of the species *Asellus aquaticus*. *PloS ONE*, **10**, e0140484.

Stanford, J.A., and Gaufin, A.R. (1974). Hyporheic communities of two Montana rivers. *Science*, **185**, 700–2.

Stanford, J.A., Ward, J.V., and Ellis, B.K. (1994). Ecology of the alluvial aquifers of the Flathead River, Montana. In J. Gibert, D.L. Danielopol, and J.A. Stanford, eds. *Groundwater Ecology*, pp. 367–90. Academic Press, San Diego, California.

Stevens, T.O., and McKinley, J.P. (1995). Lithautotrophic microbial ecosystems in deep basalt aquifers. *Science*, **270**, 450–4.

Stoch, F., Artheau, M., Brancelj, A., Galassi, D.M.P., and Malard, F. (2009). Biodiversity indicators in European ground waters: towards a predictive model of stygobiotic species richness. *Freshwater Biology*, **54**, 745–55.

Stone, F.D., Howarth, F.G., Hoch, H., and Asche, M. (2012). Root communities in lava tubes. In W.B. White and D.C. Culver, eds. *Encyclopedia of caves. Second edition*, pp. 658–64. Elsevier/Academic Press, Amsterdam, The Netherlands.

Studier, E.H. (1996). Composition of bodies of caves crickets (*Hadenoecus subterraneus*), their eggs, and their egg predator, *Neaphaenops tellkampfi*. *American Midland Naturalist*, **136**, 101–9.

Sugihara, G. (1981). S=CA^z, z=1/4: a reply to Conner and McCoy. *American Naturalist*, **117**, 790–93.

Tabin, C.J. (2016). Introduction: the emergence of the Mexican cavefish as an important model system for understanding phenotypic evolution. In A.C. Keene, M. Yoshizawa, and S.E. McGaugh, eds. *The biology and evolution of Mexican cavefish*, pp. 1–8. Academic/Elsevier, Amersterdam, The Netherlands.

Taiti, S. (2004). Crustacea: Isopoda: Oniscidea (Woodlice). In J. Gunn, ed. *Encyclopedia of caves and karst science*, pp. 265–7. Fitzroy Dearborn, New York.

Taylor, S.J., Krejca, J., and Denight, M.L. (2005). Foraging and range habitat use of *Ceuthophilus secretus* (Orthoptera: Rhaphidophoridae), a key trogloxene in central Texas cave communities. *American Midland Naturalist*, **154**, 97–114.

ter Braak, C.J.F., and Verdonschot, P.F.M. (1995). Canonical correspondence analysis and related multivariate methods in aquatic ecology. *Aquatic Sciences*, **57**, 1015–21.

Tercafs, R. (2001). *The protection of the subterranean environment. Conservation principles and management tools*. P.S. Publishers, Luxemburg.

Thibaud, M., and Deharveng, L. (1994). Collembola. In C. Juberthie and V. Decu, eds. *Encyclopaedia Biospeologia, Tome I*, pp. 267–76. Société Internationale de Biospéologie, Moulis, France.

Thinès, G. (1969). *L'evolution regressive des poissons cavernicoles et abyssaux*. Mason et Cie, Paris.

Thinès, G., and Tercafs, R. (1972). *Atlas de la vie souterraine*. Ealbert de Visscher Ed., Bruxelles.

Thomas, D.W. (1997). Oilbirds in caves. In I.D. Sasowsky, D.W. Fong, and E.L. White, eds. *Conservation and protection of the biota of karst*, pp. 105–6. Karst Waters Institute Special Publication 3, Charles Town, West Virginia.

Thomas, D.W., Bosque, C., and Arends, A. (1993). Development of thermoregulation and the energetics of nestling oilbirds (*Steatornis caripensis*). *Physiological Zoology*, **66**, 322–48.

Tobler, M., Kelley, J.L., Plath, M., and Riesch, R. (2018). Extreme environments and the origins of biodiversity: Adaptation and speciation in sulphide spring fishes. *Molecular Ecology*, **27**, 832–59.

Toomey, R.S., Hobbs III, H.H., and Olson, R.A. (2017). An orientation to Mammoth Cave and this volume. In H.H. Hobbs III, R.A. Olson, E.G. Winkler, and D.C. Culver, eds. *Mammoth Cave. A human and natural history*, pp. 1–28. Springer, Cham, Switzerland.

Trajano, E. (2001). Mapping subterranean biodiversity in Brazilian karst areas. In D.C. Culver, L. Deharveng, J. Gibert, and I.D. Sasowsky, eds. *Mapping Subterranean*

Biodiversity. Cartographie de la Biodiversité Souterraine, pp. 67–70. Karst Waters Institute Special Publication 6, Charles Town, West Virignia.

Trajano, E. (2007). The challenge of estimating the age of subterranean lineages: examples from Brazil. *Acta Carsologica*, **36**, 191–8.

Trajano, E. (2012). Ecological classification of subterranean organisms. In W.B. White and D.C. Culver, eds. *Encyclopedia of caves. Second edition*, pp. 275–7. Elsevier/ Academic Press, Amsterdam, The Netherlands.

Trajano, E., Bichuette, E.M., and Kapoor, B.G., eds. (2010). *Biology of subterranean fishes*. CRC Press, Boca Raton, Florida.

Trajano, E., and de Carvalho, M.R. (2017). Towards a biologically meaningful classi-fication of subterranean organisms: a critical analysis of the Schiner–Racovitza system from a historical perspective, difficulties of its application and implications for conservation. *Subterranean Biology*, **22**, 1–26.

Trontelj, P. (2018). Structure and genetics of cave populations. In O.T. Moldovan, L. Kovác, and S. Halse, eds. *Cave ecology*. Springer, Cham, Switzerland.

Trontelj, P., Blejec, A., and Fišer, C. (2012). Ecomorphological convergence in cave communities. *Evolution*, **66**, 3852–65.

Trontelj, P., Douady, C.J., Fišer, C., Gibert, J., Gorički, Š., Lefébure, T., Sket, B., and Zakšek, V. (2009). A molecular test for cryptic diversity in groundwater: how large are ranges of macro-stygobionts? *Freshwater Biology*, **54**, 727–44.

Trontelj, P., Gorički, Š., Polak, S., Verovnik, R., Zakšek, V., and Sket, B. (2007). Age estimates for some subterranean taxa and lineages in the Dinaric Karst. *Acta Carsologica*, **36**, 183–90.

Turquin, M.J., and Barthelemy, D. (1985). The dynamics of a population of the troglo-bitic amphipod *Niphargus virei* Chevreux. *Stygologia*, **1**, 109–17.

Tvrtković, N. (2012). Vertebrate visitors—birds and mammals. In W.B. White and D.C. Culver, eds. *Encyclopedia of caves. Second edition*, pp. 845–9. Elsevier/ Academic Press, Amsterdam, The Netherlands.

U.S. EPA (2018). *Semiannual report of UST performance measures mid-fiscal year 2018 (October, 2017 – March, 2018)*. www.epa.gov/gov/sites/production/files.2018-05/ documents/ca-18-12.pdf

Valentine, J. M. (1945). Speciation and raciation in *Pseudanophthalmus* (cavernicolous Carabidae). *Transactions of the Connecticut Academy of Arts and Sciences*, **36**, 631–72.

Vandel, A. (1964). *Biospéologie: la biologie des animaux cavernicoles*. Gauthier-Villars, Paris, France.

Vandel, A. (1965). *Biospeleology: the biology of cavernicolous animals* (B.E. Freeman, trans). Pergamon Press, New York.

Van Dover, C.L. (2000). *The ecology of deep-sea hydrothermal vents*. Princeton University Press, Princeton, New Jersey.

Venarsky, M.P., Benstead, J.P., and Huryn, A.D. (2012b). Effects of organic matter and season on leaf litter colonisation and breakdown in cave streams. *Freshwater Biology*, **57**, 773–86.

Venarsky, M.P., Benstead, J.P., Huryn, A.D., Huntsman, B.M., Edmonds, J.W., Findlay, R.H., and Wallace, J.B. (2017). Experimental detritus manipulations unite surface and cave stream ecosystems along a common energy gradient. *Ecosystems*, **21**, 629–42. DOI: 10.1007/s10021-017-0174-4

Venarsky, M.P., Huntsman, B.M., Huryn, A.D., Benstead, J.P., and Kuhajda, B.R. (2014). Quantitative food web analysis supports the energy-limitation hypothesis in cave stream ecosystems. *Oecologia*, **176**, 859–69.

Venarsky, M.P., Huryn, A.D., and Benstead, J.P. (2012a). Re-examining extreme longevity of the cave crayfish *Orconectes australis* using new mark–recapture data: a lesson on the limitations of iterative size-at-age models. *Freshwater Biology*, **57**, 1471–81.

Vergnon, R., Leijs, R., van Nes, E.G., and Scheffer, M. (2013). Repeated parallel evolution reveals limiting similarity in subterranean diving beetles. *American Naturalist*, **182**, 67–75.

Vermeulen, J.J., and Whitten, T. (1999). *Biodiversity and cultural property in the management of limestone resources—lessons from East Asia*. World Bank, Washington, DC.

Verovnik, R. (2012). *Asellus aquaticus:* a model system for historical biogeography. In W.B. White and D.C. Culver, eds. *Encyclopedia of caves. Second edition*, pp. 30–6. Elsevier/Academic Press, Amsterdam, The Netherlands.

Verovnik, R., Sket, B., Prevorčnik, S., and Trontelj, P. (2003). Random amplified polymorphic DNA diversity among surface and subterranean populations of *Asellus aquaticus* (Crustacea: Isopoda). *Genetica*, **119**, 155–65.

Verovnik, R., Sket, B., and Trontelj, P. (2004). Phylogeography of subterranean and surface populations of water lice *Asellus aquaticus* (Crustacea: Isopoda). *Molecular Ecology*, **13**, 1519–32.

Verovnik, R., Sket, B., and Trontelj, P. (2005). The colonization of Europe by the freshwater crustacean *Asellus aquaticus* (Crustacea: Isopoda) proceeded from ancient refugia and directed by habitat connectivity. *Molecular Ecology*, **14**, 4355–69.

Vesper, D.J. (2012). Contamination of cave waters by heavy metals. In W.B. White and D.C. Culver, eds. *Encyclopedia of caves. Second edition*, pp. 160–6. Elsevier/ Academic Press, Amsterdam, The Netherlands.

Viele, D.P., and Studier, E.H. (1990). Use of a localized food source by *Peromyscus leucopus*, determined with an hexagonal grid. *Bulletin of the National Speleological Society*, **52**, 52–3.

Villacorta, C., Jaume, D., Oromí, P., and Juan, C. (2008). Under the volcano: phylo-geography and evolution of the cave-dwelling *Palmorchestia hypogaea* (Amphipoda, Crustacea) at La Palma (Canary Islands). *BMC Biology*, **6**, doi:10.1186/1741-7007-6-7.

Vlăsceanu, L., Popa, R., and Kinkle, B. (1997). Characterization of *Thiobacillus thioparus* LV43 and its distribution in a chemoautotrophically based groundwater ecosystem. *Applied and Environmental Microbiology*, **63**, 3112–27.

Vuilleumier, F. (1973). Insular biogeography in continental regions. II. Cave faunas from Tessin, southern Switzerland. *Systematic Zoology*, **22**, 64–76.

Waltham, T. (2004). Mulu, Sarawak. In J. Gunn, ed. *Encyclopedia of caves and karst science*, pp. 531–3. Fitzroy Dearborn, New York.

Ward, J.V., and Palmer, M.A. (1994). Groundwater copepods: diversity patterns over ecological and evolutionary scales. *Hydrobiologia*, **453**, 227–53.

Ward, J.V., and Voelz, N.J. (1994). Groundwater fauna of the South Platte River System, Colorado. In J. Gibert, D.L. Danielopol, and J.A. Stanford, eds. *Groundwater Ecology*, pp. 391–423. Academic Press, San Diego.

Watts, C.H.S., and Humphreys, W.F. (2009). Fourteen new Dytiscidae (Coleoptera) of the genera *Limbodessus* Guignot, *Paroster* Sharp, and *Exocelina* Broun from underground waters in Australia. *Transactions of the Royal Society of South Australia*, **133**, 62–107.

Weber, A. (2004). Amphibia. In J. Gunn, ed. *Encyclopedia of caves and karst science*, pp. 61–2. Fitzroy Dearborn, New York.

Weigand, A.M., Jochum, A., Pfenninger, M., Steinke, D., and Klussmann-Kolb, A. (2011). A new approach to an old conundrum—DNA barcoding sheds new light on phenotypic plasticity and morphological stasis in microsnails (Gastropoda, Pulmonata, Carychiidae). *Molecular Ecology Resources*, **11**, 255–65.

White, W.B. (1988). *Geomorphology and hydrology of karst terrains*. Oxford University Press, New York.

White, W.B., and Culver, D.C., eds. (2012). *Encyclopedia of caves. Second edition.* Elsevier/Academic Press, Amsterdam, The Netherlands.

White, W.B., Culver, D.C., Herman, J.S., Kane, T.C., and Mylroie, J.E. (1995). Karst lands. *American Scientist*, **83**, 450–59.

Whitman, W.B., Coleman, D.C., and Wiebe, W.J. (1998). Prokaryotes: the unseen majority. *Proceedings of the National Academy of Sciences*, **95**, 6578–83.

Wiens, J.J., Chippendale, P.T., and Hillis, D.M. (2003). When are phylogenetic analyses misled by convergence? A case study in Texas cave salamanders. *Systematic Biology*, **52**, 501–14.

Wilkens, H. (1971). Genetic interpretation of regressive evolutionary processes: studies on hybrid eyes of two *Astyanax* cave populations (Characidae, Pisces). *Evolution*, **25**, 530–44.

Wilkens, H. (1988). Evolution and genetics of epigean and cave *Astyanax mexicanus* (Characidae, Pisces): support for the neutral mutation theory. *Evolutionary Biology*, **23**, 271–367.

Wilkens, H., and Strecker, U. (2017). *Evolution in the dark: Darwin's loss without selection*. Springer, Cham, Switzerland.

Williams, P.W. (2008). The role of the epikarst in karst and cave hydrogeology: a review. *International Journal of Speleology*, **37**, 1–10.

Woloszyn, B.W. (1998). Chiroptera. In C. Juberthie and V. Decu, eds. *Encyclopaedia Biospeologica, Tome II*, pp. 1267–96. Société Internationale de Biospéologie, Moulis, France.

Woods, L.P., and Inger, R.F. (1957). The cave, spring, and swamp fishes of the family Amblyopsidae of central and eastern United States. *American Midland Naturalist*, **58**, 232–56.

Wright, S. (1964). Pleiotropy in the evolution of structural reduction and of dominance. *American Naturalist*, **98**, 65–70.

Xiao, H., Chen, S.Y., Liu, Z.M., Zhang, R.D., Li, W.X., Zan, R.G., and Zhang, Y.P. (2005). Molecular phylogeny of *Sinocyclocheilus* (Cypriniformes: Cyprinidae) inferred from mitochondrial DNA sequences. *Molecular Phylogenetics and Evolution*, **36**, 67–77.

Yager, J. (1981). Remipedia—a new class of Crustacea from a marine cave in the Bahamas. *Journal of Crustacean Biology*, **1**, 328–33.

Yager, J. (1994). Remipedia. In C. Juberthie and V. Decu, eds. *Encyclopaedia Biospeologica, Tome I*, pp. 87–90. Société Internationale de Biospéologie, Moulis, France.

Yamamoto, Y., and Jeffery, W.R. (2000). Central role for the lens in cavefish eye degeneration. *Science*, **289**, 631–3.

Yamamoto, Y., Stock, D.W., and Jeffery, W.R. (2004). Hedgehog signalling controls eye degeneration in blind cavefish. *Nature*, **431**, 844–7.

Yoshizawa, M., and Jeffery, W.R. (2008). Shadow response in the blind cavefish *Astyanax* reveals conservation of a functional pineal eye. *The Journal of Experimental Biology*, **211**, 292–9.

Zagmajster, M., Culver, D.C., Christman, M.C., and Sket, B. (2010). Evaluating the sampling bias in pattern of subterranean species richness—combining approaches. *Biodiversity and Conservation*, **19**, 3035–8.

Zagmajster, M., Culver, D.C., and Sket, B. (2008). Species richness patterns of obligate subterranean beetles in a global biodiversity hotspot—effect of scale and sampling intensity. *Diversity and Distributions*, **14**, 95–105.

Zagmajster, M., Eme, D., Fišer, C., Galassi, D., Marmonier, P., Stoch, F., Cornu, J., and Malard, F. (2014). Geographic variation in range size and beta diversity of groundwater crustaceans: insights from habitats with low thermal seasonality. *Global ecology and biogeography*, **23**, 1135–45.

Zakšek, V., Sket, B., Gottstein, S., and Trontelj, P. (2009). The limits of cryptic diversity in groundwater: phylogeography of the cave shrimp *Troglocaris anophthalmus* (Crustacea: Decapoda: Atyidae). *Molecular Ecology*, **18**, 931–46.

Zhao, Y., and Zhang, C. (2009). *Endemic fishes of* Sinocyclocheilus *(Cypriniformes, Cyprinidae) in China. Species diversity, cave adaptation, systematics, and zoogeography*. Science Press, Beijing, China (in Chinese).

Index